Wild Product Governance

Wild Product Governance

Finding Policies that Work for Non-timber Forest Products

*Edited by Sarah A. Laird, Rebecca J. McLain
and Rachel P. Wynberg*

publishing for a sustainable future

London • New York

First published in 2010 by Earthscan

Earthscan
2 Park Square, Milton Park, Abingdon, Oxon OX14 4RN
Simultaneously published in the USA and Canada by Earthscan
711 Third Avenue, New York, NY 10017
Earthscan is an imprint of the Taylor & Francis Group, an informa business

First issued in paperback 2011

ISBN: 978-1-84407-500-3 hardback
ISBN: 978-0-415-50713-4 paperback

Typeset by Composition and Design services
Cover design by Susanne Harris

A catalogue record for this book is available from the British Library

Library of Congress Cataloging-in-Publication Data

Wild product governance: finding policies that work for non-timber forest products/edited by Sarah A. Laird, Rebecca McLain, and Rachel P. Wynberg.
    p. cm.
    Includes bibliographical references and index.
    ISBN 978-1-84407-500-3 (hardback)
  1. Non-timber forest products–Government policy. 2. Non-timber forest resources–Management. I. Laird, Sarah A. II. McLain, Rebecca J. (Rebecca Jean) III. Wynberg, Rachel.
    SD543.W55   2010
    338.1'74980973–dc22   2009053715

At Earthscan we strive to minimize our environmental impacts and carbon footprint through reducing waste, recycling and offsetting our CO_2 emissions, including those created through publication of this book.

Contents

List of Figures, Tables and Boxes   *ix*
About the Contributors   *xiii*
Acknowledgements   *xxi*
Foreword by J. E. Michael Arnold   *xxiii*
List of Acronyms and Abbreviations   *xxv*

Introduction   **1**
Sarah A. Laird, Rebecca J. McLain and Rachel P. Wynberg

Chapter 1   **Changing Policy Trends in the Emergence of
                Bolivia's Brazil Nut Sector**   **15**
                Peter Cronkleton and Pablo Pacheco

Case Study A   **In Search of Regulations to Promote the Sustainable
                Use of NTFPs in Brazil**   **43**
                *Marina Pinheiro Klüppel, Júlio César Raposo Ferreira,
                José Humberto Chaves and Antônio Carlos Hummel*

Chapter 2   **Integrating Customary and Statutory Systems:
                The Struggle to Develop a Legal and Policy Framework
                for NTFPs in Cameroon**   **53**
                *Sarah A. Laird, Verina Ingram, Abdon Awono, Ousseynou Ndoye,
                Terry Sunderland, Estherine Lisinge Fotabong and Robert Nkuinkeu*

Case Study B   **Policies for *Gnetum* spp. Trade in Cameroon: Overcoming
                Constraints that Reduce Benefits and Discourage Sustainability**   **71**
                Ousseynou Ndoye and Abdon Awono

Case Study C   **Regulatory Issues for Bush Mango (*Irvingia* spp.)
                Trade in South-west Cameroon and South-east Nigeria**   **77**
                Terry Sunderland, Stella Asaha, Michael Balinga and Okon Isoni

Chapter 3   **NTFPs in India: Rhetoric and Reality**   **85**
                Sharachchandra Lele, Manoj Pattanaik and Nitin D. Rai

Chapter 4   **Policy Gaps and Invisible Elbows: NTFPs in British Columbia**   **113**
                *Darcy A. Mitchell, Sinclair Tedder, Tim Brigham, Wendy Cocksedge
                and Tom Hobby*

Chapter 5   **NTFPs in Scotland: Changing Attitudes to Access Rights in a
                Reforesting Land**   **135**
                Alison Dyke and Marla Emery

Chapter 6   **From Barter Trade to Brad Pitt's Bed: NTFPs and Ancestral Domains in the Philippines**   155
Yasmin D. Arquiza, Maria Cristina S. Guerrero, Augusto B. Gatmaytan and Arlynn C. Aquino

Chapter 7   **From Indigenous Customary Practices to Policy Interventions: The Ecological and Sociocultural Underpinnings of the Non-timber Forest Trade on Palawan Island, the Philippines**   183
Dario Novellino

Case Study D   **Overregulation and Complex Bureaucratic Procedure: A Disincentive for Compliance? The Case of a Valuable Carving Wood in Bushbuckridge, South Africa**   199
Sheona Shackleton

Chapter 8   **Overcoming Barriers in Collectively Managed NTFPs in Mexico**   205
Catarina Illsley Granich, Silvia E. Purata, Fabrice Edouard, Maria Fernanda Sánchez Pardo and Citlali Tovar

Chapter 9   **Fiji: Commerce, Carving and Customary Tenure**   229
Francis Areki and Anthony B. Cunningham

Chapter 10   **One Eye on the Forest, One Eye on the Market: Multi-tiered Regulation of Matsutake Harvesting, Conservation and Trade in North-western Yunnan Province**   243
Nicholas K. Menzies and Chun Li

Chapter 11   **Managing Floral Greens in a Globalized Economy: Resource Tenure, Labour Relations and Immigration Policy in the Pacific Northwest, USA**   265
Rebecca J. McLain and Kathryn Lynch

Chapter 12   **NTFP Policy, Access to Markets and Labour Issues in Finland: Impacts of Regionalization and Globalization on the Wild Berry Industry**   287
Rebecca T. Richards and Olli Saastamoinen

Chapter 13   **Navigating a Way through Regulatory Frameworks for *Hoodia* Use, Conservation, Trade and Benefit Sharing**   309
Rachel P. Wynberg

Chapter 14   **Laws and Policies Impacting Trade in NTFPs**   327
Alan Pierce and Markus Bürgener

Chapter 15   **The State of NTFP Policy and Law**   343
Sarah A. Laird, Rachel P. Wynberg and Rebecca J. McLain

Chapter 16   **Recommendations**   **367**
Sarah A. Laird, Rachel P. Wynberg and Rebecca J. McLain

Appendix: **NTFP Law and Policy Literature: Lie of the Land and**
                  Areas for Further Research   **375**
Alan Pierce

Index

List of Figures, Tables and Boxes

FIGURES

1.1	Bolivia's northern Amazon region	17
1.2	Woman breaking Brazil nut fruits to extract nuts	18
2.1	Map of Cameroon and neighbouring countries in Africa	54
2.2	Eru (*Gnetum* spp.) loaded onto taxi in transition from Cameroon to Nigeria	62
B.1	Selling *Gnetum* in the market, Cameroon	72
C.1	Bush mango storage by sticking the fruits to the outer walls of the house for later use. This image was taken in the village of Ekuri, Cross River State, Nigeria	79
3.1	Location of central dry forest region and Western Ghats region in India	86
3.2	Collection of Kalabhalia seed by the villagers of Kalasulia, Boudh	88
3.3	Population structure of uppage individuals in low-intensity and high-intensity harvest sites	105
4.1	Map of British Columbia, Canada	114
4.2	Salal (*Gaultheria shallon*) harvest	117
5.1	Map of United Kingdom, including Scotland	136
5.2	Cep (*Boletus edulis*), one of Scotland's highest value mushrooms	137
5.3	Native bluebell (*Hyacinthoides non-scripta*), a controversial wild harvest	142
6.1	Alangan Mangyan rattan gatherers hauling bundles of split rattan	156
6.2	Map of the Philippines	157
7.1	Map of Palawan, the Philippines	184
7.2	Tapping resin from an Agathis tree, Palawan	185
D.1	Harvesting *P. angolensis* in Bushbuckridge – procedure and problems	200
8.1	*Agave cupreata* in its natural habitat	206
8.2	Map of Guerrero State in Mexico	213
9.1	Map of Fiji	230
9.2	Cultural value and production of kava bowls (tanoa)	234
10.1	Jiedi: Matsutake buyers and sellers at the official village matsutake market waiting for the signal to begin trading	244
10.2	Distribution of matsutake production in Yunnan (data based on year 2005)	246
11.1	Map of the Pacific Northwest	266
11.2	Weighing salal on the Olympic Peninsula	267
12.1	Map of Finland	288
12.2	Mushroom pickers waiting their turn to sell their weekend crop on a Sunday afternoon in 1995 outside 'Tuote ja Vihannes Ky', the oldest firm in the Joensuu region in the wild berry and mushroom business	291

13.1   Flowering *Hoodia gordonii*, Ceres (Karoo), Western Cape, South Africa   310
13.2   The intersection of regulatory frameworks for *Hoodia*   310
13.3   Distribution ranges of six southern African *Hoodia* species   311
13.4   Production cycle of *Hoodia* spp.   323

TABLES

1.1   Appendix: Principal legal mechanisms affecting Brazil nuts   42
A.1   Regulations for the use and trade of NTFPs in Brazil   44
B.1   Amount paid in informal taxes by traders (for a car of about
       1.5 tonnes) from Lékié Division to Idenau market, Cameroon   74
3.1   NTFP categories and relevant legislation   93
3.2   Percentage share of collectors in final KL auction price   94
3.3   Collector share in final sale price of tendu leaf in Madhya Pradesh   96
3.4   Prices offered to collectors for select NTFPs and LAMPS margins   100
3.5   Revenues, margins and profit/loss for all LAMPS
       in Karnataka, 1994–95   101
6.1   Major NTFPs in the Philippines and their uses   158
6.2   Timeline: NTFP utilization and ancestral domain legislation
       in the Philippines   160
6.3   Institutional aspects of rattan operations   164
6.4   Rattan value chain: Custody and benefits   165
6.5   Almaciga resin prices   170
6.6   Unofficial payments in the NTFP Trade (for checkpoints only)   174
8.1   Examples of important NTFPs for market and subsistence
       economies in Mexico   207
8.2   Legal procedures for extraction of NTFPs in Mexico   210
8.3   Use of wild mushrooms in Mexico   211
8.4   Role of local institutions in regulating the agave-mezcal production
       chain in Guerrero   217
8.5   Appendix: Barriers within the national legal framework   226
11.1A Common and scientific names of plants mentioned in the text   286
13.1   Key laws and policies pertaining to the use, trade and conservation
       of *Hoodia* in South Africa, Namibia and Botswana   315
14.1   Estimated value of select NTFPs in international trade   328
14.2   The 12 leading countries for imports and exports of the commodity
       group 'pharmaceutical plants' (SITC.3: 292.4 = commodity group
       HS 1211), average annual trade volumes, 1991–2003   329

BOXES

A.1   Legislation of NTFPs: The case of Acre   47
A.2   Palm heart and Normative Instruction 05 of 1999: The challenges
       of accommodating different species, extracting practices and
       commercial demand through legislation   50
2.1   Forestry laws in Cameroon   57
2.2   Special Forest Products regulated by the 1994 Forestry Law   59
5.1   Mushroom gathering and the 'right of user'   141
5.2   Bluebells   143
5.3   The Scottish Wild Mushroom Code   149
6.1   Draft administrative order for NTFP utilization in ancestral domains 176
6.2   Sustainable harvest of wild honey in Bukidnon: The role of
       customary law and resource management   178
D.1   Carvers' and furniture makers' comments on the harvesting system   202
8.1   Wild mushrooms in Mexico   210
11.1   Federal and Washington State labour law requirements   277
11.2   Excerpt from a permit for harvesting on a brush shed lease   279
11.3   Five-step checklist for determining whether a brush shed
       is an employer   279
11.4   Comparison of floral greens and wild mushroom industries   281
12.1   The rise of the cep (*Boletus* spp.) export industry in Finland   289
12.2   Finnish participation rates in wild berry harvesting   289
12.3   Wild berry yields in Finland, Sweden and Russia   290
12.4   Everyman's rights restrictions on cloudberry harvesting in Lapland 295
12.5   Everyman's rights do not apply to lichen harvesting   296
13.1   Overview of the ecology of *Hoodia* spp.   311
13.2   *Hoodia* permit requirements in Namibia   319
14.1   The importance of tracking and statistics for NTFP policy   337

List of Contributors

Arlynn C. Aquino has 17 years of programme management experience in areas of rural economic development, natural resource economics, agriculture, microfinance, social research, policy advocacy, capacity building especially for the rural sector, and gender awareness in the Philippines, Cambodia and Indonesia. She is currently the Country Program Advisor (Household Economic Security) for Plan International-Philippines. She holds an MSc in natural resource economics from The University of Queensland and an MBA from Ateneo De Manila University.

Francis Areki studied at the University of the South Pacific in Suva. He studied trade and development issues related to use of *Intsia bijuga* (vesi) for several years while working for the WWF South Pacific Programme, based in Suva. He has family connections to the Lau group of islands and continues his commitment to environment and development in that remote part of Fiji.

Yasmin D. Arquiza is the founding director of *Bandillo ng Palawan*, a multi-award-winning environmental weekly based in Puerto Princesa City, Philippines. An environmental journalist, three of her articles have won prizes from the Developing Asia Journalism Awards, Jaime V. Ongpin Awards for Investigative Journalism, and Louie V. Prieto Journalism Awards. Ms Arquiza graduated from the University of the Philippines in Diliman and obtained her Master's degree in development management from the Asian Institute of Management in 2003. She received a Reuters Foundation fellowship grant at the Oxford University in 2005 and the US-Asia Environmental Partnership grant from the Asia Foundation in 1995.

Stella Asaha is a founder of Forests, Resources and People (FOREP), an NGO devoted to improving livelihoods and promoting sustainable natural resource management in Cameroon. She was a research assistant at the Limbe Botanic Gardens before accepting the post of Social Research Officer with the African Rattan Research Programme, a DFID-funded project which ran 2000–2004 in Cameroon, Nigeria and Ghana. Asaha has been Assistant Coordinator at FOREP since 2004. She holds a BSc (Hons) in Botany from the University of Calabar, Nigeria.

Abdon Awono is a research officer working with the Center for International Forestry Research (CIFOR) in the Central Africa Regional Office. He specializes in non-timber forest products markets, forest policy and farmer enterprise development. He holds a Master's degree in political science and a Bachelor's degree in Business law from the University of Yaoundé II in Cameroon.

Michael Balinga began his career as a research assistant with the African Rattan Research Programme. He then helped set up and coordinate the NGO Forests, Resources and People until 2007, when he joined CIFOR's Forests and Livelihoods Programme as a Regional Scientist based in Conakry-Guinea. He also worked as a researcher and consultant to the Smithsonian Institution and the Wildlife Conservation Society in Cameroon and much of Central Africa. He has a degree in Forestry Engineering from the University of Dschang in Cameroon.

Tim Brigham, MA, is the Coordinator, Capacity Building and Extension at the Centre for Non-Timber Resources at Royal Roads University. He specializes in consulting and training on NTFP business development for the benefit of rural communities.

Markus Bürgener is a Senior Programme Officer with TRAFFIC, the wildlife trade monitoring organization, which focuses on the trade in wild plants and animals. He has been based with the East/Southern African regional programme of TRAFFIC for the past ten years and has worked on national, regional and international policy and legislation related to biodiversity conservation. Markus has also conducted research into and carried out advocacy and training initiatives on many species and broad resource issues concerning wildlife trade. Markus is a qualified attorney and has a Master's degree in International Environmental Law.

José Humberto Chaves is the general coordinator for the Management of Forest Resources and Deputy Director of Sustainable Use of Biodiversity and Forests of IBAMA in Brazil. He has worked in the Directorate of Forests at IBAMA since 2001, where he held the position of Coordinator for Forest Monitoring and Control and coordinated the development of the Electronic Forest Control System. He has published texts on forest restoration, the forest sector and forest control mechanisms. He has a degree in Forest Engineering and an MSc in Forestry from the Federal University of Viçosa-UFV.

Wendy Cocksedge, MSc, is the Coordinator for Research and Extension at the Centre for Non-Timber Resources at Royal Roads University. She specializes in NTFP resource management, focusing on ethnobotanical participatory research with rural communities.

Peter Cronkleton is an anthropologist with the Center for International Forestry Research (CIFOR). Dr Cronkleton is a specialist in community forestry development, forest social movements and participatory approaches for research and capacity building. Currently based in Bolivia, he has worked as a researcher and development practitioner in Latin America for more than 15 years. A graduate of the University of Florida (MA 1993, PhD 1998) he has recently focused on institutional change in forest communities during periods of policy change.

A. B. (Tony) Cunningham works for People and Plants International and is an Adjunct Professor at the School for Environmental Research (SER), Charles Darwin University, Australia. He has worked in the interdisciplinary field of plant use by people for 30 years, mainly in Africa and Asia.

Alison Dyke is an independent researcher, consultant, activist and gatherer who over the last ten years has been working to gain greater understanding and recognition in policy for the importance of wild harvests in the lives of Scottish people. Her recent work has focused on bringing together groups of gatherers, scientists and government agency staff to share knowledge and understanding. One objective of this is to develop guidance on sustainable harvesting practice that is informed by the perspectives of all involved.

Fabrice Edouard is a French agro-economist and founder of the Mexican NGO Methodus and the RAISES network. He works with farmer organizations on the integration of NTFP chains of production in tropical and temperate ecosystems in Mexico, particularly Oaxaca. He also studies the factors that contribute to the commercial success of NTFPs in Mexico.

Marla R. Emery, PhD, is Research Geographer with the United States Forest Service, where her research focuses on contemporary NTFP use in the global North. She is particularly interested in documenting and theorizing the role of NTFPs in the lives and livelihoods of the people who gather them. Dr Emery led the team reporting on subsistence uses of US forests for the Montréal Process. Recently she has conducted research on women's uses of wild plants in a Maya village (Mexico) and wild harvests in Scottish woodlands.

Augusto B. Gatmaytan is an anthropologist, lawyer and advocate of indigenous peoples' rights, who has been working with Philippine indigenous peoples since 1985. He has been working more intensively in the Agusan del Sur province, north-eastern Mindanao since 1993, with particular focus on state-indigenous peoples' relations, as it affects issues of tenure, resource management and constructions of ethnicity. He is a co-founder and board-member of the Legal Rights and Natural Resources Center (FOE-Philippines), a Darrell Posey Fellow, and is currently undertaking research for a doctorate with the London School of Economics and Political Science, through a scholarship of the Ford Foundation.

Maria Cristina S. Guerrero has worked on NTFPs and indigenous peoples concerns for over 15 years. Themes of focus have been community enterprise development, policy advocacy and recently traditional ecological knowledge and participatory resource monitoring. Her training is in business management and international development with long-term field experience on the island of Palawan in the Philippines. She is currently the Deputy Director of the Non-Timber Forest Products Exchange Programme for South and Southeast Asia.

Tom Hobby holds an MA in Economics from the University of Victoria, British Columbia, an MSc in Agriculture and an MBA from California Polytechnic State University, San Luis Obispo, California. His academic research work has been focused primarily in British Columbia where he has researched many aspects of NTFP commercialization and the associated economic aspects of these emerging markets.

Antônio Carlos Hummel, a forest engineer with a postgraduate certificate in Environmental Law and an MSc in Tropical Forestry, is currently Director of Sustainable Use of Biodiversity and Forests of IBAMA in Brazil. He previously worked at IBAMA in the State of Amazonas and coordinated the ProManejo project. He has published more than 20 texts on the environment, especially on issues of the Amazon forest.

Catarina Illsley Granich, biologist, is coordinator of the Peasant Management of Natural Resources programme for the Group for Environmental Studies in Mexico. She has over 20 years' experience working on different aspects of traditional knowledge, local governance, NTFP and land management in several regions of Mexico. For the past 15 years she has worked in the NGO sector developing and implementing participatory methods for sustainable community management of NTFPs, water and land.

Verina Ingram has worked in the Central Africa office of CIFOR since 2008 researching the relations between forests and livelihoods. She is also currently working on her PhD on the governance aspects of sustainable livelihoods from NTFPs with the University of Amsterdam. Prior to this she worked for four years with the Netherlands Development Organization in Cameroon on natural resources and poverty reduction. She has a BSc in Management Sciences, and two Masters, in Human Resources and Environmental Management.

Okon Isoni is a professional forester working for the Cross River State Forestry Commission in Nigeria. His interests are related to illegal logging and NTFP promotion and development. He is a graduate of the University of Calabar, Nigeria.

Marina Pinheiro Klüppel, a biologist with a postgraduate certificate in Medicinal Plants and an MSc in Tropical Biology from the National Institute for Amazonian Research (INPA), joined the Brazilian Institute of Environment and Renewable Natural Resources (IBAMA) in 2002, where she worked in the management of protected areas and use of natural resources. In 2007 IBAMA became part of the Chico Mendes Institute for Conservation of Biodiversity (ICMBio), where she works in the Directorate of Biodiversity Conservation on the conservation of Brazilian plant diversity.

Sarah A. Laird is the Director of People and Plants International. Her work focuses on NTFPs used for subsistence, and those found in local and international trade. Her work in Cameroon includes research on indigenous use and management of NTFPs around Mt Cameroon, and laws and policies relating to the commercial use of biodiversity. She has also worked for the last 20 years on legal frameworks relating to the commercial use of genetic resources, including those growing from the Convention on Biological Diversity. Sarah holds an MSc in Forestry from Oxford University.

Sharachchandra Lele, Senior Fellow at the Centre for Environment and Development, Ashoka Trust for Research in Ecology and the Environment, Bangalore, has a PhD in Energy and Resources from the University of California, Berkeley. His research interests include conceptual issues in sustainable development and sustainability, and analyses of institutional, economic, ecological and technological issues in forest, energy and water resource management. He has a strong interest in understanding and facilitating interdisciplinary research and teaching on the environment.

Chun Li is Executive Director of the Kunming Division of China's CITES (Convention on International Trade in Endangered Species) Administration Authority. He has an MSc in Forest Resources Management from the University of the Philippines at Los Baños and has spent over 20 years working on issues concerning wildlife conservation and CITES enforcement in Yunnan Province. He is the author of the Handbook of Import and Export of Endangered Species of Fauna and Flora (China Forestry Press, 2003), which is widely used as a reference work by the customs authorities and forest police.

Estherine Lisinge Fotabong is an environmental lawyer and policy analyst and is the Head of the Environment Sector at the NEPAD Secretariat. Prior to this she served in various capacities including UNEP Country Programme Coordinator for South Africa. Her main interest of work has been in understanding the links between international policy processes and domestic policy/regulatory development, and implementation in the field of environment and sustainable use of natural resources, with a particular interest in institutional capacity and the implementation and applicability of laws and policies

Kathryn Lynch is co-coordinator of the Environmental Leadership Program at the University of Oregon. Her research has focused on the relationships between the conservation of biodiversity, the cultural uses of wild plants, and community-based natural resource management. Other research interests include critical pedagogy, engaged environmental education and how gender impacts natural resource issues. She holds a PhD in Environmental Anthropology from the University of Florida.

Rebecca J. McLain is a senior social scientist at the Institute for Culture and Ecology. Her work includes articles on wild mushroom and floral greens policy, a country-wide assessment of the US Forest Service's non-timber forest management programme, a review of the literature examining the links between informal economic activity and natural resource policy, and two books on nontimber policy in the US. She is currently co-leading a collaborative effort to document gathering practices and their social and environmental impacts in urban ecosystems in the USA.

Nicholas K. Menzies (Nick) is Executive Director of the Asia Institute at the University of California. He has a PhD in Wildland Resource Science from the University of California at Berkeley and has spent some 20 years working on issues concerning communities and forest resources management in China, Southeast Asia and East Africa. He is the author of *Our Forest, Your Ecosystem, Their Timber: Communities and the State in Community-Based Forest Management* (Columbia University Press, 2007), which includes case studies from China, India, Zanzibar and Brazil.

Darcy A. Mitchell established the Centre for Non-Timber Resources at Royal Roads University in Victoria, British Columbia, Canada in 2004. Since retiring from the position of Director, she remains active in research, curriculum development and teaching, with a focus on comparative international studies in NTFPs and services. She holds a Master's and PhD in public administration, an MA in political science and a BA in sociology.

Ousseynou Ndoye is an agricultural and resource economist currently working with the Food and Agriculture Organization of the United Nations (FAO) in Yaoundé, Cameroon. He specializes in NTFP markets, forest-based enterprise development, gender and non-timber forest products, and forest policy. He has a Licence and Maîtrise in Economics from the University of Dakar, an Msc from Kansas State University and a PhD from Michigan State University in the USA.

Robert Nkuinkeu is an ethnobotanist who has worked for more than 15 years in the medicinal plant industry including for Plantecam (Cameroon branch of the pharmaceutical firm Fournier – France) and Cexpro (business partner of Starlight Products – France). He helped establish a reference herbarium for medicinal and other useful plants. He is currently the executive director of World Botanical Exchange and Services. Robert has a BSc degree in botany from Yaoundé University and is finishing his MSc studies in seed technology at the University of Yaoundé 1.

Dario Novellino is an affiliate of Peoples and Plants International (PPI) and a Researcher at the Centre for Biocultural Diversity at the University of Kent at Canterbury (UK). Dario received his Master's in social anthropology from the School of Oriental and African Studies (University of London) and his doctorate in environmental anthropology from the University of Kent. Since 1986 Southeast Asia has become the focus of his research activities, and he has spent a total period of seven years with the Batak and Palawan communities.

Pablo Pacheco holds a PhD from the Graduate School of Geography at Clark University, MA, USA. He is a senior researcher at the Center for International Forestry Research (CIFOR), Bogor, Indonesia in the Forests and Livelihoods Programme. His work focuses primarily on the Amazon. His interests embrace issues related to the implications of public policy and regional development on natural resources management, human dimensions of land-use/cover change, rural development and people livelihoods, and institutional change and decentralization.

Manoj Pattananaik has been working in the areas of community-based natural resource management with a focus on community forestry for more than 15 years. Manoj has designed and implemented several programmes and projects on poverty reduction and sustainable livelihood using natural resources. He has worked with different models of community-based natural resource management and has been actively engaged in policy research and advocacy on different development and environmental issues.

Alan Pierce is a consultant specializing in NTFP research and forest policy analysis. Pierce has over 15 years' experience working with environmental NGOs and is co-editor of *Tapping the Green Market: Certification and Management of Non-Timber Forest Products* (2002). He is currently a PhD candidate at Antioch New England University.

Silvia E. Purata is an ecologist working for People and Plants International. She supports communities to manage their forest resources sustainably, in order to conserve Mexico's rich cultural and biological diversity. She writes resource-use manuals directed to peasants and forest technicians, and collaborates with NGOs to organize training workshops. She has also been involved in certification of good forest management under Forest Stewardship Council (FSC) standards.

Nitin D. Rai is a Fellow at the Ashoka Trust for Research in Ecology and the Environment (ATREE), Bangalore, India. He received a PhD from Pennsylvania State University for his study on the social, economic and ecological aspects of NTFP harvest in a wet tropical forest in India. His current interests include interdisciplinary approaches for decentralized biodiversity conservation and the political ecology of protected areas.

Júlio César Raposo Ferreira has been working in the state of Acre, Brazil since 2000 on non-timber forest products (NTFP) management and is participating in the development of state standards for NTFPs. Due to his wide experience in various areas of the forest sector in Acre, he has become a resource on NTFPs at IBAMA (Brazilian Institute of Environment and Renewable Natural Resources) in Acre, where he has been working since 2005 on sustainable use of forest resources. He has a degree in Forest Engineering from the Universidade Estadual Paulista – UNESP (Botucatu-SP).

Rebecca T. Richards is Professor of Sociology at the University of Montana in Missoula where she teaches environmental and rural sociology courses. She has conducted field research on wild mushroom and wild berry harvesting in the Pacific Northwest since 1993. Her collaborative research on wild berry harvesting in Finland with Olli Saastamoinen and colleagues at the University of Joensuu was supported by a Fulbright Senior Scholar award in 2005.

Olli Saastamoinen is a Professor of Forest Economics at the Faculty of Forest Sciences, University of Joensuu, Finland. His major research interest has been economics of multiple-use forestry, and he has been involved in wild berry studies since 1970s. Other research interests include forest policy and governance. He is also a coordinator of IUFRO unit 6.05. on forest ethics.

Maria Fernanda Sánchez-Pardo is an environmental lawyer with ENLACE legislación, ambiente y sociedad S.C. in Guanajuato, Mexico. She conducts legal research and capacity building in environmental law and public policy and specializes in NTFPs and sustainable community management.

Sheona Shackleton is a lecturer in the Department of Environmental Science at Rhodes University, South Africa and a Research Associate at the Center for International Forestry Research, Indonesia. Her areas of research specialization include: natural resource utilization and valuation in communal tenure systems; theoretical, policy and practical aspects of community-based natural resource management; community forestry and woodland management; rural livelihood systems and commercialization of wild resources/NTFPs.

Terry Sunderland is a Senior Scientist with CIFOR's Forests and Livelihoods programme, and leads the research domain 'Managing trade-offs between conservation and development at landscape scale'. Prior to joining CIFOR Terry was based in central Africa for many years and worked there for the UK's Department for International Development, University College London, and more recently the Wildlife Conservation Society. He has extensive consultancy experience with various international organizations, including the United States Forest Service, GTZ and the Smithsonian Institution. Sunderland holds a PhD from University College London.

Sinclair Tedder has been an economist with the British Columbia Ministry of Forests and Range for 15 years. For the past ten years he has participated in a variety of projects, publications and conferences related to NTFPs, mainly in the areas of property rights and commercial production. Sinclair is also a PhD Candidate in the Department of Forest Resources Management, Faculty of Forestry at the University of BC, where he is undertaking a study of institutional change associated with common pool resources.

Citlali Tovar is an environmental lawyer with ENLACE legislación, ambiente y sociedad S.C. in Guanajuato, Mexico. She conducts legal research and capacity building in environmental law and public policy and specializes in NTFPs and sustainable community management.

Rachel Wynberg, a natural scientist and environmental policy analyst, specializes in the commercialization and trade of biodiversity, and the integration of social justice into biodiversity concerns. She is based at the Environmental Evaluation Unit, University of Cape Town, South Africa and holds a PhD from the University of Strathclyde focused on pro-poor approaches for NTFP commercialization. Over the past 15 years Rachel has worked closely with governments and NGOs, focused in particular on southern Africa, to formulate appropriate policy frameworks for biodiversity conservation; NTFPs; access and benefit-sharing; intellectual property rights and traditional knowledge; and community-based natural resource management. She is a trustee and founding member of two South African NGOs: the Environmental Monitoring Group and Biowatch South Africa.

Acknowledgements

Many individuals contributed insight and support to this process, helping to shape the book as a comprehensive resource for policy-makers, academics and practitioners. In particular, we would like to thank Patricia Shanley, Sam Johnston and Miguel Alexiades for their guidance, review and support throughout. Others offered important comments and analysis during the process, including the many authors who contributed to this book, and in particular Alan Pierce, Marla Emery and Darcy Mitchell.

Excellent editing of the manuscript and pieces along the way was provided by Paul Wise. Leilan Greer also provided editorial assistance at various points in the process, which is gratefully acknowledged. At People and Plants International, Elizabeth Skinner expertly and patiently supported the authors in compiling the book, while Andre Almeida assisted with translation. Fahdelah Hartley provided ever-efficient secretarial support at the University of Cape Town.

Financial and institutional support was provided by the project partners: People and Plants International, the Environmental Evaluation Unit (University of Cape Town), the Institute for Culture and Ecology, the United Nations University, and the Center for International Forestry Research. The Christensen Fund also provided invaluable support to the research and writing process, as did the Northern Territory Government, Australia. Additional financial support was provided to Rachel Wynberg by South Africa's National Research Foundation (NRF) although any opinion, findings and conclusions or recommendations expressed in this book are those of the authors and the NRF does not accept any liability for them.

Finally, we thank our partners and families for their endurance of intercontinental Skype calls at all hours, and their support throughout.

Foreword

Products other than timber and fibre have always constituted a large part of the overall economic outputs of forests. But until quite recently most attracted relatively little attention; often being referred to as 'minor' forest products – reflecting prevailing perceptions that they were largely low value goods primarily consumed locally or confined to limited market niches.

As more information has accumulated, it has become apparent that such products and trading activities are much more important and prevalent than had been recognized previously. Wild products feature prominently in the health and nutrition of most rural populations – adding balance to diets and fuels with which to cook food, and products that help maintain or restore health. They also provide sources of the income that even the poorest need in order to keep subsistence livelihoods from deteriorating further, fill seasonal gaps in food supply or income flows, and provide 'safety nets' in periods of shortage or more extreme poverty. In addition, they are often important in providing the basis for local production and trading activities that can generate income on a scale that can help people to escape from subsistence and poverty. Many non-timber forest product production and trading activities have low capital and skill entry thresholds, and widely constitute one of the largest components of the non-farm part of the rural economy.

However, using a forest resource to generate income can conflict with the needs of those who still depend upon it to meet their subsistence needs. Harvesting of the tradable components may so change the composition of a resource as to deplete or remove products used locally. Or control over local resources containing tradable products may be captured by local elites or outsiders engaged in commercial production, so depriving gatherers of access to products they need for their own use. Growth in non-timber forest product production and use has consequently frequently featured problems arising from lack of or ineffective regulation and control, or poor management of the resource or the trade, or both.

The growing use of non-timber forest products can also raise other issues requiring increased attention to regulation and control. Harvesting these products may need to be consistent with a primary objective of producing timber. Or it may be necessary to ensure that they are not harvested unsustainably, or on a scale that adversely affects ecological or environmental values and functions of the forest, or that their use is consistent with other uses and claims on the lands where they occur. There can also be important issues of quality and composition of the product that need to be monitored and controlled, notably where it is to be eaten or used for medicinal purposes, or needs to meet industrial or international trade standards. Where production of a product for sale is on public land, issues are likely to arise relating to fees and taxes to be paid, and about their distribution among the different agencies involved.

Effective governance is thus likely to be important at several different levels, and in a number of different ways, throughout the management, harvesting, trade and use of most non-timber forest products. However, by comparison with the recent attention paid to other aspects of these products, the governance dimension has received relatively little coverage. Though a number of countries have by now taken steps to set in place more relevant policies and regulations governing production and trade of such forest products, there is very little that has attempted to draw this information together and assess what common patterns emerge, or what lessons of broader application might be learned.

This publication is intended to help address this gap. An introductory review of the literature is followed by 15 chapters and case studies, each of which provides in-depth description and analysis of particular non-timber forest product situations. In the concluding two chapters, the editors first identify a number of features that seem to be common to policies, regulations and governance in most of the case situations examined; and then draw upon this analysis to provide a wide-ranging set of recommendations that could prove valuable to those concerned with understanding and improving governance of non-timber forest product situations. The result is a compendium that adds substantially to the existing literature on this subject. By bringing together the experiences and analysis of many of those who, as researchers or managers, have been at the forefront of research into governance of non-timber forest products, it creates a valuable source of information about what is known. At the same time it provides practical guidance to those seeking to improve governance in this area, and identifies key areas where further research and experimentation could be needed.

J. E. Michael Arnold
April 2010

List of Acronyms and Abbreviations

AAC	Annual Allowable Cut
ADMP	Ancestral Domain Management Plan
ADSDPP	Ancestral Domain Sustainable Development and Protection Plan
AHPA	American Herbal Products Association
AMCS	Agency Marketing Cooperative Society
ANAFOR	National Forest Development Agency
AP	Andhra Pradesh
ATPF	Autorização de Transporte de Produtos Florestais (Transport of Forest Products Licence, Brazil)
BC	British Columbia
BETP	Bureau of Export Trade Promotions
BIND	Broad Initiatives for Negros Development
BOLICERT	Boliviana de Certificación, the main organic certification body in Bolivia
CADC	Certificate of Ancestral Domain Claim
CADT	Certificate of Ancestral Domain Title
CBD	Convention for Biological Diversity
CBFMA	community-based forest management agreement
CBSRM	Community Based Sustainable Resource Management
CCARC	Comisión de Conciliación, Arbitraje y Resolución de Conflictos (Commission for Reconciliation, Arbitration and Resolution of Conflicts)
CDC	Cameroon Development Corporation
CEMAC	Economic and Monetary Community of Central Africa
CFA	community forest agreement
CFIF	Cebu Furniture Industries Foundation
CFS	Canadian Forest Service
CFV	Consejo Boliviano para la Certificación Forestal Voluntaria (Council for Voluntary Forest Certification)
CIFOR	Center for International Forestry Research
CITES	Convention on International Trade in Endangered Species of Wild Flora and Fauna
CNTR	Centre for Non-Timber Resources
COINACAPA	Cooperativa Integral Agroextractivistas Campesinos de Pando (Integral Agro-extractive Farmers' Cooperative of Pando, Bolivia)
COMERCAM	Mexican Regulatory Council for the Quality of Mezcal
CONABIO	Mexican National Commission for the Knowledge and Use of Biodiversity
CONAFOR	Comisión Nacional Forestal (National Forestry Commission)
CPR	Common-pool resource

CSIR	Council for Scientific and Industrial Research (S. Africa)
DAO 2	DENR Administrative Order No. 2, 1993
DENR	Department of Environment and Natural Resources
DFPCU	District Forest Produce Cooperative Union
DLI	Washington State Department of Labor and Industries
DNR	Department of Natural Resources, Washington State
DO	Denominaton of Origin (Mexico)
DOF	Documento de Origem Florestal (Document of Forest Origin, Brazil)
DPT	Directorate of Promotion and Transformation of Forest Products (Cameroon)
DWAF	Department of Water Affairs and Forestry
ECAN	Environmentally Critical Areas Network
EU	European Union
FDA	US Food and Drug Administration
FES	Función económica-social (economic and social function)
FIDA	Fiber Industry Development Authority
FRPA	Forest and Range Practices Act
FSC	Forest Stewardship Council
FSP	forest stewardship plan
FTC	US Federal Trade Commission
GDP	gross domestic product
GEA	Grupo de Estudios Ambientales, AC (Mexican NGO)
GI	Geographical Indication (Mexico)
HCA	hydroxycitric acid
HOGRAN	*Hoodia* Growers Association of Namibia
IBAMA	Instituto Brasileiro do Meio Ambiente e dos Recursos Naturais Renováveis (Brazilian Institute for the Environment and Renewable Natural Resources)
ICC	indigenous cultural community
ICDP	integrated conservation and development project
IMAC	Instituto do Meio Ambiente do Acre (Environmental Institute of Acre, Brazil)
IMPI	Instituto Mexicano de la Propiedad Industrial (Mexican Institute for Industrial Property)
INRA	Instituto Nacional de Reforma Agraria (National Institute for Agrarian Reform)
IP	Indigenous people
IPR	Intellectual Property Rights
IPRA	Indigenous Peoples Rights Act
ITTO	International Tropical Timber Organization
IUCN	International Union for Conservation of Nature
JFM	Joint Forest Management
KFD	Karnataka Forest Department
KL	Kendu Leaf (country cigar wrappers)
LAMPS	Large Area Multi-Purpose Society/Large-Scale Adivasi Multi-Purpose Society

LGEEPA	Ley General del Equilibrio Ecológico y Protección al Ambiente (General Law for Ecological Balance and Environmental Protection)
LRMP	Land and Resource Management Plan
MFP	Minor Forest Produce
MINEPIA	Ministry of Livestock Fisheries and Animal Husbandary
MINFOF	Ministry of Forestry and Wildlife (Cameroon)
MOEF	Ministry of Environment and Forests
MP	Madhya Pradesh
NAFTA	North American Free Trade Agreement
NATRIPAL	Nagkakaisang Mga Tribu ng Palawan (Indigenous Peoples Federation of Palawan, Philippines)
NCIP	National Commission on Indigenous Peoples
NFPP	Natural Forest Protection Programme
NGO	Non-Governmental Organization
NIPAS	National Integrated Protected Areas System
NIPAS Act	National Integrated Protected Areas System Act
NLTB	Native Lands Trust Board
NOM	Norma Oficial Mexicana (Mexican Official Norm)
NRC	Natural Resources Canada
NTFP	non-timber forest product
NTFP-EP	Non-Timber Forest Products Exchange Programme
NTFP-TF	Non-Timber Forest Products Task Force
NWFP	non-wood forest product
OFDC	Orissa Forest Development Corporation
OGMA	old growth management area
OM	Ordinary Minor (a type of forest licence)
PAWB	Protected Areas and Wildlife Bureau
PCS	Primary Cooperative Society
PCSD	Palawan Council for Sustainable Development
PENRO	Provincial Environment and Natural Resources Office
PESA	Provisions of Panchayat (Extension to Scheduled Areas) Act, 1996
PFPCS	Primary Forest Produce Cooperative Society
PROMAB	Programa Manejo de Bosques de la Amazonía Boliviana) (Programme of Forest Management in the Bolivian Amazon)
RCDC	Regional Centre for Development Cooperation
RCW	Revised Code of Washington State
REDD	Reduce Emissions from Deforestation and Forest Degradation
SAHGA	Southern African Hoodia Growers Association
SAMAKA	Samahan sa Maoyon ng mga Katutubo (Association of Indigenous people in Maoyon, the Philippines)
SE	Secretaría de Economía (Ministry of Economy)
SEMARNAT	Secretaría de Medio Ambiente y Recursos Naturales, Mexico (Federal Ministry of the Environment and Natural Resources)
SEP	Strategic Environmental Plan
SMFPCF	State Minor Forest Produce (Trading & Development) Co-operative Federation

SOP	standard operating procedure
SSSI	sites of special scientific interest
STAR	Science to Achieve Results
TDCC	Tribal Development Cooperative Corporation
TRIFED	Tribal Cooperative Marketing Development Federation
TRIPS	Trade-Related Aspects of Intellectual Property Rights
UFPL	Utkal Forest Products Ltd.
UK	United Kingdom
UN	United Nations
USDA	US Department of Agriculture
VASS-NZAID	Voluntary Agency Support Scheme, New Zealand Agency for International Development
WHA	wildlife habitat areas
WIMSA	Working Group of Indigenous Minorities in Southern Africa
WTO	World Trade Organization
WWF	Worldwide Fund for Nature

Introduction
Wild Product Governance

Sarah A. Laird, Rebecca J. McLain and Rachel P. Wynberg

People have long developed and depended upon useful species from diverse ecosystems. Even today, wild products provide critical subsistence and trade goods for forest and other communities, as the chapters in this book and many studies undertaken in the last few decades attest.[1] In many areas, NTFPs are the main source of cash to pay school fees, buy medicines, purchase equipment and supplies, and even buy food.

However, wild products – or non-timber forest products (NTFPs) – have been both overlooked and poorly regulated by governments. As the following chapters describe, with a few notable exceptions NTFP measures instituted in recent decades were tagged onto timber-centric forestry laws, were neither strategic nor well-informed, and inadequate resources were allocated for oversight and implementation. Regulations rarely followed from careful analysis of the complex factors involved in NTFP management, use and trade, or from consultations with producers, who are often on the political and economic margins. These and other experiences are remarkably similar around the world.

In the end NTFP law and policy has often created new opportunities for corruption and exploitation and, in conjunction with other bodies of law like agriculture and land tenure, provided perverse incentives to over harvest NTFPs. In many cases policy interventions also criminalized NTFP extraction, further marginalizing harvesters while generating new forms of inequity (Alexiades and Shanley, 2005). Customary law and local institutions better suited to regulating many species were also often undermined by efforts to establish statutory control over NTFPs (Arnold and Ruiz-Pérez, 2001; Michon, 2005).

With greater information, effective consultations with stakeholders, and strategic approaches to policy-making, however, laws and policies can promote ecological sustainability, equity in trade, and improved rural livelihoods. Many governments today are revisiting NTFP laws and policies, and this book was produced in order to support and assist this important and timely process.

NTFPS DESCEND INTO AND EMERGE FROM 'INVISIBILITY'

Over the course of the past century, the meaning of the term 'forest products' has been narrowed to the point where it essentially has come to include only timber and wood fibres harvested on an industrial scale for use in the manufacture of products such as lumber, paper, cardboard and particle board. This has occurred even in regions where commercial NTFPs such as rattans, medicinal plants and wild foods are far more valuable than the 'forest products' that have traditionally fallen within the scope of this limited definition. Also entirely lost from this view are the substantial subsistence values of plants used for medicine, food, building materials, grazing and dozens of other purposes (for example, see Anderson et al, 2000; Neumann and Hirsch, 2000; Cunningham, 2001; Campbell and Luckert, 2002; Shanley et al, 2002; Carroll et al, 2003; Shackleton and Shackleton, 2004; Alexiades and Shanley, 2005; Emery and Pierce, 2005; McLain, 2008).

As forest products became equated in the minds of natural resource professionals and policy-makers with industrial forms of wood and their by-products, the majority of useful species present in forests and other ecosystems, and other values such as watershed protection and recreation, became largely invisible in natural resource management and policy in most parts of the world (Westoby, 1989; Jones and Lynch, 2007; Hurley et al, 2008). A shift began to occur in the late 1980s, however, as scientists, natural resource managers and policy-makers increasingly recognized the non-timber values of forests (e.g. McNeely, 1988; Pearce, 1991; Swanson and Barbier, 1992), including the socioeconomic and cultural importance of non-timber forest products (e.g. Peters et al, 1989; Balick and Mendelssohn, 1992; Nepstad and Schwartzman, 1992).

This shift resulted from a range of factors, including a dramatic change in the focus of conservation agencies away from a purely protectionist approach to one that also incorporated sustainable use, and viewed equity and social justice as integral to conservation. Originally articulated by the Brundtland Commission in 1987 (WCED, 1987), this view culminated in the various agreements that emerged from the 1992 United Nations Conference on Environment and Development in Rio de Janeiro, including the legally binding Convention on Biological Diversity (CBD). The CBD explicitly made the link between conservation, sustainable use and equity, as did national laws drafted to implement the CBD, and a range of protected area, forestry and conservation measures that grew from these trends in conservation thinking. In many countries, governments were also forced to accept a broader view of conservation in exchange for World Bank and other loans.

At the same time, pressure on policy-makers to recognize the land, resource, human, cultural and intellectual property rights of indigenous peoples influenced a suite of global instruments and institutions (Posey and Dutfield, 1996; Posey, 1999) including the CBD, the International Labour Organization's Convention No. 169, the United Nations Declaration on the Rights of Indigenous Peoples (UN, 2007), the United Nations Permanent Forum on Indigenous Issues and processes within the World Intellectual Property Organization and the World Trade Organization. Implementation challenges remain, and in many countries indigenous peoples' rights are often little more than window dressing (e.g. Chapters 6 and 7; Castillo and Castillo, 2009). However, national laws increasingly recognize the rights of indigenous peoples to control the use of their resources and associated traditional knowledge (e.g. Chapters 3 and 4).

As a result of these trends, small-scale producers and NTFPs have emerged from 'invisibility' in recent decades. For a time, conservation and development groups experimented with NTFP-based projects as ecologically benign and socially just income-generating activities. The commercial use of a handful of NTFPs was promoted as a way to help people live well with minimal damage to the environment (e.g. Clay, 1992; Nepstad and Schwartzman, 1992; Plotkin and Famolare, 1992; Freese, 1997). More recently, international agencies have worked to devolve, or decentralize, natural resource governance, and promote greater local participation (Chapter 3; Case Study A).

Conservation and development gains from these efforts have often proved elusive, however. This should not be surprising given the complex and multidimensional nature of forest production systems and livelihoods (Lynch and Alcorn, 1994; Arnold and Ruiz-Pérez, 1996, 2001; Clay, 1996; Neumann and Hirsch, 2000; Alexiades and Shanley, 2005). However, some have come to see NTFPs as a 'poverty trap' rather than a 'golden egg', a livelihood of last resort that locks producers into hardship (Belcher and Ruiz Pérez, 2001; Wunder, 2001; Shiel and Wunder, 2002). NTFPs continue to receive attention at the national level, however, because in many countries they are important parts of the forest economy and are central to rural livelihoods and cultures. This includes commercial species with large international markets like Brazil nuts (Chapter 1), rattan (Chapter 6), wild berries (Chapter 12), and *Hoodia* (Chapter 13); important national and regional products like *Gnetum* spp (Case study B), *Instia bijuga* (Chapter 9), and tendu (Chapter 3); and the thousands of species used around the world for subsistence and in local trade.

THE BOOK: *WILD PRODUCT GOVERNANCE*

Numerous works have been published about NTFPs over the past two decades, including descriptions of their use, harvest and/or conservation, analyses of the factors influencing successful commercialization (e.g. Neumann and Hirsch, 2000; Sunderland and Ndoye, 2004; Belcher and Kusters, 2004; Alexiades and Shanley, 2005; Marshall et al, 2006) and 'how-to' manuals for inventorying and monitoring NTFPs or measuring their economic value (e.g. Peters, 1996; Cunningham, 2001; Shanley and Medina, 2005; Stockdale, 2006). While many of these works touch on policy issues, NTFP policy is not their primary focus. Other publications have a strong focus on policy (Dewees and Scherr, 1996; Jones et al, 2002; Shanley et al, 2002; Michon, 2005; McManis, 2007; Wynberg and Laird, 2007; Cunningham et al, 2009), but tend to be either geographically or topically narrow (e.g. dealing with a single or few species or types of products).

This book complements the existing NTFP literature by providing a comparative analysis of a broad spectrum of experiences with NTFP policy and law from around the world. By including cases from post-industrial contexts as well as the more commonly studied context of developing economies, it emphasizes the truly global importance of these products, and highlights similarities in issues and lessons that emerge with NTFP regulation. The majority of cases in this book focus on species found in formal, usually regional or international, trade. Many of these also have extensive local markets and are used for subsistence. However, numerous products, situations and regions are not covered in this book, which is far from comprehensive, in part reflecting the vast and diverse nature of NTFP use.[2]

The subject of this book is the vast array of botanical materials other than industrial timber and wood fibre. Although a number of terms have long been used to describe such products (e.g. minor forest products, specialty forest products, alternative forest products), in the last decade the somewhat vague and catch-all term 'non-timber forest products' (NTFPs) has become the most widely accepted among policy-makers and natural resource managers in many parts of the world (Arnold and Ruiz-Pérez, 1996; Belcher, 2003). Despite its ambiguities, 'NTFP' is a useful term for our purposes because it emphasizes a departure from a century of professional forestry in which timber production dominated research, management and policy-making. Additionally, it is a more inclusive term than 'non-wood forest product' (NWFP), the most common alternative used in international policy arenas. NTFPs include products such as firewood, artisanal woodcarvings and Christmas trees, all of which are wood-based products, but whose use, processing and trade patterns bear little resemblance to those of industrial timber and wood fibre.

The book excludes game, fish and insects, on the grounds that policies for mobile organisms are likely to differ greatly from those regulating the use, harvest and trade of organisms rooted in place. We acknowledge, however, that these organisms are extremely important parts of complex livelihood and production systems that also include botanical NTFPs (e.g. DeFoliart, 1995; Dounias, 2000; Bennett et al, 2006), and that for other purposes such as community forest management plans, and conservation and livelihood strategies, they may fall naturally under the heading of NTFPs.

While the term 'non-timber forest products' suggests species harvested from forests alone, we interpret the term 'forest' broadly, including in our definition a variety of ecosystems ranging from densely treed 'natural' forests and managed forests to agroforests and agrosilvopastoral systems, to the southern African veld and desert, and to the Arctic tundra. Likewise the term may seem to apply only to 'wild' products, but we include products derived from species relatively unmodified by humans as well as those that humans have managed with varying degrees of intensity, but that fall short of intensive domestication.

The concept of 'governance' is currently fashionable but, as Weiss (2000) remarks, is as old as human history. Although the term has wide interpretation, it commonly refers both to the political dimension of policy formulation as well as to the 'system of rules that shape the actions of social actors' (cited by Mayntz in Treib et al, 2007, p3). In this book we embrace a wide definition of governance, recognizing that it refers not only to government regulation and law enforcement, but also to the 'political, institutional, and cultural frameworks through which diverse interests in natural and cultural resources are coordinated and controlled' (Cronkleton et al, 2008, p1). Thus, 'governance' incorporates the rules adopted to organize and manage activities to serve larger social objectives and solve conflicts between different groups; the functioning of institutions and their acceptance by the public; and the broader efficacy of government.

Policy outputs may be legally binding, such as laws, regulations and directives, or may represent softer approaches that include guidelines, policy statements or more general norms. By 'policies' we thus mean both legally enforceable instruments that are rigid and have clear sanctions, and approaches that adopt a more flexible approach to implementation and enforcement. Institutions are a vital component of governance and refer both to the formal public institutions that are typically involved in law and policy-making and implementation, as well as the myriad of so-called informal (but often very formalized) institutions and systems that exist among rural communities. In fact, as several chapters in this book demonstrate, institutions rooted in customary law often lead to extremely effective regulation of natural resources. Increasingly, such systems co-exist with statutory systems of governance in a context of legal pluralism (Griffiths, 1986).

The book includes 13 chapters and four shorter case studies drawn from specific countries or regions: Bolivia, Brazil, Cameroon, Canada, China, Fiji, Finland, India, Mexico, the Philippines, southern Africa, the United Kingdom and the United States. Then follow two synthesis chapters, one on laws and policies relating to NTFP trade, and the other providing an overview of the state of NTFP policy and law which integrates findings drawn from the preceding chapters and case studies. The final chapter is recommendations for policy-makers, NGOs, producer groups, industry and others with an interest in NTFP policy. An appendix to the book introduces an annotated bibliography on NTFP policy and law that is found on the People and Plants International website (www.peopleandplants.org).

The chapters in this book cover a range of overlapping themes and issues, and are difficult to separate into distinct categories. The similarities in experience around the world are striking, and the chapters evoke many common features of NTFP law and policy.

In Chapter 1, Peter Cronkleton and Pablo Pacheco describe how a consistent, comprehensive policy strategy has never been developed to guide the management, use and trade of Brazil nuts in Bolivia, even though this product is the primary motor for the economy of the country's northern Amazon. What laws do exist have been created in a piecemeal fashion over decades, and the policies with the greatest impact on the sector were not intended to address Brazil nut management at all, including property rights' formalization through land reform and macroeconomic policies seeking to promote non-traditional exports. The authors argue that, while Bolivia developed progressive forestry legislation in the mid-1990s, the focus was on timber production, and a strategic approach to governing NTFPs, including promotion and maintenance of the Brazil nut sector, has yet to emerge.

In Case Study A, Marina Pinheiro Klüppel, Júlio César Raposo Ferreira, José Humberto Chaves and Antônio Carlos Hummel describe recent efforts by the Brazilian government to regulate NTFPs more effectively and to correct inappropriate approaches adopted in earlier years that were lifted directly from the timber sector, including requirements for management plans. The authors also describe the increasing role of states in regulating NTFPs, including the progressive and comprehensive work of the State of Acre.

In Chapter 2, Sarah A. Laird, Verina Ingram, Abdon Awono, Ousseynou Ndoye, Terry Sunderland, Estherine Lisinge Fotabong, and Robert Nkuinkeu examine NTFP policies drafted in Cameroon, largely at the behest of foreign donors, and as part of wider revision of the country's forestry and environment laws in the 1990s. A lack of clarity and consistency have made implementing these new provisions difficult, and although skeletal institutional structures were created to address NTFPs, they have little power, few resources and limited capacity. The result is a legal environment in which the bulk of NTFPs – those used for subsistence or traded locally – remain under the control of customary law, and high-value products in national or international trade are regulated in inconsistent and confusing ways that create burdens on harvesters and traders. In Case Study B, dealing with the trade in *Gnetum* spp. in Cameroon, Ousseynou Ndoye and Abdon Awono demonstrate the very real cost of allowing bribes to serve as an informal tax on traders, and the way this erodes benefits and promotes overharvesting. In Case Study C, which focuses on the bush mango trade, Terry Sunderland, Stella Asaha, Michael Balinga and Okon Isoni describe the use and trade of this valuable NTFP in parts of Cameroon and Nigeria, some of the customary regulations in effect in local communities, and the poor knowledge of statutory laws, which remain vague and ambiguous both to communities and to most traders.

The inadequate and often negative role of the state when it seeks to control all aspects of the NTFP trade and channel revenues into government coffers has been demonstrated in the forest regions of India. In Chapter 3, Sharachchandra Lele, Manoj Pattanaik and Nitin D. Rai give an account of different policy and institutional approaches to the governance of NTFPs. India has a long history of strong state inter-vention in forestry, initiated by the British colonial government and continuing today. The rationale for continued government involvement in high-value NTFPs is to allow collectors to benefit from guaranteed prices, but in practice the government profits to a far greater extent than collectors. The government also regularly collects taxes (royalties) on the trade of NTFPs, but there is no reinvestment of these funds in the NTFP sector, or in sustainable management. Recent progressive laws have devolved control over lower-value NTFPs ('minor forest produce') to villages, but high-value NTFPs, like tendu (*Diospyros melanoxylon*) remain under state control. Government intervention in most aspects of the NTFP trade, coupled with a lack of secure resource rights, has limited communities' ability to control and benefit from the trade in NTFPs.

In Chapter 4, Darcy A. Mitchell, Sinclair Tedder, Tim Brigham, Wendy Cocksedge and Tom Hobby explore key issues associated with NTFP policies in the Canadian province of British Columbia. They describe the ways in which forestry and other policy initiatives have, largely unintentionally, created problems in the NTFP sector. When deliberate policies have been crafted for NTFPs, they have generally been created species by species, in response to sudden increases in demand. Although tenure arrangements exist under which multi-species management would be possible (e.g. treaties with First Nations and community forest agreements), at present none of these mechanisms have been widely implemented on the ground. Recent interest in NTFPs on the part of the provincial government has grown out of court decisions affirming the rights of First Nation peoples to share in benefits from, and decision-making about, land and resources, and the decline in timber-related employment due to large-scale mortality of the province's pine forests. However, the authors identify the ongoing lack of political and economic organization in the NTFP sector as a key challenge to gener-ating long-term and large-scale government support for NTFP industries.

In Chapter 5, Alison Dyke and Marla Emery identify similar problems in the United Kingdom. This chapter examines NTFP policy development in the context of a post-industrial economy that is undergoing a revival of interest in NTFP harvesting as the nation's forest cover expands and consumer demand for wild foods increases. The commercial NTFP sector, though growing, is still very small, but, as in other countries, access to NTFPs is fraught with tension. In Scotland, much of this tension is linked to recent codification of the customary 'right to roam' and questions about the rights of gatherers to harvest from land under different forms of ownership. A lack of knowledge about NTFPs on the part of land managers and large private landowners, as well as the limited participation of NTFP gatherers in policy-making, has created obstacles to the development of workable NTFP policies.

In the Philippines, a quite different legal, economic, and cultural environment produces similar problems with regards to clarity and consistency in government regulation. In Chapter 6, Yasmin D. Arquiza, Maria Cristina S. Guerrero, Augusto B. Gatmaytan, and Arlynn C. Aquino describe through the lens of two products – rattan and almaciga resin (*Agathis philippinensis*) – how rights granted to indigenous forest communities in the Philippines have not improved the benefits they receive from NTFPs, and the role inappropriate and onerous government procedures have played in this. The authors explain how overlapping laws and institutional mandates have resulted in confusing policies that leave communities uncertain as to which procedures to follow. This is compounded by unnecessarily bureaucratic requirements for detailed inventories and management plans, which remain too costly and difficult for most harvesters to implement. To overcome these constraints, harvesters resort either to higher levels of extraction or to the bribery of government personnel, which remains a major problem for NTFP enterprises in the Philippines. Throughout the value chain for both rattan and almaciga resin, high levels of inequality are the norm, and bribes, forest taxes and transportation costs make it difficult for harvesters to benefit from the trade.

Further fleshing out what might be a promising picture in the Philippines, but is in fact very bleak, Dario Novellino addresses many of these same issues in Chapter 7, but through the experience of one indigenous group, the Batak on Palawan Island. The Batak trade NTFPs to support themselves following decades of extreme pressure on their livelihoods, lands and culture. In the past century, migrants have been granted title to Batak land, logging concessions overlay their territory, and more recently traditional forms of swidden cultivation were prohibited and mining claims made on their lands. Driven into the least hospitable areas, the Batak have now been informed that these last remaining 'wild' places are biologically diverse protected areas, and that their use of the forest is to be strictly controlled. Today, the Batak are recipients of poorly formulated and bureaucratic programmes instituted by the government and non-governmental organizations (NGOs) to promote the commercial sale of NTFPs as a form of sustainable and 'traditional' land use. Novellino presents the many interlocked events and policies that over the past 100 years have made NTFPs one of the few possible income sources for the impoverished and disempowered Batak.

In Case Study D, Sheona Shackleton describes how *Pterocarpus angolensis*, commonly known as kiaat (or African teak or wild teak), is the basis of the local woodcraft industry and a mainstay for several hundred entrepreneurs. Concerns about overexploitation prompted a profusion of legislation in South Africa, at all levels of government and among a variety of different government institutions. The system has been highly problematic, relying heavily on law enforcement, excluding producers from resource management decisions, and having little regard for the ecological management of the wild resource. The ultimate effect is that producers bypass the regulations because they are so costly and complex.

Government intervention from 'on high' often has unintended consequences and leads to policies that do not function or, worse, undermine producers and local systems of governance. In Chapter 8, Catarina Illsley Granich, Silvia E. Purata, Fabrice Edouard, Maria Fernanda Sánchez Pardo and Citlali Tovar examine the extraction of wild agave by poor communities in Mexico, and the conflict between inappropriate national laws and policies on the one hand and local and state-led resource management on the other. Agave extraction has been managed for hundreds of years through local institutions, with sophisticated systems developed to regulate access to the resource, to manage it sustainably and to distribute its benefits. A unique value chain has also developed around the production of mezcal, spirits distilled from agave, and this generates much local employment. Independently of local institutions and systems, however, the national government has developed a regulatory system for the commercial use of NTFPs, including agave, that has proven largely ineffectual. At the same time, the establishment of an appellation of origin for mezcal duplicates procedures required by the Ministry of the Environment, makes the process of legal extraction difficult and expensive, and leads to a classic array of 'unintended consequences': encouraging monoculture plantations, discouraging consideration of the many species of agave, and undermining local management, production and regulation of agave.

In some cases, however, local and customary laws are not sufficient to deal with increased commercial demand for a species, or with changed circumstances, and a combination of statutory and customary law is needed. Writing about Fiji in Chapter 9, Francis Areki and Anthony B. Cunningham advocate support for and the renewed enforcement of customary laws, complemented by stronger and better-coordinated national laws for NTFPs, and in particular the woodcarving species *Intsia bijuga*. Most land in Fiji is under customary tenure, but in many areas customary law has become weak. At the same time, increased commercial pressure from the tourist trade means that *Intsia bijuga* – which is slow-growing, occurs in low densities and is subject to new technology that significantly increases harvesting rates – is now endangered. A combination of factors has led to a crisis in sustainability that many are working to address at a legal and policy level. The authors suggest that state regulation is needed to bolster weakened customary law, but resources must be allocated for implementation and building capacity within government and producer groups.

In Chapter 10, Nicholas K. Menzies and Chun Li describe a multi-tiered system of regulation developed in China's Yunnan province that encourages sustainable matsutake production in the absence of a formal national or provincial policy on NTFPs. Although there have been no systematic scientific surveys to assess the impacts of commercial harvesting, matsutake is a protected species under national law, and exports are controlled at the national level. Harvesting is allowed only with the approval of the provincial forestry and agricultural departments. Counties control the marketing networks, and village rules define rights of access and monitoring procedures. The authors argue that better conservation and equity outcomes are likely to result from a multi-tiered system of this kind, which allows each level of government to control those aspects of harvesting and marketing that it is best suited to manage. They also warn against moves to impose all-encompassing policy under one level of government.

NTFPs are not only subject to laws and events in the countries where they are found. In many parts of the world, global forces have immediate and significant impacts on the local harvest, use and trade of NTFPs. In Chapter 11, Rebecca J. McLain and Kathryn Lynch examine the evolution of the relationship between labour relations, land tenure and immigration policy in the floral greens industry in the Pacific Northwest of the United States during the 1990s and early 2000s. At the beginning of the 1990s, many floral greens harvesters participated in the industry as self-employed workers; by the end of the 1990s, a significant number were working as de facto wage labourers for a handful of economically powerful floral greens wholesale companies. A combination of factors contributed to this transformation: the development of floral greens resource allocation systems that favoured larger companies, the state's inability to control floral greens poaching, the entry of large numbers of undocumented Latino immigrants into the workforce, and an immigration policy that encouraged the development of abusive labour conditions. The authors highlight the need for NTFP managers and policy-makers to develop understandings of how policies on seemingly unrelated matters, such as immigration and labour, can affect the ecological health of NTFPs, as well as the ways in which economic costs and benefits associated with NTFP industries are distributed.

Regional and global forces have similarly created significant changes in the NTFP sector in Finland. In Chapter 12, Rebecca T. Richards and Olli Saastamoinen examine how the collapse of the Soviet Union, European Union (EU) regionalization, and globalization have combined to alter the competitive advantage of the Finnish wild berry industry, despite favourable income tax policies and resource tenure arrangements. The case is unusual in that Finland is one of the few countries to have adopted a comprehensive and proactive approach to NTFP market and policy development. Until very recently, many rural Finns participated extensively in NTFP harvesting for economic and subsistence reasons, and even today more than half the Finnish population harvests wild berries and mushrooms for recreation and home use. However, entry into the EU led to weaker farm price support policies and thereby accelerated the migration of many Finns from rural berry-producing north-eastern parts of the country to the south-western cities. This out-migration has resulted in a shortage of local labour. Wild berry companies have turned to seasonal immigrant labour, often from Eastern Europe and Thailand, to fill the gap. Additionally, with the globalization of wild berry markets, high labour costs mean Finnish companies have lost their competitive advantage, despite ample supplies of berries. The Finnish case serves as an important reminder that NTFP policies need to address labour availability as well as resource supply.

In Chapter 13, Rachel Wynberg describes the bewildering complexity of laws that have emerged to regulate the harvesting, trade, intellectual property and commercial development of *Hoodia* in southern Africa, and the way a lack of regional cooperation can have as much impact on NTFPs as regionalization. *Hoodia* is a succulent plant sold as an appetite suppressant, based on traditional knowledge of the indigenous San peoples. Not only is the plant shared across national borders, but so too is traditional knowledge about its properties. Ideally, common regional policies should govern strategic resources such as *Hoodia*, but in practice, the complexity and diversity of legal and institutional mechanisms across countries, and the multiple jurisdictions and cross-cutting nature of conservation, trade, intellectual property and benefit sharing, mean that governments have found it difficult to fully streamline policies. Some steps have been put in place by southern African countries to collaborate more strongly on *Hoodia* poaching and trade and the transport of illegally harvested material, but the more slippery political issues of benefit sharing and indigenous peoples' rights remain disconnected and incoherent between countries.

In Chapter 14, Alan Pierce and Markus Bürgener review the impacts of trade policies on NTFPs. They describe the range of laws and policies that affect NTFP harvesters, producers and manufacturers as resources move from forest to local market or shop shelf, including the importance of access rights, customary oversight and local control, and compatibility with other natural resource laws. The relevance of transport regulations, as well as those governing manufacturing, quality control and safety, is also reviewed, as are legal requirements that regulate the trade of species between countries, including tariffs, taxes and licences, as well as customs and health and sanitation inspections.

An analysis of findings from the chapters and case studies, and from the broader literature, is presented in Chapter 15. In this chapter we review how and why NTFP laws and policies are developed; the content and nature of laws and policies affecting NTFPs; policy implementation and impacts; and global and regional trends that underlie and influence NTFP law and policy. Recommendations for policy-makers, NGOs, producer groups and others working on NTFP law and policy are found in Chapter 16.

Finally, in an appendix, Alan Pierce comments on the state of the literature on NTFP law and policy. It is linked to an annotated bibliography found on the People and Plants International website (www.peopleandplants.org) that has been created as a resource for publications in this area. The literature is diverse, mixed and not always easy to access, so the bibliography is intended to help guide the first stages of research on this subject.

NOTE

1   For example, see De Beer and McDermott, 1989; Falconer, 1990; Redford and Padoch, 1992; Dounias, 1993; Alexiades, 1999; Neumann and Hirsch, 2000; Boa, 2004; Shackleton and Shackleton, 2004; McLain and Jones, 2005; Alexiades and Shanley, 2005; Wynberg and Laird, 2007.
2   For example, the fuel wood and charcoal situation in Africa has generated many years of evolving and invaluable discussions around governance of wild products (e.g. Hofstad, 1997; World Bank, 2002; SEI, 2002; Arnold et al, 2006).

REFERENCES

Alexiades, M. N. (1999) 'Ethnobotany of the Ese Eja: Plants, health and change in an Amazonian society', PhD dissertation, City University of New York

Alexiades, M. N. and Shanley, P. (2005) *Forest Products, Livelihoods and Conservation: Case Studies of Non-timber Forest Product Systems*. Volume 3 – Latin America, Center for International Forestry Research, Bogor, Indonesia

Anderson, J. A., Blahna, D. J. and Chavez, D. J. (2000) 'Fern gathering on the San Bernadino National Forest: Cultural versus commercial values among Korean and Japanese participants', *Society and Natural Resources*, vol 12, pp747–762

Arnold, J. E. M., Kohlin, G. and Persson, R. (2006) 'Wood fuels, livelihoods and policy interventions: Changing perspectives', *World Development*, vol 34, no 3, pp596–611

Arnold, J. E. M. and Ruiz-Pérez, M. (1996) *Current Issues in Non-Timber Forest Products Research*, Center for International Forestry Research, Bogor, Indonesia

Arnold, J. E. M. and Ruiz-Pérez, M. (2001) 'Can non-timber forest products match tropical forest conservation and development objectives?', *Ecological Economics*, vol 39, no 3, pp437–447

Balick, M. J. and Mendelssohn, R. (1992) 'Assessing the economic value of traditional medicines from tropical rain forests', *Conservation Biology*, vol 6, no 1, pp128–130

Belcher, B. (2003) 'What isn't an NTFP?', *International Forestry Review*, vol 5, no 2, pp161–167

Belcher, B. and Kusters, K. (eds) (2004) *Forest Products, Livelihoods and Cconservation: Case Sstudies of Non-timber Forest Product Systems*. Volume 1 – Asia. Center for International Forestry Research, Bogor, Indonesia

Belcher, B. and Ruiz-Pérez, M. (2001) 'An international comparison of forest product development: An overview, description and data requirements', Working Paper no 23, Center for International Forestry Research, Bogor, Indonesia

Bennett, E. L., Blencowe, E., Brandon, K., Brown, D., Burn, R. W., Cowlish, G., Davies, G., Dublin, H., Fa, J. E., Milner-Gulland, E. J., Robinson, J. G., Rowcliff, J. M., Underwood, F. M. and Wilke, F.M. (2006) 'Hunting for consensus: Reconciling bushmeat harvest, conservation, and development policy in West and Central Africa', *Conservation Biology*, vol 21, no 3, pp884–887

Boa, E. (2004) 'Wild edible fungi: A global overview of their use and importance to people', *Non-Wood Forest Products 17*, Food and Agriculture Organization, Rome

Campbell, B. M. and Luckert, M. K. (2002) *Uncovering the Hidden Harvest: Valuation Methods for Woodland and Forest Resources*, Earthscan, London

Carroll, M. S., Blatner, K. A. and Cohn, P. J. (2003) 'Somewhere between: Social embeddedness and the spectrum of wild edible huckleberry harvest and use', *Rural Sociology*, vol 68, no 3, pp319–342

Castillo, R. C. A. and Castillo, F. A. (2009) 'The law is not enough: Protecting indigenous peoples' rights against mining interests in the Philippines', in R. Wynberg, R. Chennells and D. Schroeder (eds) *Indigenous Peoples, Consent and Benefit Sharing: Learning from the San-Hoodia Case*, Springer, Berlin

Clay, J. (1992) 'Some general principles and strategies for developing markets in North America and Europe for non-timber forest products', in M. Plotkin and L. Famolare (eds) *Sustainable Harvest and Marketing of Rainforest Products*, Island Press, Washington, DC

Clay, J. (1996) *Generating Income and Conserving Resources: Twenty Lessons from the Field*, World Wildlife Fund, Washington, DC

Cronkleton, P., Taylor, P.L., Barry, D., Stone-Jovicich, S. and Schmink, M. (2008). Environmental Governance and the Emergence of Forest-Based Social Movements, CIFOR Occasional Paper 49, Center for International Forestry Research, Bogor, Indonesia.

Cunningham, A. (2001) *Applied Ethnobotany: People, Wild Plant Use and Conservation*, Earthscan, London

Cunningham, A. B., Garnett, S., Gorman, J., Courtenay, K. and Boehme, D. (2009) 'Eco-Enterprises and *Terminalia ferdinandiana*: "Best-laid plans" and Australian policy lessons', *Economic Botany*, vol 63, no 1, pp16–28

De Beer, J. H. and McDermott, M. (1989) *The Economic Value of Non-Timber Forest Products in South East Asia*, Netherlands Committee for IUCN (World Conservation Union), Amsterdam

DeFoliart, G. R. (1995) 'Edible insects as minilivestock', *Biodiversity and Conservation*, vol 4, no 3, pp306–321

Dewees, P. A. and Scherr, S. J. (1996) 'Policies and Markets for Non-Timber Tree Products', EPTD Discussion Paper no 16, Environment and Production Technology Division, International Food Policy Research Institute

Dounias, E. (1993) *Perception and Use of Wild Yams by the Baka Hunter-Gatherers in South Cameroon*, Man and the Biosphere series, UNESCO and Parthenon, Paris

Dounias, E. (2000) *Review of the Ethnobotanical Literature for Central and West Africa*. The African Ethnobotany Network, Bulletin No 2, August 2000

Emery, M. R. and Pierce, A. R. (2005) 'Interrupting the telos: Locating subsistence in contemporary forests', *Environment and Planning A*, vol 37, pp981–993

Falconer, J. (1990) *The Major Significance of 'Minor' Forest Products: The Local Use and Value of Forests in the West African Humid Zone*, Community Forestry Note No. 6, Food and Agriculture Organization of the United Nations, Rome

Freese, C. H. (1997) *Harvesting Wild Species: Implications for Biodiversity Conservation*, Johns Hopkins University Press, Baltimore and London, pp283–314

Griffiths, J. (1986) 'What is legal pluralism?', *Journal of Legal Pluralism and Unofficial Law* vol 24, pp1–56, available at www.law.gsu.edu/jjuergensmeyer/spring08/bonilla_session1_Griffiths.pdf

Hoftsad, O. (1997) 'Woodland deforestation by charcoal supply to Dar es Salaam', *Journal of Environmental Economics and Management*, vol 33, pp17–32

Hurley, P. T., Halfacre, A., Levine, N. and Burke, M. (2008) 'Finding a "disappearing" non-timber forest resource: Using grounded visualization to explore urbanization impacts on sweetgrass basket-making in greater Mt Pleasant, SC', *The Professional Geographer*, vol 60, no 4, pp1–23

Jones, E. T. and Lynch, K. A. (2007) 'Nontimber forest products and biodiversity management in the Pacific Northwest', *Forest Ecology and Management*, vol 246, pp29–37

Jones, E. T., McLain, R. J. and Weigand, J. F. (2002) *NTFP Management and Policy in the United States*, University of Kansas Press, Lawrence, KS

Lynch, O. J. and Alcorn, J. B. (1994) 'Tenurial rights and community-based conservation', in D. Western and M. Wright (eds) *Natural Connections: Perspectives in Community-Based Conservation*, Island Press, Washington, DC, pp373–392

Marshall, E., Schreckenberg, K. and Newton, A. C. (eds) (2006) *Commercialization of Non-Timber Forest Products: Factors Influencing Success. Lessons Learned from Mexico and Bolivia and Policy Implications for Decision-Makers*, United Nations Environment Programme World Conservation Monitoring Centre, Cambridge

McLain, R. J. (2008) 'Constructing a wild mushroom panopticon: The extension of nation-state control over the forest understory in Oregon, USA', *Economic Botany*, vol 62, no 3, pp343–355

McLain. R. J. and Jones, E. T. (2005) *Nontimber Forest Products Management on National Forests in the United States*, General Technical Report PNW-GTR-655, US Department of Agriculture, Forest Service, Pacific Northwest Research Station, Portland, OR

McManis, C. R. (2007) *Biodiversity and the Law: Intellectual Property, Biotechnology and Traditional Knowledge*, Earthscan, London

McNeely, J. A. (1988) *Economics and Biological Diversity: Developing and Using Economic Incentives to Conserve Biological Resources*, IUCN (World Conservation Union), Gland, Switzerland

Michon, G. (2005) 'NTFP development and poverty alleviation: Is the policy context favourable?', in J. L. Pfund and P. Robinson (eds) *Non-Timber Forest Products: Between Poverty Alleviation and Market Forces*, Intercooperation, Berne, Switzerland

Nepstad, D. C. and Schwartzman, S. (eds) (1992) 'Non-timber products from tropical forests: Evaluation of a conservation and development strategy', *Advances in Economic Botany*, vol 9, New York Botanical Garden, Bronx, NY

Neumann, R. and Hirsch, E. (2000) *Commercialisation of Non-Timber Forest Products: Review and Analysis of Research*, Center for International Forestry Research, Bogor, Indonesia

Pearce, D. (1991) *Assessing the Returns to the Economy and to Society from Investments in Forestry*, Occasional Paper 47, Forestry Commission, Edinburgh

Peters, C. M. (1996) *The Ecology and Management of Non-Timber Forest Resources*, World Bank Technical Paper No 322, World Bank, Washington, DC

Peters, C. M., Gentry, A. H. and Mendelssohn, R. O. (1989) 'Valuation of an Amazonian rainforest', *Nature*, vol 339, pp655–656

Plotkin, M. and Famolare, L. (1992) (eds) *Sustainable Harvest and Marketing of Rainforest Products*, Island Press, Washington, DC

Posey, D. A. (ed) (1999) *The Cultural and Spiritual Value of Biodiversity*, United Nations Environment Programme, Nairobi

Posey, D. and Dutfield, G. (1996) *Beyond Intellectual Property: Toward Traditional Resource Rights for Indigenous Peoples and Local Communities*, International Development Research Centre, Ottawa

Redford, K. and Padoch, C. (eds) (1992) *Conservation of Neotropical Forests: Working from Traditional Resource Use*, Columbia University Press, New York, NY

Shackleton, C. and Shackleton, S. (2004) 'The importance of non-timber forest products in rural livelihood security and as safety nets: A review of evidence from South Africa', *South African Journal of Science*, vol 100, pp658–664

SEI (2002) *Charcoal Potential in Southern Africa, CHAPOSA Final Report*, INCO-DEV, Stockholm Environment Institute, Stockholm

Shanley, P and Medina, G. (eds) (2005) *Fruitiferas e Plantas Uteis na Vida Amazonica*, CIFOR/IMAZON, Belem, Brazil

Shanley, P., Pierce, A. R., Laird, S. A. and Guillen A. (eds) (2002) *Tapping the Green Market: Certification and Management of Non-Timber Forest Products*, Earthscan, London

Shiel, D and Wunder, S. (2002) 'The value of tropical forest to local communities: Complications, caveats, and cautions', *Conservation Ecology* vol 6, no 2, p9

Stockdale, M. (2006) *Steps to Sustainable and Community-Based NTFP Management: A Manual Written with Special Reference to South and Southeast Asia*, Non-Timber Forest Products Exchange Programme for South and Southeast Asia, Quezon City, Philippines

Sunderland, T. and Ndoye, O. (eds) (2004) *Forest Products, Livelihoods and Conservation: Case Studies of Non-Timber Forest Product Systems*, Volume 2 – Africa, Center for International Forestry Research, Bogor, Indonesia

Swanson, T. M. and Barbier, E. B. (1992) *Economics for the Wilds: Wildlife, Wildlands, Diversity and Development*, Earthscan, London

Treib, O., Bahr, H. and Falkner, G (2007) 'Modes of governance: towards a conceptual clarification', *Journal of European Public Policy* vol 12, no 1, pp1–20

United Nations (2007) *United Nations Declaration on the Rights of Indigenous Peoples*, adopted by United Nations General Assembly Resolution 61/295 on 13 September, www.un.org/esa/socdev/unpfii/documents/DRIPS_en.pdf, accessed 29 April 2008

WCED (World Commission on Environment and Development) (1987) *Our Common Future*, Oxford University Press, Oxford

Weiss, T. G. (2000) 'Governance, good governance and global governance: conceptual and actual challenges', *Third World Quarterly* vol 21, no 5, pp795–814

Westoby, J. C. (1989) *Introduction to World Forestry*, Blackwell, New York, NY

World Bank (2002) *Report of the AFTEGI/AFTRS joint seminar on household energy and woodland management*, World Bank, Washington, DC

Wunder, S. (2001) 'Poverty alleviation and tropical forests: What scope for synergies?', *World Development* vol 29, no 11, p18

Wynberg, R. P. and Laird, S. A. (2007) 'Less is often more: Governance of a non-timber forest product, marula (*Sclerocarya birrea* subsp. *caffra*) in southern Africa', *International Forestry Review*, vol 9, no 1, pp475–90

Chapter 1

Changing Policy Trends in the Emergence of Bolivia's Brazil Nut Sector

Peter Cronkleton and Pablo Pacheco

INTRODUCTION

For more than a decade Brazil nuts (*Bertholletia excelsa*) have been one of Bolivia's most important forest exports. Paradoxically, national policies and initiatives to support the management of Brazil nuts in Bolivia's northern forests have generally taken a back seat to timber management and other economic development programmes. The Brazil nut sector emerged despite the lack of a clear policy framework defining access to the resource or specific guidelines for its management. In fact, those policies that most shaped Brazil nut production were not intended to address their management, but were instead linked to macroeconomic policy to promote non-traditional exports, to expand infrastructure, to increase Brazil nut processing capacity and to formalize property rights.

As the primary motor for economic activity in Bolivia's northern Amazon, Brazil nuts are a resource that supports much of the population directly or indirectly. They are the most recent manifestation of an extractive economy based on the exploitation of NTFPs and subjected to extreme boom–bust cycles. Historically, the benefits of the extractive economy were highly concentrated and flowed out of the region, but there are now signs that more democratic and equitable forms of resource access and distribution may be developing. To ensure that such a trend can be sustained, the Brazil nut sector needs greater attention from policy-makers.

Compared to other types of land use, Brazil nut collection is relatively benign. However, improvements in terrestrial transportation networks that have contributed to the development of industrial processing capacity have also increased pressure and competition for control of the resource base. At the same time, ambiguous or contradictory policies have provided a weak regulatory framework to mediate the competing interests of stakeholders dependent on forests and others intent on forest transformation. These disparate factors have placed this strategic resource base in a precarious position.

Although Bolivia has never had a consistent, comprehensive policy strategy to guide Brazil nut management, use and trade, some of these issues have been dealt with in a piecemeal or peripheral fashion. A key factor in understanding shifts in Bolivia's Brazil nut sector has been the lack of secure property rights in the Bolivian north. Some stakeholders have maintained strong de facto control over the resource base, but legally their claims remain ill-defined and, in recent decades, contested, and such conflict has shaped the actions and plans of forest stakeholders across the region. A second issue has been the absence of clear policies for guiding the management of this important NTFP. Historically, Bolivian forestry legislation has not addressed management practices or the regulation of access rights to Brazil nut resources in the forest. Bolivia has gained much attention for its progressive forestry legislation, but too little effort has been invested in trying to influence NTFP management, whether by maintaining benign practices, discouraging destructive ones or pursuing consistent efforts to promote NTFP management as an important strategy for regional livelihoods. Given the threats to this resource, which are likely to expand, it is important to examine how this robust Brazil nut sector developed, reconsider existing policy frameworks to identify positive or negative influence, and define adjustments that could promote good management and maintain the economic and social contribution of the sector.

The chapter is divided into three sections. Part 1 reviews the characteristics of the Brazil nut sector, the resource itself and the key stakeholders. Part 2 traces the history of the extractive forest economy in northern Bolivia and the emergence of Brazil nuts as a key resource. It illustrates how the development of the sector through the 20th century maintained earlier patterns and power relations, and how the few policy interventions by the state allowed benefits to be concentrated in the hands of a few. Part 3 examines major policy shifts in the mid-1990s that changed the playing field by recognizing the access and property rights of rural peoples and further consolidated the Brazil nut sector. In the conclusion, the authors offer recommendations for policies and actions to better support the Brazil nut sector and its stakeholders in the face of new frontier challenges.

PART 1: BRAZIL NUT PRODUCTION IN BOLIVIA

Importance of the Brazil nut economy

Bolivia's northern Amazon, a region that includes the department of Pando and the provinces of Iturralde (department of La Paz) and Vaca Diez (department of Beni), is covered by approximately nine million hectares of Brazil-nut-rich forest. Brazil nuts have been called the 'single most important pillar of the regional economy' (Stoian, 2000, p284) forming the basis for employment and livelihoods for most residents. It is estimated that the Brazil nut sector generates approximately 22,000 direct and indirect jobs[1] (Bojanic, 2001). Approximately 6000 peasant and indigenous households in agro-extractive communities depend on Brazil nut extraction as their main source of income (Stoian, 2000). In fact, for many rural and urban families in the region, the income generated during the Brazil nut harvest provides most of the cash they

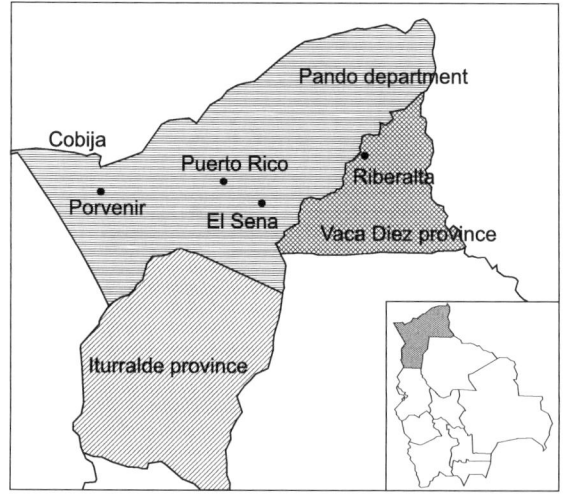

Source: Public domain base map from University of Texas Perry-Castañeda Library Map Collection, www.lib.utexas. edu/maps/cia08/bolivia_sm_2008.gif, modifications by Elizabeth Skinner.

Figure 1.1 *Bolivia's northern Amazon region*

will have throughout the year. Thousands of migrant labourers (about 5500 people) move seasonally between rural forest estates (*barracas*) and the region's urban centres (mainly Riberalta) to work in large, capital-intensive Brazil nut processing plants. These factories provide seasonal employment for approximately 8500 people (Stoian, 2004). Finally, there are extensive networks of intermediaries who organize labour groups to collect the nuts in the *barracas* and transport nuts from the forests to urban processing plants.

Brazil nut export values have become a key segment of Bolivia's forest sector, along with timber products. In 2005, shelled and unshelled Brazil nuts were Bolivia's most important forest export, worth almost US$74 million, practically 45 per cent of the value of all forest-related exports, while unprocessed and semi-processed wood accounted for just over 19 per cent of the total, with manufactured wood products at 31 per cent (Cámara Forestal, 2006). In the first five years of this century, Bolivia accounted for over 50 per cent of the world's Brazil nut exports – or over 70 per cent of processed, shelled nuts (FAOSTAT, 2007). In fact, Bolivia has led the world in the export of shelled nuts since 1992 (Stoian, 2000).

Recently international Brazil nut prices have increased dramatically, with local prices paid to producers in Bolivia jumping from around US$6 per *barrica* (a 66kg bag used as the traditional measure) during the 2002 harvest to around US$35 a *barrica* in 2004, and remaining above US$25 since then. This increase in price has meant more income for regional stakeholders, but it has also brought more competition for control of access to the resource.

Source: Kristen Evans.

Figure 1.2 *Woman breaking Brazil nut fruits to extract nuts*

Biophysical and ecological characteristics

Brazil nut trees are found throughout the Amazon, but occur in greater concentrations in Brazil, Bolivia and Peru. They grow across Bolivia's northern Amazon, although 80 per cent of the country's production comes from the department of Pando (Cámara Forestal, 2006). The Brazil nut tree is a rainforest giant reaching heights of nearly 60m and diameters of over 2m and can live for more than 1000 years (Ortiz, 2002). In Bolivian forests there are about one to five trees per hectare (DHV, 1993). Virtually all

Brazil nuts come from wild trees, possibly due to the species' reliance on pollinators found in closed canopy primary forest (Ortiz, 2002).

The tree produces large woody fruit, containing 15–25 seeds each covered by a hard shell (Ortiz, 2002). The fruit begins to drop in November and by January most are on the forest floor. Once the fruit has fallen, the harvest begins and normally runs until March.[2] Harvest practices have varied little over time. Workers collect fruit from under the trees in natural forests and, when they have gathered enough, break them open to release the seeds. The seeds are then packed in bags and carried from the forests to the roadside or river's edge for transport to urban centres for processing and export.

High Brazil nut prices have intensified the harvest, but the activity still has a low impact on the forest compared to logging or other land uses. There is debate about whether high levels of Brazil nut extraction can be sustained without negative impacts. Research has indicated 'good prospects for continued regeneration of exploited populations' (Zuidema and Boot, 2002), although other studies suggest that long histories of exploitation affect the population structure, with intensely harvested areas lacking juvenile trees less than 60cm diameter at breast height. This indicates that regeneration could be a problem (Perez et al, 2003). But the same authors agree that more immediate threats to Brazil nut populations are deforestation and forest degradation from conventional logging and an increased incidence of fire, which could be avoided through greater emphasis on sustainable forest management and the protection of primary forest (Perez et al, 2003). One way to assure such protection would be to keep Brazil-nut-rich forests securely under the control of the stakeholders who depend on them for their livelihoods.

PART 2: HISTORICAL CONTEXT: EMERGENCE OF AN EXTRACTIVE ECONOMY AT BOLIVIA'S NORTHERN FOREST FRONTIER

Understanding the current context surrounding Brazil nut production in Bolivia's northern Amazon requires a brief review of the region's boom–bust economic history and the production patterns and stakeholders involved. Bolivia's Brazil nut sector grew out of a regional extractive economy that had developed largely in a policy vacuum in the late 19th century. The limited regulatory guidance from or intervention by state authority meant that the spontaneous development of the production system favoured the interests of regional elites and foreign investors that drove the initial incursions into Bolivia's northern Amazon. Control over forest resources in the region was based on informal claims that were never converted to formal tenure rights. The pattern of traditional but informal rights persisted throughout the 20th century, although the stakeholders holding these rights changed somewhat over time.

This section will describe the production systems and power relations that supported the informal basis of the region's economy and survived the transition from a dependence on rubber to greater reliance on other products such as Brazil nuts. Near the end of the 20th century the economically powerful stakeholders that had benefited most from earlier booms were poised to corner benefits from policies related to structural adjustment and economic reform.

Property rights and the sector's shaky legal foundation

In the late 19th century, non-indigenous explorers flooded Bolivia's northern Amazon in search of forest resources, primarily rubber (*Hevea brasiliensis*), to supply world markets. This spontaneous occupation established property claims and commercial networks that set a pattern that persists more than a century later (Fifer, 1970). From the outset they were tied to global markets and relied on foreign financing advanced over complex networks that extended across the Amazon to Europe and North America in the form of credit for harvesting forest products. During the early period (1894–1913), commercial activity focused on rubber extraction, gradually spreading along rivers in the region due to the comparative difficulty of penetrating the upland forests (Riviere, 1900).

The explorers and businessmen who entered the region established forest estates called *barracas* and they became known as *barraqueros*. To harvest rubber, the *barraqueros* relied on rural workers, initially drafted from the region's indigenous population and later imported from the departments of Beni and Santa Cruz, and even migrants from abroad (Gamarra, 2007). During the rubber boom, workers responsible for the harvest were assigned homesteads with rubber trees on two or three forest trails. They were not allowed to farm, but were instead extended supplies on credit that could be paid off with their production, although this usually bound them through perpetual debt peonage to the *barraca* estates. As they found themselves in remote forests, rural workers were dependent on this relationship and had little choice but to accept the terms dictated by the *barraqueros* (Pacheco, 1992).

The *barraqueros* did not have legal title to their *barracas* but established claims through the de facto occupation of forest lands and mutual agreement among their *barraquero* neighbours. Their economic power allowed them to enforce their claims, so they had a great deal of local legitimacy. The central government in the Bolivian capital, La Paz, did not exert much influence over the remote region. From 1880 to 1910 some regulations were established to facilitate the purchase of vacant land in the region (*Law No. 1096*, 26 October 1905 and *Law No. 1141*, 3 December 1907), for land measurement (*Resoluciones*, 1912) and for the awarding of property rights under registered title with the payment of taxes (*Law No. 1166*, 11 September 1915 and *Law No. 1223*, 26 September 1917) based on the number of rubber trees (Tambs, 1966). In 1924, additional regulations were enacted that designated the region exclusively as an area of colonization (*tierra de colonias*) (Pacheco, 2008). However, given the weakness of the central government in the remote northern Amazon, this legislation apparently had little long-term effect, as the *barraquero* claims were not titled. Their claims were, however, respected by other *barraqueros*, and, given that this elite held economic and political power over the region, that was sufficient.

During the first decades of the 20th century, plantation rubber from Southeast Asia entered the world market and the price of rubber plummeted. The price collapse depressed the economy throughout the Amazon and the *barraqueros* lost much of the economic power they had wielded over Bolivia's northern forests. Briefly, when Southeast Asia was under Japanese occupation during World War II, the demand for Amazonian rubber increased, but by the end of the 1940s economic stagnation had returned. In the process other intermediaries began competing to purchase forest products, denying *barraqueros* the ability to monopolize access to trade in the region, so they had less power to control workers through debt (Ormachea and Fernández, 1989). However, the general economic and social patterns framing the extractive forest economy did not substantially change, since the remaining *barraqueros* continued to control markets and trade (Pacheco, 1992). In 1958, under the terms of the Treaty of Roboré, Brazil extended rubber subsidies to Bolivian producers, artificially propping up the price and allowing *barraqueros* to continue their rubber operations with certain margins of profit (Pacheco, 1992).

As the rubber economy faltered so-called 'independent communities' not controlled by *barraqueros* began to appear (Ormachea and Fernández, 1989). Some evolved from small *barracas* as the descendants of the original owners began working the forest individually. Others were settled by rural labourers occupying forests abandoned by their former patrons or that were otherwise unoccupied. From the outset, they were ethnically mixed, combining indigenous people from the region and others who had migrated from other departments. Rather than enterprises, they claimed territory collectively but worked the forests as individual households establishing trade and credit relationships with buyers on their own. Though independent from earlier patrons, residents were still held in debt relationships, only now, with the arrival of itinerant merchants, there was a wider array of patrons.

These settlements were agro-extractive communities with diversified livelihoods, combining the collection of forest products with swidden agriculture. The communities have varying levels of forest dependence and organization depending on their proximity to urban centres, difficulty of access, and relations with former landlords (Stoian and Henkemans, 2000). In general, the form taken by rural settlements has been driven by the demands of forest extraction. The basic production unit is the household and rural families were initially dispersed throughout the forest to facilitate the daily extraction of wild rubber. Later, after the collapse of rubber prices and greater dependence on Brazil nuts, communities began shifting to more nucleated settlements with seasonal occupation of forest holdings during the harvest from January to March (Ormachea and Fernández, 1989; Pacheco, 1992).

Historically, customary property rights claimed in the region have been based on 'tree tenure' (Fortmann et al, 1985). These traditional property rights have evolved over time and are based on de facto control over the forests without legal title to back up their claims. The system does not emphasize control of contiguous territory but instead recognizes the right to access individual trees and related infrastructure (previously rubber trees and trails, and now more commonly Brazil nut trees and connecting trail networks). In agro-extractive communities access rights to Brazil nut trees are organized by clusters of trees, called *castañales*. Typically, a *castañal* can have anywhere from a few dozen to several hundred trees, spread over hundreds of hectares (Cronkleton et al, 2010). In newer communities the system may be less defined, but in established communities the customary tree tenure is well developed and quite specific, even though no formal written record of these rights exists. For example, in a recent participatory mapping exercise in Pando, one community mapped approximately 11,000ha of their forest, revealing 38 distinct *castañales* with 8366 Brazil nut trees (Cronkleton et al, 2008). Though lacking a clear legal foundation, the system has been sufficiently resilient to allow NTFPs to drive the regional economy and to allow people to sort out forest property issues and maintain a very lucrative and important forest industry.

The emergence of Brazil nuts

With the collapse of rubber prices during the first decades of the 20th century, producers began searching for alternative products that could fill gaps in the regional economy. Brazil nut production first grew in importance during the 1920s, shortly after the decline in rubber prices (CIDOB, 1979). The Brazil nut could be easily integrated into the production systems of both *barracas* and agro-extractive communities. Since the harvest occurred during the rainy season, when rubber trees were less productive, labour could be shifted to Brazil nut collection. To harvest nuts in the extensive areas they controlled, *barraqueros* began to rely on migrant labour (*zafreros*) contracted in regional urban centres, mainly Riberalta, to augment the rural workforce.

Income generated from Brazil nuts was initially marginal compared with rubber. Most nuts were exported in the shell, but some *barraqueros* began experimenting with machinery to remove shells from the seeds in the 1920s (Fifer, 1970). However, Brazil nuts gradually grew in economic importance, and by the 1950s they accounted for a larger share of exports than rubber (Stoian, 2000). Processing plants began to appear in the late 1970s adding value by shelling large volumes of nuts. In 1986 the region's economy reached a crossroads, as Brazil nuts overtook rubber in terms of export income (Stoian, 2000).

Like the earlier pattern with rubber, the emergence of the Brazil nut sector was largely spontaneous, with little policy guidance. One exception is seen in efforts to influence the treatment of the sector's labour force on *barracas* and in processing plants. The notorious history of exploitative debt peonage and miserable working conditions had plagued the region's economy from the outset. In response a series of governmental resolutions attempted to regulate payment and assure coverage for the region's workforce under existing labour law (*Resoluciones Supremas No. 158242, No. 158243*, and *No. 158244*, 15 June 1971, as well as *Resolucion Ministerial No. 135/80*, 21 April 1980). However, given the government's weak influence in the region, it is unclear to what extent the resolutions brought change for workers. Regardless of these mechanisms, the perceived exploitation of workers continued to be an issue in later conflicts between *barraqueros* and rural people in the region.

In the early 1980s, an economic crisis struck Bolivia and hit the northern Amazon region particularly hard. Bolivia's overvalued exchange rate created disincentives for exports. Hyperinflation increased the debt of *barraqueros* to foreign banks, causing many to abandon their *barracas* (Stoian, 2000). The economic downturn further weakened the ability of *barraqueros* to provide for and control their rural workforce. More independent communities were formed as workers took over abandoned *barracas*, but there was also a surge in rural-to-urban migration. The crisis also had an impact on the Brazil nut economy, as the two remaining Brazil nut processing plants closed down due to the lack of liquidity (Stoian, 2000). Finally, in 1986, Brazil cancelled its subsidy for rubber, which had supported Bolivian producers since 1958, and without this incentive rubber production in the region ground to a halt. This marked the definitive collapse of the rubber economy (Pacheco, 1992).

In less than 100 years, the region had gone through a series of economic booms and busts around forest resources. Throughout this process, the forest basis of the extractive economy persisted, but it remained precarious even as it shifted from rubber to Brazil nuts. The *barraqueros* and communities that occupied the land lacked legal title and continued to depend on customary rights and tree tenure concepts that did not fit easily into existing tenure frameworks focused on land rights. In such a weak economy, forest extraction was not very lucrative, but the lack of infrastructure sheltered the region from frontier change and competing land uses. Under these conditions, however, government programmes designed to respond to the economic crisis could bring dramatic change. Fortunately for the region's forests and forest-dependent peoples, macroeconomic policies adopted by the government breathed new life into the Brazil nut trade.

Neoliberal reforms and the growth of the Brazil nut sector

The region's economy, having reached a low point in the mid-1980s, began to turn around in response to national policies. However, the policies that had the greatest influence on the Brazil nut sector at this point were macroeconomic policies, rather than those developed expressly for the sector, NTFPs or even forest management. In 1985, the government of Victor Paz Estenssoro implemented structural adjustment policies, contained in the *Decreto Supremo No. 21060* (29 August 1985), to address the economic crisis through market liberalization and non-traditional export promotion, under which Brazil nuts were prioritized along with other products, mainly soybeans (Pacheco, 1998). The structural adjustment policies created conditions favourable for exporters by devaluing the Bolivian currency against the US dollar, eliminating licensing requirements for the export of goods and services, freeing non-traditional exports (like Brazil nuts) from taxes and fees, and approving a tax refund for consumption and import tariffs that were incorporated in the costs of export goods (Morales, 1994).

In addition, there have been several attempts to articulate regional development strategies that include Brazil nut extraction as a crucial component. The most significant has been a World Bank project implemented by an international consultancy group DHV in the early 1990s (DHV, 1993) to develop the competitiveness of several non-traditional exports, including Brazil nuts. This project provided credit to national investors for expanding the Brazil nut processing plants.

However, probably the most important factor driving the Brazil nut economic expansion was the opening of the road from La Paz to the northern Amazon city of Riberalta in the early 1990s, which attracted investors from outside the region to establish Brazil nut processing plants. Soon after, in response to the road and increases in world market price, the number of processing plants soared to 20, allowing the Brazil nut industry to experience its 'first outright boom' (Stoian, 2000).

During this time of transition, the regional elite, joined by investors from other departments, attempted to maintain their dominance of the economy. With the boom in the Brazil nut industry, the owners of processing plants began buying up *barracas* to secure sources of raw materials (Assies, 1997; Stoian, 2000). These were dubious transactions, taking place in a juridical vacuum due to the absence of an effective legal system governing land tenure. However, because the power relations arising from the economic system had not changed, the new owners maintained a level of de facto legitimacy and control over the resource base.

Furthermore, in 1995, the *Fundación Bolivia Exporta* – an entity created to promote exports with the help of international cooperation – shifted from an initial strategy aimed at the 'democratization of the Brazil nut economy', which emphasized the role played by small-scale producers. Instead it promoted the vertical integration through pilot projects using joint venture mechanisms to improve the quality of Brazil nuts produced and the infrastructure of *barracas* (Assies, 1997, p40). At this point, some predicted that the insertion of Brazil nut exports into a neoliberal strategy of non-traditional export promotion would exclude small producers as processing plants gained control over supplies of raw materials, and community participation would eventually become 'residual at best' (Assies, 1997, p40). This may have been the case had there not been growing pressure from rural social movements to change the rules of the game.

Although the government promoted exports and investments in Brazil nut processing, it initially did little to resolve the underlying problem of ill-defined forest property rights in the region. In fact, government actions complicated the issue by adding another layer of 'rights' over the region's forests when it encouraged the entrance of timber companies. In 1995 the government implemented the *Plan Soberanía* (Sovereignty Plan), which allocated long-term timber harvest contracts (the mechanism pre-dating Bolivia's current timber concession system) to 17 timber companies from other regions (primarily Santa Cruz) to occupy forests near the borders with Brazil and Peru (Pacheco, 1998). Despite the existing property rights claims of communities and *barracas*, the government superimposed additional rights under the logic that it was necessary for national security and for the economic development of the region.

PART 3: NATIONAL REFORMS IN THE 1990s

The policy context framing the Brazil nut sector took a new course in the mid-1990s as a series of wide-ranging reforms swept the country. Two of the most significant policy changes during this period were the new Tenure Reform Law (*Ley No. 1715, Ley del Servicio Nacional de Reforma Agraria*) – called the 'INRA Law' after the land agency it created, the National Institute for Agrarian Reform[3] (INRA) – and the new Forestry Law (*Ley No. 1700, Ley Forestal*, 12 July 1996). Neither emphasized the specific issues related to Brazil nut management or the forest property rights system in the northern Amazon context, nor did they immediately affect conditions in the region. When Bolivia's agrarian reform and forestry laws were negotiated in the mid-1990s, the region's *barraquero* elite felt that their traditional forest holdings were relatively secure and did not push to have the laws address the unique peculiarities of the region's production system by legalizing their forest properties and validating their management practices (Ruiz, 2005). Over time, these new laws set the stage for a power struggle between *barraqueros* and agro-extractive communities over claims to forest resources in the region, a struggle that continues.

The struggles for property rights

Although the 1996 INRA law initially had little impact on property rights in the northern Amazon, as the Brazil nut sector became more lucrative, pressure to clarify or formalize access rights grew. The informal and frequently overlapping property claims of *barraqueros* and agro-extractive communities were increasingly contested and insecure. Representing the two extremes on the land issue, they struggled to gain the upper hand, often using distinctive channels of influence to produce decisions in their favour (Ruiz, 2005). The economically and politically powerful *barraqueros* hoped to maintain their claims to large forest estates and attempted to exploit their influence in the national government to solidify their traditional land holdings. For their part, the peasants relied on collective action to defend their access to the forests they occupied.

The INRA law was an ambitious attempt to organize the complex and contradictory property rights system in Bolivia, to resolve competing claims and to distribute and title undocumented or unclaimed lands throughout the country. The law defined several size classes of private property, but also recognized collective community lands. The standard limit for peasant landholdings was 50ha per family – based on previous definitions in agrarian law – regardless of whether the lot was individual or part of a collective area. Implementing this law was a challenge, given that records were often vague, claims overlapping and documentation at times fraudulent. In Bolivia's northern Amazon, the widespread lack of formal land title meant that traditional claims had to be documented and the status of contested areas resolved as part of the process of implementing the INRA law.

To evaluate whether a property was legitimate and justified, INRA required property owners to demonstrate that their land use served an 'economic and social function' (known as FES, the Spanish acronym for *'función económica-social'*). If a property did not meet FES criteria, the 'owner' could lose the claim or have it reduced. The INRA law defined FES as 'the sustainable use of the land for the development of agriculture, ranching, forestry or other productive activities, as well as the conservation and protection of biodiversity, research and ecotourism based on the land's capacity, for the benefit of society, the common good and that of the property'[4] (*Ley INRA*, article 2, II). It is hard to conceive of activities that could not be included under these vague terms. As a result, INRA employees lacked clear guidance, which increased the likelihood of arbitrary decisions and manipulation by outsiders attempting to influence decisions. Usually forest uses like Brazil nut extraction were not deemed to meet FES criteria, a determination that prevented *barraqueros* from justifying their extensive land claims and provided an incentive for land clearing, as they could comply with FES if they found pasture for ranching. Forestry could be considered as a productive use meeting the FES requirement, but only if the owner was able to present an approved forest management plan.

INRA only began work in the northern Amazon in 2000. Since the legislation did not support *barraquero* demands for extensive private properties, which they would have preferred, they attempted to use political influence to move reforms in their favour. Relying on powerful senators and congressmen from the region, the *barraqueros* tried to press their interests using a legal instrument that became infamously known as the '*Barraquero* Decree' (*Decreto Supremo No. 25532*, 5 October 1999). The decree built on articles in the Forestry Law (described below) that prioritized the claims of 'traditional users' in NTFP-rich forests. The decree identified the *barraqueros* as the 'traditional forest users' and the socioeconomic base of the region, and defined steps for converting their holdings into forest concessions, although it also stated that the *barraqueros* were not renouncing their claims to titled private property. It set a six-month time limit for *barraqueros* to present documentation supporting their concession claims to INRA, later extended an additional three months by a second decree (*Decreto Supremo No. 25783*, 19 May 2000). While the concessions were not supposed to be superimposed on community forests, in practice the *barraquero* demands largely ignored community claims (Ruiz, 2005).

These decrees would have created 3–3.5 million hectares of concessions benefiting about 200 *barraqueros* (Aramayo, 2004). The new concessions would have re-established *barraquero* control over contested areas that were in the de facto possession of communities. Opponents interpreted the decrees as the first steps towards granting the *barraqueros* private title. News of this decree catalysed activist opposition among peasant and indigenous organizations that represented the interests of agro-extractive communities and rural workers in the region.

Because of the absence of public consultation surrounding the decrees and the apparent lack of channels for public dialogue and mediation, small producers in the region began to organize collective action to pressure the government to address their complaints. Mass marches by indigenous people and peasants to La Paz had been successful in forcing the government to consider their demands during the debates leading up to the ratification of the INRA law, so in 2000 the Third National March for Land Territory and Natural Resources[5] was organized by a coalition of regional peasant and indigenous organizations supported by NGOs. The goal was to force an official response from a national government that was increasingly interested in populist measures to appease rural tension. The strategy worked and on 10 July 2000 the central government annulled the *barraquero* decree (Ruiz, 2005). Although still a powerful regional player, the *barraquero* bloc had overplayed its hand and the pendulum was swinging in favour of small-scale rural producers.

After the march, the government went further in addressing the demands of the region's rural population. It issued another decree that became known as the '500-hectare decree' (*Decreto Supremo No. 25848*, 18 July 2000) because it defined 500ha per family as the appropriate area for peasant and indigenous communities, rather than the 50 hectare standard. Such areas corresponded roughly to the territory used by individual households to harvest NTFPs in agro-extractive communities. However, rather than each family being given title to an individual property, the communities would receive communal titles covering areas more or less equivalent to 500ha per family. This decree effectively recognized the de facto hold over expanses of forest by peasant households, allowing them to maintain their NTFP-based livelihoods.

Implementation of this decree was slow owing to continued political opposition by *barraqueros* and regional tension generated by disputes between communities and *barraqueros.* The decree confirmed peasant and indigenous rights to contested lands but provided no mechanism for implementing Brazil nut concessions, leaving the possibility of further cuts in *barraquero* claims as INRA sorted out the boundaries. At this point, *barraqueros* were only receiving title to 50 hectare plots around their homesteads in the claimed forests.

Finally, the stalemate was broken with a decree by the Carlos Mesa government (*Decreto Supremo No. 27572*, 17 June 2004) that confirmed the 500-hectare per family measure for agro-extractive communities and reaffirmed the eventual implementation of Brazil nut concessions. According to the decree, all communities that had existed five years prior to the 1996 reforms, that were legally incorporated and that had officially registered organizations were eligible. The 500-hectare measure was defined as the minimum (with 1000ha per family as the maximum) and INRA was required to find additional land to compensate communities that had not received enough. The decree also created a Commission for Reconciliation, Arbitration and Resolution of Conflicts[6] (CCARC) to mediate the process. Once community titling was resolved, the government would move on to define NTFP concessions for *barraqueros* on the remaining state lands not titled to agro-extractive communities. With this decree, *barracas* could be voluntarily converted into concessions with a maximum size of 15,000ha, if they developed the required management plans, paid the required forest fees and did not have conflicts with neighbouring properties or communities.

Although modifications to the INRA Law have allowed a regional consensus on tenure rights for Brazil-nut-rich forest lands, there is still uncertainty about how the process will run its course. One variable is new agrarian legislation introduced by the government of Evo Morales. In November 2006 the government issued the Community Redirection of Agrarian Reform Law (*Ley No. 3545, Reconducción Comunitaria de la Reforma Agraria*) and then the corresponding by-laws (*Decreto Supremo No. 29215*, 2 August 2007). The law maintains the concept of FES to validate justifications for property claims, but modifies it to allow some projections for growth in land use, to require field inspections of properties, and to prohibit the use of illegal land clearing to justify compliance with FES. These changes could diminish the use of land clearing driven by efforts to establish property claims. The law also states that at the conclusion of the agrarian reform process, all remaining state lands should be granted exclusively to landless indigenous and peasant communities (*Disposición Transitoria Décimo Primera*).

The 2004 decree by the Mesa government defined NTFP concessions as part of the agrarian reform process. Therefore the new law should not affect concession claims, but to date none have been issued. Given the country's political instability, the central government has not made a concerted effort to begin implementing this law. In the meantime, more analysis is needed to identify the implications of these policy changes for Brazil nuts and the people who depend on them.

Agrarian reform in the region accelerated following the July 2004 decree. By late 2008, 139 of the 245 peasant communities in the region had been titled, receiving 1,807,320ha. An additional 106 are having their claims processed, which will add another 567,638ha of forest. Once INRA finalizes the titling of community lands it will be able demarcate areas for NTFP concessions. According to Ruiz (2005), in 2000 there were 221 *barracas*, whose owners claimed over 3 million hectares of forest, although 71 per cent of this area was controlled by just 44 *barracas*. However, by 2008

INRA has received 237 *barraquero* claims for NTFP concessions with 'expected rights' (*derechos expectaticios*) although the total surface area had decreased to 1,535,790ha, all in the department of Pando. In February 2007 INRA announced that 68 areas had been demarcated and could be classified as concessions. However, without Brazil nut management plans the concessions have not been finalized (PrensaBolivia, 2007).

The titling process has dramatically improved the position of agro-extractive communities in the region, but it has not resolved problems of insecurity and property conflicts (Cronkleton et al, 2009). Frequently the boundaries drawn by INRA do not fully reflect the patterns of traditional forest used, and consequently generate or accentuate boundary conflicts between communities and other property owners. The process has also introduced a competing agrarian model of property access that could undercut existing tree tenure systems. Some residents erroneously believe that the state plans to impose internal divisions on the communal property by demarcating each family's 500ha. Such divisions would be difficult to reconcile with the mosaic of customary tree tenure rights, and the redistribution of resources that would be required, for such change has increased tension among neighbours in some communities.

Regulating and promoting sustainable forest use

Interestingly, Bolivia's new Forestry Law, which one would have expected to directly address policy issues related to Brazil nut management, did not emphasize NTFPs and, in fact, mentioned Brazil nuts only in passing references to access rights and fees. It did not prescribe relatively simple actions to recognize and prioritize the protection of this valuable resource. For example, despite the rapid growth and increasing economic significance of Brazil nut production in the 1990s, the felling of Brazil nut trees was not prohibited even while this comprehensive forestry reform was taking place – at least, not until the 2004 decree addressing the property conflict between communities and *barraqueros* (*Decreto Supremo No. 27572*, article 39).

The Forestry Law was promoted by international cooperation agencies to support the development of Bolivia's timber sector. In general, it brought significant change to the forest sector, with a broad agenda to redefine not only who could legally manage forests and derive commercial benefits, but also how that management should take place, by establishing norms for sustainable forest management. Although a series of technical norms and administrative mechanisms for timber was quickly put in place in the months and years following the law's ratification, less urgency was given to NTFPs or the needs of traditional stakeholders in the northern Amazon. In particular, the norms required to regulate or guide the Brazil nut sector have been slow to develop, and to date efforts to devise policy instruments have been ineffective, as will be explained below. This section will examine how the law's emphasis on timber in defining who could manage and how management should take place relegated Brazil nuts to a secondary status.

Bolivia's 1996 Forestry Law was lauded as an innovative attempt to better regulate Bolivia's forest sector, mandate sustainable management practices and devolve forest access and management rights to a greater diversity of stakeholders including private property owners and, in particular, communities and indigenous peoples (Mancilla and Andaluz, 1996; Pavez and Bojanic, 1998). The law created the Forest Superintendence, responsible for authorizing commercial forest exploitation and policing forest use to ensure compliance with regulations. The new Forestry Law brought dramatic change to how access to forests was determined. In Bolivia, all forests are owned by the state, regardless of whether they are on state land, individual properties or communal land. With the approval of a management plan, the state grants commercial access rights to the forest, but before a management plan can be developed, the individual, group or company has to demonstrate valid property rights claims to the land. Forests on private properties or on indigenous or peasant community land could only be commercially managed by their respective owners, a right that had previously been denied.

In Bolivia's northern Amazon, the lack of clear property rights and the resulting conflicts over the control of forest areas created a bottleneck in the implementation of the Forestry Law and made it harder for most rural residents to benefit from the new rights. In an apparent reference to the northern Amazon, the law states that 'areas with Brazil nut, rubber, palm heart or similar resources shall be ceded preferentially to traditional users, peasant communities and local forest user associations (*Asociaciones Sociales del Lugar*)[7] (*Ley Forestal*, article 31, I). While these groups are not specifically defined in the law, given the resources mentioned, this is apparently a reference to *barraqueros*, indigenous people and peasant groups traditionally dependent on the northern forests for their livelihoods.

The other reference to Brazil nuts in the law occurs in the section dealing with forestry fees. Initially, the law required the same area-based patent fee system used for industrial timber concessions to NTFP producers. Area-based fees had successfully diminished extensive claims over forests made by the timber industry (Contreras and Vargas, 2001), and could possibly have had the same effect on Brazil nut holdings. The Forestry Law set the fee at 30 per cent of the minimum fee paid by timber concessionaires (i.e. US$0.30 per hectare for Brazil nut forests) (*Ley Forestal*, article 37, II). Because property rights were not defined in the region, the Forest Superintendence had no basis for determining the surface area to calculate the fee. It therefore switched to a weight-based fee system.[8] These fees are paid by the processing plants, rather than by the resource manager as is the case with timber products. Also, because the fees are charged to processing plants, efforts to record the origin of nuts have been haphazard, and a disproportionate share of the fees is paid to the municipal governments where processing plants are located (Riberalta and Cobija) rather than being channelled back to the local governments where forests are located to support resource conservation and forest development efforts. At any rate, the fee for Brazil nuts was set so low that it is largely symbolic.

Being singled out for 'preferential' rights did not result in the rapid legalization of those rights. Approval required the delimitation of the forest area, the preparation, approval and implementation of a management plan, and the submission of annual operational plans (*Ley Forestal*, article 31, IV). It would be almost a decade before the necessary norms and guidelines were issued. This has been especially important for *barraqueros* since a management plan is a key instrument for defining and justifying NTFP concessions – and, to a certain extent, the need to justify forest access rights has driven interest in defining management norms for Brazil nuts.

For the 19 timber concessions created in Pando[9], the law states that in areas where NTFPs are present, the concession rights to timber and NTFPs would subsequently be 'harmonized' through related by-laws (*Ley Forestal*, article 29, II). Industrial timber concessionaires with approved forest management plans could allow others to exploit NTFPs under 'auxiliary contracts', but guidelines as to how the management of timber and NTFPs was to be integrated were not provided. Therefore the opportunity for more secure access and formal approval for Brazil nut management proved to be ephemeral.

Management norms for Brazil nuts

Under Bolivia's forestry legislation, the approval of a management plan is a key step for gaining full legal rights to commercialize forest products. The first technical norms for developing management plans focused on timber management and defined practices such as polycyclic planning, minimum cutting diameters, the use of reduced-impact logging techniques and the protection of environmentally sensitive areas. For NTFPs the necessary technical norms that would set the standards for such management plans did not exist.

The first attempts to define management norms for Brazil nuts occurred at the outset of the region's property rights conflicts during the late 1990s. The *barraquero* decree (*Decreto Supremo No. 25532*, 5 October 1999) included a simple format for management plans consisting of the definition of the management area, the elaboration of maps showing infrastructure, and an estimation of production levels, but because the decree was nullified the mechanism was never applied. A resolution passed in late 2002 (*Resolución Ministerial No. 164*, 11 November 2002) approved the first stand-alone norms for Brazil nut management as a complement to the technical norms for timber management plans. However, because the resolution was prepared without public consultation, it was not well received. Peasant and indigenous organizations were suspicious that this was another attempt by the *barraqueros*, working through allies in La Paz, to meet FES requirements and consolidate large landholdings as private property. Two months after the resolution was announced, it was suspended by another resolution (*Resolución Ministerial No. 023*, 2 April 2003), due to the lack of involvement of indigenous people and peasant organizations in its formulation.

A more concerted effort to develop management norms then began, involving foresters from the National Forestry Direction at the Ministry of Sustainable Development in consultation with other forestry professionals, but without extensive consultations with regional stakeholders. In March 2005 the government issued the resulting Technical Norms for the Preparation of Brazil Nut Management Plans[10] (*Resolución Ministerial No. 077/2005*, 28 March 2005).

The norms take a conventional approach to forest management, apparently modelled on technical norms for timber: for example, early versions required the documentation of stem quality and basal area, which have little relevance for Brazil nut management. The norms require definition of the management area and a census of Brazil nut trees (above 30cm in diameter at breast height) that can be carried out incrementally, covering at least 10 per cent of the total management area per year. Non-harvested quadrants (*cuadrantes de seguridad*), covering 6 per cent of each census area, must be designated and set aside for three to five years. The norms also require separate maps of the census plan and of the distribution and density of Brazil nut trees, and operational plans estimating harvest volumes and identifying silvicultural practices, post-harvest storage and the placement of infrastructure. The plans must be prepared by a forestry engineer, who also must prepare annual post-harvest reports. As of mid-2008, no management plans have been approved under these norms.

The norms are intended for use by all Brazil nut producers, but they seem focused more on the needs of *barraqueros* than on those of community-level stakeholders. The norms specifically state that management plans will not recognize property rights and are only intended to authorize NTFP concessions, a clear reference to *barracas* since communities have received property rights. The fact that management plans can be approved after only 10 per cent of the territory has been covered by the census substantially lowers the start-up costs of preparing a management plan by spreading total costs over ten years. This provision would allow quick approval of large areas, which would be in the interests of *barraqueros* seeking concessions.

It is not clear how a community would benefit from preparing a management plan as they are currently being conceived other than having legal approval for the plan. Efforts to develop norms so far miss crucial issues related to internal access and boundary disputes. They impose complex and costly requirements on producers, often with dubious value for improving management decision-making. Furthermore, the state is unlikely to have either the capacity to monitor or enforce compliance with the norms, given the resources at its disposal, and the vast space to be monitored, nor will it have the political will to risk disrupting the region's economic base. It is therefore unrealistic to assume that community-level stakeholders would invest in the costly and complex application of these norms. There are, however, NGOs in the region initiating Brazil nut management plans for communities and subsidizing the costs of approval under these norms.

As mentioned earlier, customary access rights within communities comprise mosaics of individual claims to Brazil nut stands called *castañales*. Resident families should be encouraged to document the distribution and characteristics of their *castañales* to validate customary rights within their communal properties. This is particularly important when new property boundaries defined by the agrarian reform do not reflect the traditional boundaries of agro-extractive communities and hence generate confusion or uncertainty. Addressing these internal issues is key to securing access rights to Brazil nut resources for households, and could contribute to improved management planning and generate information for mediating internal resource conflicts. However, the norms do not address internal diversity in communal properties where multiple families live, claim use and management rights to specific forest resources, and practise swidden agriculture. The development of management norms for communities should begin with an explicit discussion of the interests that community-level stakeholders and governmental agencies have, as well as the range of management problems they face.

Policy initiatives and market-based mechanisms

Although policy initiatives to support Brazil nut resource managers (*barraqueros* and communities) were slow to appear, starting in the mid-1990s, the state supported the interests of processing plants and exporters in a range of ways. For example, when strict health and quality-control regulations relating to acceptable levels of aflatoxins threatened to close European and US markets to Brazil nuts, especially after the European Commission passed regulations lowering the acceptable levels from 20 to 4 parts per billion (Regulation 1525/98 EC, cited in Newing and Harrop, 2000), the government took action. In 1997, it established rules for certifying the quality of Brazil nuts for export under the norms of the Bolivian Standards, Metrology, Accreditation and Certification System[11] (*Decreto Supremo No. 24498*, 17 February 1997). The following year another decree created an oversight board called the National Brazil Nut Council[12] that required health and quality certification for all Brazil nut exports (*Decreto Supremo No. 25200*, 16 October 1998). In 2000, a new law created the National Agricultural Health and Food Safety Service,[13] responsible for overseeing the quality of Brazil nut exports (*Ley No. 2061, Ley del Servicio Nacional de Sanidad Agropecuaria e Inocuidad Alimentaria*, 16 March 2000). That same year the Bolivian Quality and Normalization Institute[14] established norms for Brazil nut classification, sanitation practices and aflatoxin sampling, drawing heavily on the Food and Agriculture Organization's *Codex Alimentarius* (Soldán, 2003). Finally, in 2001, a governmental decree created general norms for regulating the certification of Brazil nut safety and quality (*Decreto Supremo No. 26081*, 23 February 2001).

The government's agility and relative speed in developing this regulatory framework illustrates the importance accorded to the processing industry and, probably more significantly, the rapid and effective lobbying efforts on the part of that industry to protect its access to international markets. It also illustrates how governments can regulate aspects of NTFP trade efficiently and strategically, if they choose – given the right incentives to do so, a high-value product, and organized and outspoken constituents, in this case the processing companies. Much can be learned from this experience. Brazil nut exports faced a clear threat of lost markets because of aflatoxin contamination and there were significant benefits in addressing that threat. Mechanisms were proposed, stakeholders agreed they were worthwhile, and they were implemented. So far these measures have been successful in protecting the country's Brazil nut exports.

While the processing industry was adjusting to market forces, requiring compliance with international standards for health and quality control, Brazil nut producers controlling forest resources also sought to secure market share and other advantages such as access to specialized markets. Initially, efforts by technical assistance agencies were channelled through forest certification standards, under the umbrella of the Forest Stewardship Council (FSC), but over time greater opportunities and a better fit were found with organic and fair-trade certification.

Over a three-year period starting in 1998, a collaborative effort led by the Bolivian Council for Voluntary Forest Certification[15] (CFV), with strong support from the Programme of Forest Management in the Bolivian Amazon[16] (PROMAB), brought together the owners of processing plants, representatives of producer organizations and other experts to develop certification standards for Brazil nut forests. In 2001, after eight different drafts, the effort resulted in the Bolivian Standards for Forest Management Certification of Brazil Nuts (*Bertholletia excelsa*),[17] which were then submitted to the FSC for accreditation (Soldán, 2003). The Bolivian standards were conditionally accredited in 2002, although it was not until 2006 that all conditions were met (CFV, 2006).[18]

Initially there were high expectations that certification would bring financial benefits to the region and would encourage the owners of *barracas* to comply with social and ecological standards in return for the formal recognition and validation of their forest management practices by certifying bodies. Unfortunately, these standards have had little impact. Until recently, compliance has presented difficult, if not insurmountable, hurdles, particularly those requiring the actor seeking certification to have secure property rights and an approved Brazil nut management plan. Furthermore, investing in a costly and complex field assessment for certification would not offer a clear benefit such as a premium price or market access. As of mid-2008, no Bolivian stakeholder has attempted to certify their forest management under these standards.

Although FSC certification has not been successful, other arrangements like fair trade – which addresses social and equity issues associated with production – and organic certification have had an impact on the sector. A growing number of producers are gaining important advantages and market access in Europe through organic certification and compliance with fair-trade standards. At least five processing plants: Manutata Tahuamanu, Lourdes, Harold Claure Lens and *El Campesino*, have been organically certified as has the small Integral Agro-extractive Farmers' Cooperative of Pando[19] (COINACAPA). The obvious advantage of organic certification is that it opens specialized organic markets to those producers. Organically certified Brazil nuts demand premium prices, but, more importantly, prices for organic nuts are reportedly less volatile (personal communication with Casildo Quispe, president of COINACAPA, September 2006). Fair-trade markets offer small producers a clear advantage by providing a premium payment bonus for groups that qualify.

A good example of a small producer group gaining access to the fair-trade market is COINACAPA. Their strategy is to subcontract one of the region's processing plants to shell their members' Brazil nuts, which are then exported to fair-trade brokers in Europe. This was a key step, since the market is supposed to support small producers rather than the processing plants that usually act as the intermediaries exporting the nuts. By selling to overseas buyers since achieving fair-trade status in 2001, COINACAPA members have received almost twice the local market price for Brazil nuts they deliver to the cooperative. In addition, COINACAPA has used its fair-trade premium to provide health care and other services for members. As a result, COINACAPA has grown from 41 families in 2001 to 454 families in 47 agro-extractive communities in 2008.

COINACAPA leaders claim that organic certification and fair-trade arrangements that use market mechanisms to achieve benefits for producers have had more influence on their management and production practices than any norms or forest policies issued by the government. For example, to qualify for these programmes COINACAPA members have to maintain their forest holding and keep high quality-control standards relating to the cleanliness and humidity of their product and the safe post-harvest storage and transport of nuts to ensure that they are free of chemicals, fuels and other contaminants. The members are organized into groups of four or five producers at the community level to ensure compliance. If nuts spot-checked at delivery fail inspection, the entire group's lot is rejected, which creates a strong incentive for self-regulation. To demonstrate that they are small producers each member must map and document the location and size of their *castañal* (measured in number of trees), which also allows better planning. Finally, certification norms developed by the main organic certification body in Bolivia, BOLICERT (*Boliviana de Certificación*), which follow international standards (Council Regulation [EEC] No. 209/91; SIPPO, 2005), are currently interpreted to prohibit mechanized timber harvesting where Brazil nuts are collected (personal communication with Gróver Bustillos, BOLICERT general manager, June 2007). The experience has led to increased concern for maintaining forest quality and more specifically opposition by COINACAPA to logging in Brazil nut forests. The adoption of these practices was embraced by the members of COINACAPA because they offered clear advantages (in terms of income) and were not overly burdensome, allowing members to adopt them with little outside support.

While the market-related mechanisms discussed above have positively influenced Brazil nut producers, market forces and the region's greater integration into national and transnational market networks could greatly increase pressure on the forests and provide incentives for changes in behaviour. Infrastructural improvement in the region will certainly remove substantial barriers that have sheltered the region from frontier change. Since 2005 the Inter-American Development Bank has funded a major project to improve regional transportation infrastructure called the Northern Corridor (DHV, 2006) with the goal of integrating the region with the rest of Bolivia and linking it to international transportation corridors through Peru and Brazil. On 18 July 2008, the presidents of Bolivia, Brazil and Venezuela signed an agreement according to which Venezuela would contribute US$300 million and Brazil US$230 million to pave highways across northern Bolivia (IIRSA, 2008).

Road improvements will increase competition for land and resources as the region becomes more accessible to other stakeholders and markets. It is not clear how existing patterns of land and forest use will be affected. A broad discussion of the potential impacts of these changes on the region in general and on the Brazil nut sector more specifically is needed. In parallel, efforts should be initiated to engage representatives of agro-extractive communities, indigenous organizations and policy-makers, as well as representatives of *barraquero* and processing industry groups, to define the common interests of the sector and identify potential policies and strategies for mitigating the impact on forests and forest-dependent people.

RECOMMENDATIONS

Given the importance of Brazil nut production to the regional economy and Bolivia's forest exports, the Bolivian government should formulate policies to address a series of key issues.

Delineation of Brazil nut production forests

The government should accurately define and map the extent and location of Brazil nut production forests so that they can be monitored and maintained. Brazil-nut-rich forests in accessible areas are under threat of conversion for other uses or from fire. These forests should be given a special status that strengthens existing restrictions on deforestation and transformation for other uses. Recent modifications of agrarian law prohibiting illegal deforestation as a justification for FES compliance are a positive step. There should be maps at regional, departmental and municipal scale to facilitate decision-making. They should be made available to the public and used to stimulate the participation of local people and communities in the definition of future land uses.

Additional efforts to strengthen forest property rights in the region

Major efforts to complete land regularization have already occurred in Bolivia's northern Amazon. However, the government should move forward with the allocation of access rights to Brazil nut production forests (for peasant communities, indigenous people and *barraqueros*) to empower forest users and, more importantly, assign responsibility for forest stewardship. The titling of community lands has been an important advance in this direction, but extensive areas of Brazil nut forests claimed as NTFP concessions are still not legally demarcated. The government should accelerate efforts to define the legal status of these lands and ensure that they are held by actors committed to the sustainable management of Brazil nuts and forests in general. This should include NTFP management concessions for *barraqueros* and others. Furthermore, where forests traditionally managed by communities have been left outside their communally titled properties on state land, they should be granted similar rights to the forests to ensure that rural households are not separated from their livelihoods.

Greater transparency in agrarian reform process and results

Legal rights will gain greater legitimacy if they are available in publically accessible registries and maps that identify the holders of access rights to production forests. Similarly, policy-makers should promote local working groups or commissions (as the CCARC created in the 2004 Mesa decree) to mediate boundary disputes or adjustments.

Adapted guidelines for regulating land transactions and restricting land markets

Land transactions in communal territories are prohibited but occur nonetheless. There are many reasons why a family may wish to sell or transfer its forest access rights to other families, so this should be allowed, but in ways that do not violate the intended goals of supporting rural households. The government should acknowledge this situation by defining rules that permit land transactions but prevent abuse, as a way to avoid further land concentration. Under what circumstances should transactions be permitted and reflected in the legal documentation? How can the government ensure that transactions do not lead to land concentration or permit outside speculators to take over community land? These regulations should be defined in collaboration with representatives of community organizations that have first-hand knowledge of the context of these transactions and the problems that could arise with modifications.

Equitable development of the Brazil nut sector

The government should take proactive steps to develop the sector, including efforts to level the playing field. For example, providing sources of credit to small producers would help break cycles of debt peonage and provide market access independent of traditional networks. Monitoring should also ensure that industry and *barraqueros* comply with labour regulations and support the sector's workforce. In the past, some stakeholders were able to use debt relations to gain almost monopoly control over commercial networks. Government programmes (subsidies or export support) should guard against the concentration of capital or the monopolization of market access, and should instead be made available to the diverse groups dependent on the forests. Initiatives to facilitate the formation of cooperatives like COINACAPA could give small producers greater bargaining power. Such measures should be coupled with efforts to promote collaborative partnerships between community-level groups and industry, where horizontal relationships are possible, with potential benefits for both groups.

Guidelines or norms for integrated management

Households in agro-extractive communities have diverse livelihoods and Brazil nut production forests are multi-use forests for most stakeholders. It may be especially important to develop specific norms for timber management in Brazil-nut-rich forests. How can other commercial uses be reconciled in communal properties demarcated for sustainable forestry? Some communities and *barraqueros* extract timber from Brazil nut production forests, and this trend will probably expand in future. How can logging and NTFP extraction be integrated, and what measures are needed to ensure that the resource is maintained and not adversely affected? Also, agriculture plays an important role in community livelihoods, but what should happen if residents wish to invest in cattle, agroforestry or other products for commercial purposes? Answering these questions is crucial to the future of Brazil nut management in the region.

Promoting self-regulation and local governance with proper tools and guidelines

A key principle of Brazil nut management policy should be to promote self-regulation mechanisms and local governance to strengthen community decision-making in the management of land and forest resources. Tools and approaches are needed to help communities mediate conflicts, define rules for the inheritance of resource rights and ensure gender equity in access to resources. Policy should encourage the documentation and mapping of Brazil nut stands in communal territories to assist with management decisions and land-use planning. Approaches should assist residents to maintain traditional forest access systems or at least help them adapt to changing conditions.

Encouraging research in Brazil nut management

The government should encourage and fund research into best management practices for Brazil nuts and promote the dissemination of research findings to forest users to help them adapt management practices. Although Brazil nuts are an important resource and have been for decades, there are still many unknowns about their management. What is the state of regeneration within the country's Brazil nut stocks? What silvicultural practices should be promoted (for example, vine cutting or enrichment planting)? What post-harvest practices best reduce the incidence of aflatoxins?

Including Brazil nut production forests and stakeholders in emerging REDD schemes

Brazil nut production forests may be ideal for inclusion in schemes to reduce emissions from deforestation and forest degradation (REDD). The communities and *barraqueros* have so far been good stewards of the forest, but there are emerging drivers that may increase pressures on forests and lead to deforestation and forest degradation in the region. Criteria are needed for evaluating the efficiency and fairness of REDD schemes in contributing to forest conservation and improving incomes from forest management without generating conflict. It is necessary to give local stakeholders a role in defining mechanisms suitable for the region, and to make sure that benefits are not captured elsewhere.

Facilitating alliances across the Brazil nut sector

Policy should promote the formation of a 'Brazil nut bloc' representing the diverse stakeholders in the Brazil nut sector, and this effort should be shared among the organizations of stakeholders in the region through meetings and dialogue. Despite frequent bouts of conflict and antagonism, the stakeholders that comprise the Brazil nut sector share common interests and threats, so they may be better able to defend collective interests and lobby decision-makers by forming alliances around issues they identify as crucial. Such collaboration would give the sector greater power to lobby for supportive policies and could reassure them that the voices of local actors are heard by decision-makers.

CONCLUSION

Through much of its history, Bolivia has lacked a strategy or policy to promote the sound management and conservation of Brazil-nut-rich forests. At the same time, a number of policies originating outside the forest sector – from agrarian reforms to macroeconomic structural reforms – have strongly influenced Brazil nut management and trade. In the absence of a clear policy agenda for Brazil nuts, the interaction and competition between *barraqueros*, processing industries and the indigenous and peasant communities have given these stakeholders strong roles to play in developing the sector as it is today. For example, the region's remoteness allowed the remnants of the extractive economy to remain in place long into the 20th century, when they had disappeared elsewhere, and let groups harvesting Brazil nuts respond as economic, social and political conditions changed without being immediately overrun by other actors seeking land and forest resources. Competition between *barraqueros* and small producers led to instability, but at the same time discouraged others from trying to claim land in the north.

The inaccessibility of the area also meant that the stakeholders in the region had few alternatives to forest extraction, at least until the first decade of the 21st century, with improved infrastructural links to the rest of the country. As the frontier has opened over the past 20 years, some of these conditions have changed, but at the same time the Brazil nut processing industry has become stronger and more politically and economically powerful, and community-level stakeholders have gained more secure control over their forest resources as the agro-extractive communities have been given title to their land.

In general, the policy initiatives that were implemented responded to the needs of the region's economic elite. However, unprecedented activism by small producers and indigenous people turned reforms in their favour. Their success illustrates the profound change in power relationships in the region. *Barraqueros* and the business interests behind the processing industry still retain great power and influence, and the economic benefits from Brazil nuts are still concentrated in their hands, but community-level stakeholders and their representative organizations have gained notable strength.

Government policies have favoured processing plants, and while this has not directly addressed the needs of small producers, it has had a positive impact on the sector as a whole and brought benefits to most stakeholders, as Bolivia's processing industry controls a significant share of the international market in processed Brazil nuts. The demands of processors have led to a vibrant market and high demand for Brazil nuts from the region. If community-level stakeholders are able to build on their success with activist grassroots organization and the formation of cooperatives, it seems probable that they will further change the terms of trade and build alternative market networks to gain a more equitable share of the benefits.

With many actors in the region sharing an interest in maintaining Brazil nut production, there should be ample opportunity for collaboration in the sector to identify common strategies for responding to change from regional integration. Although it would be a break with precedent, a proactive effort to promote, coordinate and facilitate such collaboration by government agencies through well-thought-out policy decisions and actions would be welcome.

NOTES

1   To put this figure into context, the last census put the entire population of Pando at only 52,525 (INE, 2001).
2   With the high prices of recent years, the harvest continues for three or four months longer, an extension locally called the *zafrilla*, or little harvest.
3   Instituto Nacional de Reforma Agraria.
4   Translation by authors.
5   Tercera Marcha Indígena-Campesina por la Tierra, el Territorio y los Recursos Naturales.
6   Comisión de Conciliación, Arbitraje y Resolución de Conflictos.

7   Translation by authors.
8   Originally these fees were 0.30 bolivianos (US$0.05) per 20-kg box for unshelled nuts and 0.75 bolivianos (US$0.11) per 20-kg box for shelled nuts (*Instructivo Técnico* No. 003/97, 3 June 1997). However, two years later the fees were converted into dollars, with a US$0.005 per kg charge for unshelled and US$0.013 per kg for shelled nuts (*Instructivo Técnico* No. 003/98, 27 February 1999).
9   One of the companies subdivided its area into three separate concessions.
10   Norma Técnica para Elaboración de Planes de Manejo de Castaña (*Bertholletia excelsa Humb. & Bonpl.*).
11   Sistema Boliviano de Normalización, Metrología, Acreditación y Certificación.
12   Consejo Nacional de la Castaña.
13   Servicio Nacional de Sanidad Agropecuaria e Inocuidad Alimentaria.
14   Instituto Boliviano de Normalización y Calidad.
15   Consejo Boliviano para la Certificación Forestal Voluntaria.
16   Programa Manejo de Bosques de la Amazonía Boliviana.
17   Estándares Bolivianos para la Certificación del Manejo Forestal de castaña (*Bertholletia excelsa*).
18   The conditions required the development of a mechanism for periodic review and of indicators for evaluating the status of high-conservation-value forests, responses to requests for textual clarification and the restructuring of the format to fit guidelines for the FSC's national initiatives.
19   Cooperativa Integral Agroextractivistas Campesinos de Pando.

REFERENCES

Aramayo Caballero, J. (2004) *La reconstitución del sistema barraquero en el norte amazónico: Análisis jurídico del Decreto Supremo No. 27572*, CEJIS, Santa Cruz, Bolivia

Assies, W. (1997) 'The extraction of non-timber forest products as a conservation strategy in Amazonia', *European Review of Latin American and Caribbean Studies*, vol 62, pp33–52

Bojanic, A. (2001) *Balance is Beautiful: Assessing Sustainable Development in the Rainforests of the Bolivian Amazon*, PROMAB scientific series 4. Riberalta, Beni, Bolivia

Cámara Forestal (2006) *Anuario estadístico sector forestal de Bolivia*, Cámara Forestal de Bolivia

CFV (2006) *Estándares Bolivianos para la Certificación Forestal de la Castaña (Bertholletia excelsa)*, Certificación Forestal Voluntaria, Santa Cruz, Bolivia

CIDOB (1979) *Diagnóstico social del norte boliviano – Volumen II*, Centro de Información y Documentación de Bolivia, La Paz, Bolivia

Contreras, A. and Vargas, M. T. (2001) *Social, Environmental and Economic Dimensions of Forest Policy Reforms in Bolivia*, Forest Trends, Center for International Forestry Research, Washington, DC

Cronkleton, P., Gönner, C., Evans, K., Haug, M., Albornoz, M. A. and De Jong, W. (2008) 'Supporting forest communities in times of tenure uncertainty: Participatory mapping experiences from Bolivia and Indonesia', Thailand. International Conference on Poverty Reduction and Forests: Tenure, Market and Policy Reforms Proceedings, R. Fisher, C. Veer, and S. Mahanty (eds), Regional Community Forestry Training Center for Asia and the Pacific (RECOFTC), Bangkok and Rights and Resources Initiative (RRI), Washington

Cronkleton, P., Albornoz, M. A., Barnes, G., Evans, K. and de Jong, W. (2010) 'Social Geomatics: Participatory forest mapping to mediate resource conflict in the Bolivian Amazon', *Human Ecology*, vol 38, pp65–76

Cronkleton, P., P. Pacheco, R. Ibargüen, and M. Albornoz (2009) *Reformas en la tenencia forestal en Bolivia: La gestión comunal en las tierras bajas.* La Paz: CIFOR/CEDLA.

DHV (1993) 'Desarrollo de la Amazonía boliviana: De la actividad extractiva hacia un desarrollo integral sostenible', *Estudios agro-ecológicos, forestales y socio-económicos en la región de la castaña de la Amazonía Boliviana*, resumen ejecutivo, DHV, La Paz, Bolivia

DHV (2006) *Plan de Sección Estratégica: Evaluación ambiental estratégica del Corredor Norte de Bolivia*, Servicio Nacional de Caminos, Banco Interamericano de Desarrollo. La Paz, Bolivia

FAOSTAT (2007) Food and Agricultural Organization statistical database, www.faostat.fao.org/site/535/DesktopDefault.aspx?PageID=535, accessed 2 October 2008

Fifer, J. V. (1970) 'The empire builders: A history of the Bolivian rubber boom and the rise of the house of Suárez', *Journal of Latin American Studies*, vol 2, no 2, pp113–146

Fortmann, L., Riddell, J., Brick, S., Bruce, J. and Fraser, A. (1985) *Trees and Tenure: An Annotated Bibliography for Agroforesters and Others*, Land Tenure Center, University of Wisconsin, Madison, WI

Gamarra, M. (2007) *Amazonía norte de Bolivia. Economía gomera (1870–1940). Bases económicas de un poder regional. La Casa Suárez*, CIMA, La Paz, Bolivia

IIRSA (2008) 'Presidents sign agreement to boost "Northern Corridor", Initiative for the Integration of Regional Infrastructure in South America, www.iirsa.org/BancoConocimiento/N/noticia_corredor_norte_bolivia/noticia_corredor_norte_bolivia.asp?CodIdioma=ENG, accessed 25 August 2008

INE (2001) *Censo de Población y Vivienda 2001*, INE, La Paz, Bolivia

Mancilla, R. and Andaluz, A. (1996) 'Cambios sustanciales en la legislación forestal nacional', *Boletín BOLFOR*, vol 7, pp6–8

Morales, J. (1994) *Ajuste macroeconómico y reformas estructurales en Bolivia, 1985–1994*, Documento de Trabajo No 07/94, UCB-IISEC, La Paz, Bolivia

Newing, H. and Harrop, S. (2000) 'European health regulations and Brazil nuts: Implications for biodiversity conservation and sustainable rural livelihoods in the Amazon', *Journal of International Wildlife Law and Policy*, vol 3, no 2, pp109–124

Ormachea, E. and Fernández, J. (1989) *Amazonía boliviana y campesinado*, Cooperativa Agricola Integral Campesino Ltda, Riberalta, Bolivia

Ortiz, E. (2002) 'Brazil Nut (*Bertholletia excelsa*)', in P. Shanley, A. R. Pierce, S. A. Laird and A. Guillen (eds) *Tapping the Green Market: Management and Certification of Non-Timber Forest Products*, Earthscan, London

Pacheco, P. (1992) *Integración económica y fragmentación social: El itinerario de las barracas en la amazonía boliviana*, CEDLA, La Paz

Pacheco, P. (1998) 'Pando: Barraqueros, madereros y conflictos por el uso de los recursos forestales', in P. Pacheco and D. Kaimowitz (eds), *Municipios y Gestión Forestal en el Trópico Boliviano*, Bosques y Sociedades 3, CIFOR/CEDLA/TIERRA/BOLFOR, CID/Plural Editores, La Paz, Bolivia

Pacheco, P. (2008) *Enfoques forestales homogéneos para actores diversos: La encrucijada del manejo de bosques en Bolivia*, Documento de Trabajo, Informe Temático sobre Desarrollo Humano en Bolivia 2008, PNUD, La Paz, Bolivia

Pavez, I. and Bojanic, A. (1998) *El proceso social de formulación de la Ley Forestal de Bolivia de 1996*, CID, La Paz, Bolivia

Perez, C. A., Baider, C., Zuideman, P. A., Wadt, L. H. O., Kainer, K. A., Gomes-Silva, D. A. P., Salamão, R. P., Simões, L. L., Francisiosi, E. R. N., Valverde, F. C., Gribel, R., Shepard, G. H. Jr, Kanashiro, M., Coventry, P., Yu, D. W., Watkinson, A. R. and Freckleton, R. P. (2003) 'Demographic threats to the sustainability of Brazil nut exploitation', *Science* 302, pp2112–2114

PrensaBolivia (2007) *INRA anuncia nuevas concesiones forestales en Pando*, www.bolivia.com/noticias/autonoticias/DetalleNoticia34536.asp, accessed 2 October 2008

Rivière, H. A. de (1900) 'Explorations in the rubber districts of Bolivia', *Journal of the American Geographical Society of New York*, vol 32, no 5, pp432–440

Ruiz, S. (2005) *Rentismo, conflicto y bosques en el norte amazónico boliviano*, CIFOR, Santa Cruz, Bolivia

SIPPO (Swiss Import Promotion Programme) (2005) 'Guidance manual for organic collection of wild plants', SIPPO, Zürich, www.sippo.ch/internet/osec/en/home/import/publications/food.-ContentSlot-44399-ItemList-93332-File.File.pdf/pub_food_wildplants.pdf

Soldán, M. P. (2003) *The Impact of Certification on the Sustainable Use of Brazil Nut (Bertholletia excelsa) in Bolivia*, Food and Agriculture Organization, www.fao.org/forestry/foris/pdf/NWFP/Brazilnuts.pdf, accessed 2 October 2008

Stoian, D. (2000) 'Shifts in forest production extraction: The post-rubber era in the Bolivian Amazon', *International Tree Crops Journal*, vol 10, pp277–297

Stoian, D. (2004) 'Cosechando lo que cae: La economía de la castaña (*Bertholletia excelsa* H.B.K.) en la Amazonia boliviana', in N. Alexiades and P. Shanley (eds) *Productos forestales, medios de subsistencia y conservación*, vol 3, América Latina, CIFOR, Bogor, Indonesia

Stoian, D. and Henkemans, A. (2000) 'Between extractivism and peasant agriculture: Differentiation of rural settlement in the Bolivian Amazon', *International Tree Crops Journal*, vol 10, pp299–319

Tambs, L. A. (1966) 'Rubber, rebels and Rio Branco: The contest for the Acre', *The Hispanic American Historical Review*, vol 46, no 3, pp254–273

Zuidema, P. and Boot, R. (2002) 'Demography of the Brazil nut tree (*Bertholletia excelsa*) in the Bolivian Amazon: Impact of seed extraction on recruitment and population dynamics', *Journal of Tropical Ecology*, vol 18, pp1–31

Table 1.1 Appendix *Principal legal mechanisms affecting Brazil nuts*

Agrarian reform	
Ley No.1715, Ley del Servicio Nacional de Reforma Agraria ('INRA law')	
Decreto Supremo No. 25532, 5 October 1999	Known as the '*Barraquero* Decree', intended to create mechanisms to convert *barracas* into NTFP concessions.
Decreto Supremo No. 25783, 19 May 2000	An annulment of the 1999 decree, which was perceived as attempt to solidify *barraquero* control over the region.
Decreto Supremo No. 25848, 18 July 2000	Known as the '500-hectare decree', establishing a standard of 500ha per family for titling community lands in the northern Amazon.
Decreto Supremo No. 27572, 17 July 2004	A compromise decree of the Carlos Mesa government that confirmed the mechanisms of the '500-hectare decree' but added more mechanisms to define NTFP concessions for *barraqueros* once community land issues had been resolved.
Ley No. 3545, Reconducción Comunitaria de la Reforma Agraria (Community Redirection of Agrarian Reform Law)	A law passed by the government of Evo Morales in an attempt to refocus agrarian reform to prioritize demands by community-level stakeholders.
Decreto Supremo No. 29215, 2 August 2007	The by-laws corresponding to the above law.
Forestry regulations	
Ley No. 1700, Ley Forestal (Forestry Law)	
Resolución Ministerial No. 164, 11 November 2002	A first attempt at technical norms for Brazil nuts.
Resolución Ministerial No. 023, 2 April 2003	An annulment of the 2002 technical norms because of the lack of community consultation.
Resolución Ministerial No. 077/2005, Norma Técnica para Elaboración de Planes de Manejo de Castaña (Technical Norms for the Preparation of Brazil Nut Management Plans) [*Bertholletia excelsa Humb & Bonpl.*], 28 March 2005	A serious attempt at management norms modelled on norms already created under the new Forestry Law. To date, no management plans have been approved under these norms.
Mechanisms affecting the processing industry	
Decreto Supremo No. 21060, 29 August 1985	A structural adjustment law that created investment and export incentives for the processing industry.
Decreto Supremo No. 24498, 17 February 1997	A decree creating the Bolivian Standards, Metrology, Accreditation and Certification System (*Sistema Boliviano de Normalización, Metrología, Acreditación y Certificación*).
Decreto Supremo No. 25200, 16 October 1998	A decree creating the National Brazil Nut Council (*Consejo Nacional de la Castaña*), requiring health and quality certification for all Brazil nut exports.
Decreto Supremo No. 26081, 23 February 2001	A decree creating general norms for regulating the certification of Brazil nut safety and quality.
Labour Laws	
Ley General del Trabajo, 8 December 1942 (General Labour Law)	
Resolución Suprema No. 158242, 15 June 1971	
Resolución Suprema No. 158243, 15 June 1971	
Resolución Suprema No. 158244, 15 June 1971	
Resolución Ministerial No. 135/80, 21 April 1980	

Case Study A: In Search of Regulations to Promote the Sustainable Use of NTFPs in Brazil

Marina Pinheiro Klüppel, Júlio César Raposo Ferreira,
José Humberto Chaves and Antônio Carlos Hummel

A host of uncertainties surrounds the regulation of NTFPs in Brazil owing to the complex nature of the exploitation of these resources. On the one hand, the vast majority of these products are used for subsistence and traded in local markets (Shanley et al, 2006). Notwithstanding debate on the sustainability of the extractive economy (Homma, 1993), many traditional communities in Brazil depend on NTFPs for their livelihood. (For examples, see Diegues and Viana, 2000; Shanley et al, 2002; Schmidt et al, 2007; Sampaio et al, 2008.) On the other hand, the large-scale, unregulated commercial exploitation of some species has led to their inclusion in the official list of endangered species. This is the case, notably, for rosewood (*Aniba rosaeodora*) (May and Barata, 2004), xaxim (*Dicksonia sellowiana*), jaborandi (*Pilocarpus* spp.) (Pinheiro, 1997) and palm heart from the Atlantic forest (*Euterpe edulis*) (Reis et al, 2000), as well as several ornamental species and some medicinal plants. Creating a legal framework to regulate these diverse activities and products presents a challenge to Brazilian environmental managers and policymakers. This case study presents a brief overview of some of the steps taken in recent years by the Brazilian government to address these challenges.

FOREST POLICY AND LAW REGULATING NTFPS IN BRAZIL

The dominance of timber in legal frameworks and a species-based approach for NTFPs

Historically, forest legislation in Brazil has focused on the exploitation of timber products and has not addressed NTFPs. This was in part the result of a view in government that NTFPs were exploited only for subsistence purposes and were 'minor' forest products, and that their use had only a minor impact on the forest. Despite this timber-centric approach, some laws do exist to regulate the exploitation of NTFPs (Table A.1). These measures generally deal with single species or groups of species and are both state and national in their coverage. The species regulated by these measures are generally those found on the official Brazilian list of endangered species of flora issued by the *Instituto Brasileiro do Meio Ambiente e dos Recursos Naturais Renováveis* (IBAMA – Brazilian Institute for the Environment and Renewable Natural Resources) in 1992, or those with known problems of overexploitation.

Table A.1 *Regulations for the use and trade of NTFPs in Brazil*

Rule	Year of publication	Target species	Scope	Geographic/administrative coverage
Portaria IBDF no. DC-20 (IBDF[1] Order no. DC-20)	1976	*Araucaria angustifolia* (paraná pine)	Prohibits the cutting of paraná pine and its extraction in April, May and June.	National
Portaria IBDF no. 122 (IBDF Order no. 122)	1985	*Bertholletia excelsa* (Brazil nut)	Prohibits the cutting of and trade in Brazil nut trees.	National
Portaria IBAMA no. 118-N (IBAMA Order no. 118-N)	1992	*Ilex paraguariensis* (erva-mate)	Regulates the exploitation and trade of erva-mate.	National
Instrução Normativa IBAMA no. 05 (IBAMA Normative Instruction no. 05)	1999	Palm heart	Regulates the exploitation, processing, transportation and trade of palm heart.	National
Resolução CONAMA[2] no. 294 (CONAMA[2] Resolution no. 294)	2001	*Euterpe edulis* (palm heart from Atlantic forest)	Regulates the management plan for palm heart from Atlantic forest.	State of Santa Catarina
Lei Estadual no. 631 (State Law no. 631)	2001	*Heteropsis* spp (titica vine) and *Clusia* spp (cebolão vine)	Regulates the exploitation, transport, and trade of the titica and cebolão vines.	State of Amapá
Portaria IBAMA no. 04 (IBAMA Order no. 04)	2001	*Banisteriopsis caapi* and *Psychotria viridis* (components of ayahuasca)	Regulates transport permits and other instructions for the transport and conservation of ayahuasca.	State of Acre
Portaria IBAMA/IMAC³ no.0 01 (IBAMA/IMAC³ Order no. 001)	2004	NTFPs	Establishes basic procedures for the sustainable use of NTFPs by traditional populations and rural producers.	State of Acre
Portaria IBAMA/IMAC no. 01 (IBAMA/IMAC Order no. 01)	2005	*Uncaria tomentosa* and *U. guianensis* (cat's claw)	Regulates the exploitation of cat's claw.	State of Acre
Decreto Estadual no. 25.044 (State Decree no. 25.044)	2005	*Carapa guianensis* and *C. paraense* (andiroba); *Copaifera trapezifolia, C. reticulata* and *C. multijuga* (copaíba)	Prohibits licensing of the cutting, transportation or trade in andiroba and copaíba wood.	State of Amazonas
Decreto Federal no. 5.975 (Federal Decree no. 5.975)	2006	*Bertholletia excelsa* (Brazil nut) and *Hevea* spp. (rubber tree)	Prohibits the cutting of Brazil nut and rubber trees for timber.	National

Table A.1 *Regulations for the use and trade of NTFPs in Brazil (Cont'd)*

Rule	Year of publication	Target species	Scope	Geographic/administrative coverage
Instrução Normativa SDS[a] *no. 02* (Normative Instruction SDS[a] no. 02)	2006	*Aniba rosaeodora* (rosewood)	Regulates the extraction of rosewood trees in sustainable management areas and cultivated areas.	State of Amazonas
Instrução Normativa SDS no. 01 (Normative Instruction SDS no. 01)	2008	*Heteropsis flexuosa* (titica vine), *H. jenmanii* (timbó-açu and titicão vines), *Philodendron sp.* (ambé vine)	Regulates the management of some species of vines for commercial purposes, based on traditional practices for sustainable extraction and on scientific research.	State of Amazonas
Instrução Normativa IBAMA no. 05 (IBAMA Normative Instruction no. 05)	2008	NTFPs	Regulates permits for the export of commercial species on the federal and state lists of endangered species.	National

Notes: [1]The *Instituto Brasileiro de Desenvolvimento Florestal* (IBDF – Brazilian Institute of Forest Development) ceased to exist when IBAMA was created in 1989.
[2]The *Conselho Nacional do Meio Ambiente* (CONAMA – National Council of the Environment), established by Law 6.938 of 1981, regulates national environmental policy. CONAMA is composed of several government bodies and representatives of civil society, and one of its main missions is to 'establish rules, criteria and standards for control and maintenance of environmental quality towards an optimal use of environmental resources'.
[3]*Instituto do Meio Ambiente do Acre* (IMAC – Environmental Institute of Acre).
[4]*Secretaria de Meio Ambiente e Desenvolvimento Sustentável do Amazonas* (SDS – Secretary of Environment and Sustainable Development of the State of Amazonas).

There are some laws that prohibit the exploitation of species with both timber and non-timber uses, in an effort to protect the species' non-timber use. This is the case with the paraná pine (*Araucaria angustifolia*), which was banned for exploitation in 1976, the Brazil nut (*Bertholletia excelsa*), protected in 1985, the rubber tree (*Hevea* spp.) and, more recently, the andiroba (*Carapa guianensis, C. paraense*) and copaíba (*Copaifera trapezifolia, C. reticulata, C. multijuga*), the exploitation of which has been prohibited in the State of Amazonas since 2005 (Table A.1).

Decentralization in recent years

Environmental management falls under federal, state and municipal governments. The federal government sets general norms; states, districts and municipalities adapt and implement laws according to regional conditions. This shared responsibility was established in 1981, in order to licence and control potentially polluting activities, by Law 6.938/81, which created the National Policy on the Environment. Until the late 1980s, however, the federal government had exclusive authority over the management and regulation of forests under the 1965 Brazilian Forest Code (Law 4.771/65).

In 1988, the federal constitution established separate competencies for forest management for federal, state and municipal governments, but this new law contradicted the 1965 Forest Code, leading to a duplication of responsibilities and costs, as well as legal ambiguities and confusion.[1] In this context, only a few states took the initiative to regulate the use of NTFPs. These included specific norms for NTFPs developed in Santa Catarina, Amapá, São Paulo and Acre (Box A.1).

In 2006, IBAMA's Department of Forestry initiated a process to revise and update the federal regulation of forest resources. This was due to dramatic changes in forest exploitation since the Forest Code had been published, in particular the increasing rate of deforestation in the Amazon. This led to the Public Forests Management Law (Law 11.284/06), which lays down a participatory and decentralized approach to forest management, through close collaboration between federal and state governments under the umbrella of a technical cooperation agreement between each state and IBAMA. This law provides the legal foundation necessary for states to regulate NTFPs, and since 2006 IBAMA has signed technical cooperation agreements with almost all Brazilian states. These days, IBAMA and others favour a regional and species-based approach to NTFP regulation that reflects local conditions. IBAMA also recommends the use of good-practice guidelines for key species, rather than restrictive regulations.

Activities in other areas

In addition to these developments, other recent complementary and positive steps by the government with an impact on NTFP harvesting, management and trade include a move by the Ministry of Health to implement the National Policy on Medicinal Plants and Phytotherapies, as approved by Decree 5.813 in June 2006. This policy establishes guidelines for government intervention in the area of medicinal plants and phytotherapies, with the goal, among others, of promoting and disseminating popular practices for their use. It also seeks to promote the sustainable use of biodiversity and a fair distribution of benefits derived from traditional knowledge and genetic resources (Ministério da Saúde, 2006).

The Ministry of Environment is also developing a plan to improve livelihoods and promote sustainability in indigenous and local communities through measures including the commercialization of forest products. The first priority is to set a minimum price for some NTFPs. A price has already been set by the National Monetary Council for four products: rubber, açaí, pequi (*Caryocar brasiliense*), and babaçu (*Orbignya phalerata*). In late 2008, studies by the Ministry of Environment and the National Monetary Council were concluded, and prices fixed for four other products: copaíba oil, andiroba, brazil nut and carnaúba (*Copernicia prunifera*).

Box A.1 Legislation of NTFPs: The case of Acre

The State of Acre is the only Brazilian state with comprehensive laws regulating NTFPs. In 2000, the government of Acre formed a task force comprising representatives of several governmental departments, non-governmental organizations, civil society and the scientific community. Its goal was to advance knowledge on the extraction, processing and trade of NTFPs, in order to organize the sector and promote the sustainable use of these products. The task force achieved this in several crucial ways, among them:

* definition of the most important species;
* development of terms of reference for the study of the production chains and biological and ecological characteristics of 14 key species: açaí (*Euterpe* spp.), andiroba (*Carapa guianensis*), bacaba (*Oenocarpus mapoa*), buriti (*Mauritia flexuosa*), Brazil nut (*Bertholletia excelsa*), copaíba (*Copaifera* spp.), ipê-roxo (*Tabebuia* spp.), jagube (*Banisteriopsis caapi*), jarina (*Phytelephas macrocarpa*), jatobá (*Hymenaea courbaril*), patoá *(Oenocarpus bataua),* murmuru (*Astrocaryum* spp.), rubber tree (*Hevea brasiliensis*) and cat's claw (*Uncaria* spp.);
* preparation of technical guides on the management and harvesting of the 14 key species;
* development of legal instruments regulating the use of NTFPs (Table A.1);
* development of species-based legal instruments to regulate the use of cat's claw (*Uncaria tomentosa* and *U. guianensis*), jagube/mariri (*Banisteriopsis caapi*) and rainha/chacrona (*Psychotria viridis*), the latter used for religious purposes (Table A.1).

Under *Portaria Interinstitucional IBAMA/IMAC no. 001* (IBAMA/IMAC Order no. 001) of 12 August 2004, duties relating to the extraction and transportation of NTFPs were divided between IBAMA and IMAC, the state's environmental agency. However, this role was taken over entirely by IBAMA, and activities by IMAC were sidelined. With the decentralization of forest management under Law 11.284/06, some responsibilities were transferred from IBAMA to IMAC. By early 2008, all documentation and affairs related to NTFPs had been transferred. As of mid-2008, IMAC was revising its own legal instruments regulating the extraction and transportation of NTFPs, both in general and for the 14 key species in particular.

TOWARDS AN INSTITUTIONAL FRAMEWORK FOR NTFPS: IBAMA'S 2006 WORKSHOP

In 2006, the Department of Forestry organized a workshop called 'Limitations and opportunities for the regulation of NTFPs' to address issues associated with the regulation of NTFPs. These included whether a complete institutional restructuring was needed, or whether NTFP regulation could be based on the timber model; how to differentiate between commercial and subsistence use; and whether to establish laws only for endangered species. The workshop assembled several key actors involved in this topic – extractors, companies, researchers, natural resources managers, legislators and NGOs – and at the end, three task forces were created to focus on three specific issues and came up with the recommendations set out below.

Should NTFPs be regulated? If so, how?

1   Subsistence use and local trade of NTFPs should be excluded from regulations. The first task force found unanimously that using NTFPs for subsistence and selling them in local markets – the practice with the vast majority of such products in Brazil – did not generally compromise the conservation of the species exploited, and therefore this category of products and uses should not be included in legislation regulating NTFPs. Rather than promoting conservation, such inclusion would jeopardize the livelihoods of groups depending on NTFPs that were usually poorly organized and had limited purchasing power.
2   Regulations should be species-specific and modified to reflect local conditions. Another recommendation made by the task force was that regulations on NTFP exploitation should not be national and centralized, but instead drafted in a species-specific way when possible. Regulations should also vary by region and be drafted with the involvement of local stakeholders who had direct or indirect relationships with the NTFP. In this way, laws would better address the diverse environmental, social and economic aspects of the exploitation of the resource. These recommendations have been largely given effect through the decentralization of NTFP law and policy under Law 11.284/06, with control moving from IBAMA and the federal government to the states.
3   Selection of priority species for regulation should be systematic. The selection of priority species requiring the drafting of specific regulations could either be based on the official list of endangered species issued by IBAMA or be set by states, based on results from scientific research identifying species in high commercial demand and/or associated with known problems such as overexploitation or social conflicts around harvesting.

Should the extraction of NTFPs be regulated through forest management plans? Should replanting NTFPs be mandatory?

In the view of the second task force, the requirement of an official management plan for the extraction of NTFPs created obstacles for communities and small groups, and turned traditional and extractive reserve harvesting into illegal activities. For example, under Decree 750 of 1993, all palm harvesters in the Atlantic forest must work according to a management plan approved by the local environment authority. However, many extractors do not have the means to finance such plans and have either abandoned palm heart harvesting or begun to operate illegally (Sales et al, 2000). The task force also recommended that the need to replant NTFPs should be decided on a case-by-case basis, with the primary concern being the maintenance and continuity of the species under regulation, bearing in mind that for many species, replanting is not necessary.

The task force highlighted the need to differentiate between NTFP extraction for subsistence and that for commercial trade. Requirements for management plans and documentation certifying the origin of material were considered reasonable and useful for commercial trade. In any case, companies purchasing non-timber resources have to document the origin of resources used (e.g. requirements for registration in the *Cadastro Técnico* Federal),[2] and many foreign buyers are also requesting this type of documentation as the market for certified organic, sustainable and fair-trade products grows. It was suggested that companies purchasing materials from traditional communities and small producers underwrite the costs of drafting management plans for their suppliers.

Should transport permits be required for NTFPs? If so, in which circumstances?

The *Autorização de Transporte de Produtos Florestais* (ATPF – Transport of Forest Products Licence) established by IBAMA Order (*Portaria*) no 44-N of 1993 meant that a licence was required for the transport of products and by-products of timber, and also of xaxim, palm, essential oils, ornamental plants and medicinal and aromatic plants, as well as the seedlings, roots, bulbs, vines and leaves of indigenous plants. This restriction made it impossible for many NTFP harvesters to participate in commercial trade – or to do so legally, at any rate. In fact, many products that required an ATPF to be transported were not subject to any legal restrictions on harvesting, so the transport requirement created a level of control that did not reflect concerns about sustainability. The task force recommended that official documentation be required for the transporting of only those NTFPs whose exploitation was regulated by legislation owing to their being endangered or otherwise identified as suffering from overexploitation.

These recommendations had an immediate impact on the formulation of NTFP law and policy. Decree 1.282 of 1994, regulating the forests of Brazil, was in the final stage of revision by the time the workshop ended in 2006. The original decree permitted exploitation of the Amazon forest only through sustainable forest management and in accordance with prescribed general principles and technical requirements. It was replaced by Decree 5.975 of 2006, which made the exploitation of the Amazon forest subject to the approval of a sustainable forest management plan, defined in the new decree as 'the technical reference containing the guidelines and procedures for the administration of the forest, seeking to obtain economic, social and environmental benefits' [authors' translation].

The technical procedures for the preparation, presentation, implementation and technical evaluation of the sustainable forest management plans instituted by Decree

5.975 were set down in Normative Instruction (*Instrução Normativa*) no. 05 of 2006 (published soon after the decree). Guided by the document generated at the workshop, the task force drafting the Normative Instruction opted, after extensive technical and political consultations, not to include in it a model NTFP management plan, but only one for timber products. The Normative Instruction states: 'For the exploitation of non-timber products that do not require a transportation permit, as established in specific regulations, the landowner need only inform the official environmental authority on an annual basis of their activities, including species harvested, products and quantities extracted, until the publication of specific regulations for a given species' [authors' translation].

Following the workshop, Order no. 44-N, which established the ATPF, was replaced by Normative Instruction no. 162 of 2006, establishing the *Documento de Origem Florestal* (DOF – Document of Forest Origin). The result is that medicinal plants, ornamental and aromatic plants, seedlings, roots, bulbs, vines and leaves of native origin not found on the official list of endangered species or the annexes

---

Box A.2 Palm heart and Normative Instruction 05 of 1999: The challenges of accommodating different species, extracting practices and commercial demand through legislation

In 1999 IBAMA published Normative Instruction 05 of 1999 to regulate the exploitation and trade of palm heart. Brazil has at least four species of palm trees that supply the palm heart industry: *Euterpe edulis*, indigenous to the Atlantic forest; *Bactris gasipaes*, the peach palm, indigenous to the Amazon forest, but now cultivated on a large scale; and *Euterpe oleracea* and *Euterpe precatoria*, respectively açaí de touceira and açaí solitário, also indigenous to the Amazon forest. Each species has different ecological characteristics and exploitation practices, but the Normative Instruction regulates the entire production chain of palm heart in a generalized way without considering the particularities of each species. It was published with the main objective of restricting the disorderly extraction of *E. edulis* palm heart, which was Brazil's main source of palm heart at the time. Because of its great value in national and international markets, the species was declared an endangered species by IBAMA Order no. 37-N in 1992. With the decline of natural populations of *E. edulis* and the toughening of legislation on the Atlantic forest biome (Mata Atlântica),[3] the other species of palm tree have gradually been replacing the Atlantic forest palm in supplying the palm heart market.

In recent years, the pulp of the açaí fruit became popular in health and sports markets throughout Brazil, as well as internationally, and outstripped palm heart as the most valuable product obtainable from *E. oleracea*. However, *E. oleracea* harvesters cannot sell palm heart from trees cut for açaí fruit due to restrictions imposed by Normative Instruction 05, including official management plans and transportation documentation.

to CITES (the Convention on International Trade in Endangered Species of Wild Fauna and Flora) do not require a DOF. Only those NTFP species whose collection is regulated by another legal instrument require authorization of their transport by the national manager of the environment.

Workshop conclusions and recommendations

The conclusions and recommendations of the IBAMA workshop were compiled into a document of guidelines for the Department of Forests (Diretoria de Florestas, 2006), and had a wide-ranging impact, partly through the consultation entailed in revising the forestry laws. However, although the direct revision of standards as described above was important, this was not the most significant contribution of the workshop. Its main legacy was a change in the paradigm according to which IBAMA viewed NTFP exploitation. The complexity of regulating NTFPs in a country like Brazil – rich not only in species, but also in culture and customs – has become part of the policy discussion, and this has largely resulted from consulting with and listening to a range of stakeholders.

Documents and laws that have no basis in local groups' own realities will not ensure the conservation of a specific resource. It is only through a broad consultative process that laws addressing such complex areas can achieve the sustainable management of forest resources, and economic and social development.

Acknowledgements

The authors wish to thank Dr Patricia Shanley of the Center for International Forestry Research (CIFOR) for her support and a first revision of the manuscript.

NOTES

1   The federal government body in charge of forests is IBAMA, while state and municipalities are represented by their environment secretaries. Both IBAMA and the state environment secretaries fall under the Ministry of Environment, which is in charge of environmental public policy in Brazil.
2   The *Cadastro Técnico Federal* (Federal Technical Registry) is an official IBAMA database of companies and individuals that carry on potentially polluting activities and/or who are users of environmental resources. The law requires companies in these categories to register.
3   Decree 750 of 1993 prohibits the cutting, exploitation and removal of primary vegetation and vegetation in medium and advanced stages of regeneration in the Atlantic Forest.

REFERENCES

Diegues A. C. and Viana, V. M. (eds) (2000) *Comunidades tradicionais e manejo dos recursos naturais da Mata Atlântica*, NUPAUB/LASTROP, São Paulo

Diretoria de Florestas (2006) *Limites e oportunidades para a regulamentação de PFNM*, encaminhamentos dos grupos de trabalho, oficina realizada em 20 e 21 de julho, IBAMA, Brasília

Homma, A. K. O. (1993) *Extrativismo vegetal na Amazônia: Limites e oportunidades*, EMBRAPA, CPATU, Brasília

May, P. H. and Barata, L. E. S. (2004) 'Rosewood exploitation in the Brazilian Amazon: Options for sustainable production', *Economic Botany*, vol 58, no 2, pp257–265

Ministério da Saúde (2006) *Política nacional de plantas medicinais e fitoterápicos*, UNESCO/Ministério da Saúde, Brasília

Pinheiro, C. U. B. (1997) 'Jaborandi (*Pilocarpus* sp., Rutaceae): A wild species and its rapid transformation into a crop', *Economic Botany*, vol 51, no 1, pp49–58

Reis, A. S., Fantini, A. C., Nodari, R. O., Reis, A., Guerra, M. P. and Montovani A. (2000) 'Management and conservation of natural populations in Atlantic rain forest: The case study of palm heart (*Euterpe edulis* Martius)', *Biotropica*, vol 32, no 4b, *Special Issue: The Brazilian Atlantic Forest*, pp894–902

Sales, R. R., Maretti, C. C., Portilho, W. G. and Soares, S. G. (2000) 'Programa de regularização da exploração comercial do palmito juçara Euterpe edulis', in A. C. Diegues and V. M. Viana (eds) *Comunidades tradicionais e manejo dos recursos naturais da Mata Atlântica*, NUPAUB/LASTROP, São Paulo, pp81–88

Sampaio, M. B., Schmidt, I. B. and Figueiredo, I. B. (2008) 'Harvesting effects and population ecology of the Buriti Palm (*Mauritia flexuosa* L.f., Araceae) in the Jalapão region, Central Brazil', *Economic Botany*, vol 62, no 2, pp109–126

Schmidt, I. B., Figueiredo, I. B. and Scariot, A. (2007) 'Ethnobotany and effects of harvesting on the population ecology of *Syngonanthus nitens* (Bong.) Ruhland (Eriocaulaceae), a NFTP from Jalapão Region, Central Brazil', *Economic Botany*, vol 61, no 1, pp73–85

Shanley, P., Luz, L. and Swingland, I. R. (2002) 'The faint promise of a distant market: A survey of Belém's trade in non-timber forest products', *Biodiversity and Conservation*, vol 11, pp615–636

Shanley, P., Pierce, A. and Laird, S. (2006) *Além da madeira: A certificação de produtos florestais não-madeireiros*, Center for International Forestry Research, Bogor, Indonesia

Integrating Customary and Statutory Systems: The Struggle to Develop a Legal and Policy Framework for NTFPs in Cameroon

Sarah A. Laird, Verina Ingram, Abdon Awono, Ousseynou Ndoye, Terry Sunderland, Estherine Lisinge Fotabong, Robert Nkuinkeu

INTRODUCTION

Plants are used in complex and varied ways throughout Cameroon. Household compounds contain regularly used medicinal, food, ornamental and protective species, many brought from the forest. "Wrapping" leaves are harvested for use in almost every forest village, and forest spices distinct to a regional cuisine are consumed locally and traded widely, including to urban centres where demand for forest plant products and 'bushmeat' persists in the tastes and diet of city-dwellers. Medicinal bark from a few trees has found favour in international markets, and demand from people thousands of miles away for medicines to treat prostate problems (*Prunus africana*) or enhance sexual performance and provide energy (*Pausinystalia johimbe*) has created trade networks throughout the forest zone. Forest fruits, spices, wild greens, thatching and fuelwood species, medicines, protective plants and those with myriad other uses combine to form what are known as 'non-timber forest products'.

The difficulty of regulating such diverse products as a single group is evident. What can the objectives of such regulation be, and how is it possible for the government to, for example, promote the objectives of sustainability and equity in trade, without undermining layers of other important relationships between people and their environment? Indeed, the government of Cameroon has struggled with the regulation of NTFPs, beginning with the very definition of what they will regulate, and for what purpose. The will to do so has also been limited since the majority of NTFPs – unlike timber – do not have values that can easily be captured by government.

Since the 1990s, international agencies have pressured the government to pay attention to these products as part of a new approach to forest management that incorporates values beyond timber. Although NTFPs have received increased attention from

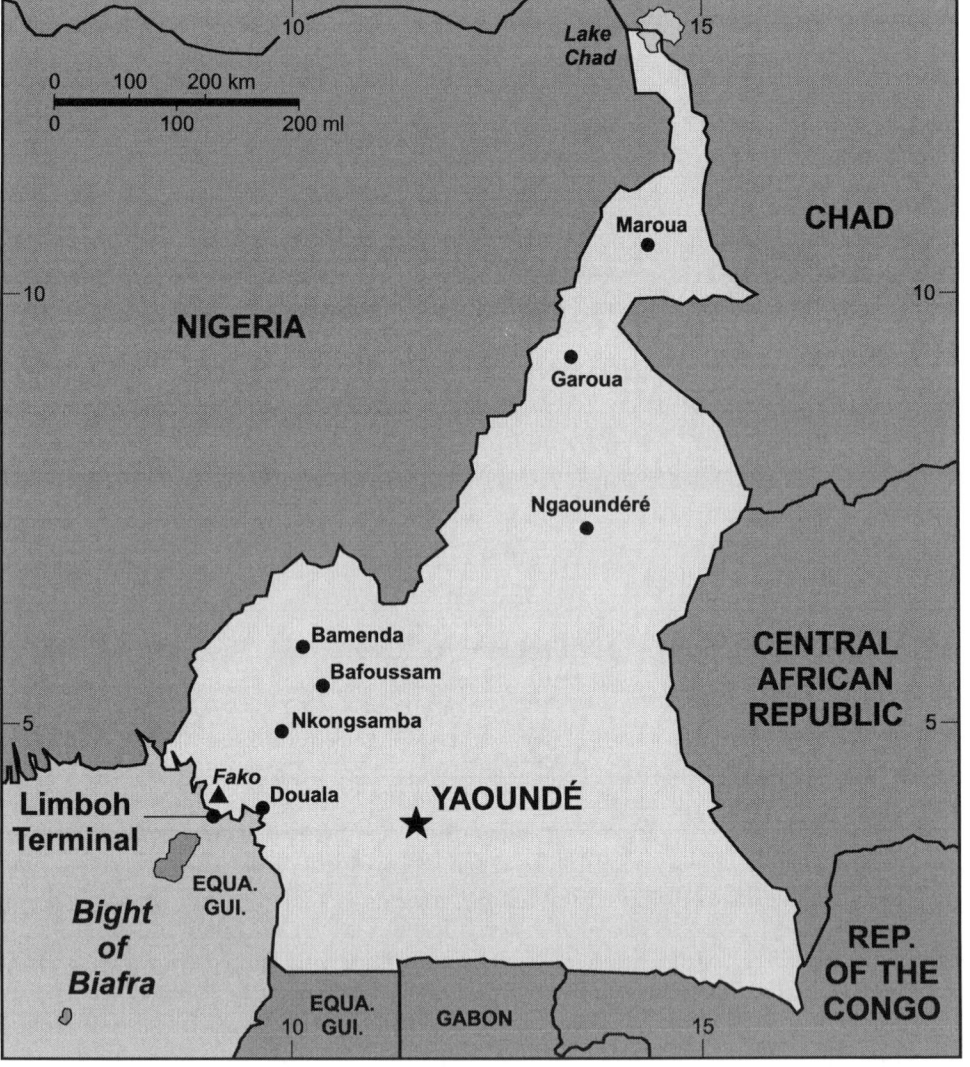

Source: CIFOR.

Figure 2.1 *Map of Cameroon and neighbouring countries in Africa*

researchers and policy-makers, this has yet to create real change in the policy framework, which is much as it has always been: the vast majority of NTFPs – those consumed on a subsistence basis or found in local trade – are regulated de facto by customary laws relating to land tenure and resource rights. On the other hand, most high-value species in national and regional trade, and internationally-traded medicinal and food plants, are subject to statutory laws that set quotas, permits and taxes, but these laws are inconsistent and confusing.

In this chapter we will review the major areas of law that impact NTFPs – land tenure and resource rights; forestry and environment law; and finance and taxation – and will discuss the institutions responsible for implementing these measures. We then offer conclusions and recommendations on ways policy-makers might address NTFP regulation more effectively.

LAND TENURE

Layers of customary and statutory laws regulate land and resource rights, reflecting the cultural, biological, political and economic diversity of the country. Statutory land rights grow from a mixed colonial heritage: Cameroon was once a German colony, subsequently divided into British and French Cameroons, and then united into a single republic. Under British colonial law, 'vacant' lands were considered the property of local communities, and were placed under the control of Native Authorities. Under French colonial law, however, all lands 'vacant and without master' belonged to the state. When the two colonial territories were unified into the Republic of Cameroon, and the legal systems subsequently merged in 1972, the British concept of communal land was replaced in favour of the French system (Burnham and Sharpe, 1997).

The 1974 Land Ordinance classified land into three major categories. *Public state land* consists of lands that prior to independence were held by foreigners, usually large plantations which after independence became state property. Some are managed by parastatal organizations such as the Cameroon Development Corporation (CDC), some have reverted to natural forest cover and others are used for public purposes. *Private land* comprises land registered by private individuals (actual persons or international organizations). *National domain land*, which is all land not registered, is divided into two categories: vacant land and land occupied and worked by indigenous populations.

Following the French model, in 1974 a large number of hitherto communally managed lands were transferred from customary control to state control. These areas include most secondary and primary forest areas, and the resources found in them. In addition, ownership over naturally growing (but not planted) trees on private land and all trees planted or naturally growing on land without a title deed are considered the property of the state. The vast majority of landholdings in rural areas, in some cases more than 90 per cent, do not have a formal title deed, largely because the process to register is expensive and bureaucratically complex (Tonye et al, 1993; Egbe, 1997; Ewane et al, 2009; Ndumbe et al, 2009). Under statutory law, therefore, the majority of NTFPs fall under government control, but in practice most NTFPs continue to be harvested and managed under better-known and respected customary laws.

RESOURCE RIGHTS

The 1994 Forestry Law addresses the issue of resource rights removed by the 1974 Land Ordinance by providing customary user rights, or *droit d'usage*, to forest communities. These allow communities to collect 'all forest, wildlife, fisheries products freely for their personal use, except protected species' (section 8, Cameroon Forestry Law, 1994). This right can be exercised in all unprotected areas, and includes subsistence fuelwood and wood for construction needs. Timber sales are not included as a user right, and instead are regulated under systems of smallholder titles or through the community forestry process created by the 1994 law.

The right of local people to exploit forest resources falls into two broad categories: 'free' access and 'paid' access. *Free access* is the usufruct right first mentioned in section 8 of the 1994 law and defined by section 4 of the Wildlife Decree of application No. 95/466-PM. *Free access* may be exercised in communal and community forests. *Paid access* refers to the right to exploit an NTFP following receipt of an exploitation permit from the government (Ngwasiri et al, 2002) and covers an assortment of 13 types of 'Special Forest Products' of interest to the government (Box 2.1).

Despite the existence of a natural resource statutory framework increasingly refined over the past 15 years, in practice most communities are unaware of statutory laws. In addition, when they are known or – as is often the case – arbitrarily enforced, statutory laws are often viewed as illegitimate and in the service of a small elite (Assembe, 2009). For the most part, government capacity is weak, and its presence is manifested primarily when community lands are allocated to outsiders for logging, mining or commercial agriculture or are included in national parks. As a result, in rural communities customary law continues to be the dominant system of governance for land and resource rights associated with NTFPs, and conflicts often erupt when statutory law intersects with customary law (Barume, 2004; Assembe, 2009).

Customary laws address – with a level of legitimacy and specificity absent in most government regulation – who owns resources, who can harvest them, where harvesting may take place and in what quantity, and who benefits and in what ways. Although this differs across Cameroon, in general harvesting NTFPs on lands held by a family may take place only with the family's permission; on communal village lands any member of the community can harvest products for subsistence use, but for higher-value products intended for sale (particularly timber, but also including some high-value NTFPs) approval is required from the chief or village council. Outsiders always require permission to harvest resources and must often provide some form of compensation before entering village lands.

Although more widely followed than statutory law, the effectiveness of customary law varies significantly. It is often weak in areas with increasing populations close to urban centres, or those characterized by cultural and social change that has undermined traditional institutions. In these cases, well-crafted and implemented statutory law could play an important role. Statutory law could also support sustainable and equitable practices when commercial pressure on resources is great and traditional structures are undermined by this pressure. For example, village chiefs and councils often receive payments or gifts to grant permission for harvesting high-value resources, even when these activities are not supported by the wider community. This has been well-documented for the sale of timber rights (Cuny et al, 2007), but it is also the case for high-value NTFPs. In the case of *Prunus africana*, for example, internal conflicts have resulted when chiefs and village councils harvest bark themselves and do not share profits, or receive payment from outsiders to harvest, often unsustainably, in village forests. The wider community, which may also earn cash from bark harvests, benefits little from these activities, and in some cases this has led to a scramble for limited resources as practices shift from the sustainable to the unsustainable, with the idea that 'if the fon [traditional ruler] can do it, so can we'.

Despite the dominance of customary law in rural communities, the legitimacy of traditional governing structures in Cameroon is often disputed. This includes chiefs, many of whom do not represent indigenous institutions, and instead were first installed by colonial governments in search of cooperative counterparts (Geschiere, 1993; Konings, 1999; Oyono, 2004, 2005). In many areas, local associations and community groups were established to improve the sustainability and equity of NTFP harvests and trade, and have lobbied for legal and institutional changes. Some have come into conflict with traditional institutions that are threatened by efforts to control these aspects of community life (WHINCONET, 2005; Cunningham, 2006; Ingram, 2008; Ingram et al, 2009).

FORESTRY AND ENVIRONMENT LAWS

Timber is the most valuable resource in Cameroon's forests and enjoys the lion's share of attention from policy-makers. However, in the 1990s international agencies, in particular the World Bank, promoted forestry laws that reflect a wider range of objectives and priorities, and emphasize sustainability and equity. In Cameroon, this resulted in the 1994 Forestry Law, which some refer to as a 'major interference of Bretton Woods experts' (Ekoko, 1999; Assembe, 2009). However well-intentioned, the 1994 law was developed without adequate or meaningful consultation with people living in the forest zone and important stakeholders such as NTFP traders and harvesters. As a result, the text is often deeply out of touch with local realities, and the law has proven largely ineffectual and in many cases undermines the very objectives it sought to achieve (Sharpe, 1998; Ekoko, 1999; Burnham, 2000; Njamnshi et al, 2008; Assembe, 2009). Revision of the 1994 law is currently under way in order to address many of its deficiencies, including the regulatory framework for NTFPs (FAO, 2009).

Box 2.1 Forestry laws in Cameroon

Since independence in 1960, Cameroon has enacted five pieces of legislation dealing with forest resources.

- Law No. 68/1/COR of 18 July 1968 regulated forest resources in the French-speaking areas of the country.
- Ordinance No. 73/18 of 22 May 1973 and its decree of application, No. 74/357 of 17 April 1974, apply to the whole of Cameroon.
- Law No. 81–13 of 27 November 1981 and three decrees of application, all issued on 12 April 1983, had a wider scope, dealing with forestry, wildlife and fisheries resources.
- Law No. 94/01 of 20 January 1994 has been followed to date by only two decrees of application (No. 95/466-PM of 20 July 1995 on wildlife and No. 95/531-PM of 23 August 1995 on forestry).
- Décision No 0336/D/MINFoF du 6 Juillet 2006 set the list of Special Forest Products.

Definitions and scope

As part of a newly expanded view of forest values, NTFPs are addressed in a number of sections of the 1994 law. However, none of the five forestry measures enacted over the last 50 years (Box 2.1) defines 'non-timber forest products'. Instead they provide lists of products referred to as 'minor forest products', 'secondary forestry products' or 'forest produce other than timber'. The 1994 law refers to 'Special Forest Products' as 'certain forest products, such as ebony, ivory, wild animals, as well as certain animal, plant and medicinal species or those which are of particular interest and shall be classified as special' (section 9(2)). The law does not give criteria or definitions of terms such as 'certain', 'interest' and 'special', and the extremely diverse collection of products included in the list of Special Forest Products elaborated more than ten years later in 2006, and each year after that, does little to clarify the wider intentions behind the law (Box 2.2; Décision No 0336/D/MINFoF du 6 Juillet 2006, fixant la liste des produits forestiers spéciaux présentant un intérêt particulier au Cameroun).

For example, the annual Special Forest Products lists include species that are native and introduced; widely cultivated and wild harvested; industrial (primarily exported) and consumed locally; and timber and non-timber. Numerous high-value NTFPs in trade – such as *Ricinodendron heudelottii* (njangsang) and *Dacryodes edulis* (bush plum) (Ndoye and Kaimowitz, 2000; Pérez et al, 2000; Awono et al, 2002b; Tajoacha, 2008) – are not mentioned in the 1994 law or the 2006 list of Special Forest Products. Some native forest species grown primarily on farms or in fallows – e.g. *Cola acuminata* and *Cola nitida* – are classified as agricultural crops and not Special Forest Products. However, introduced and cultivated Eucalyptus is included on the Special Forest Products list. According to the 2006 decision (Box 2.2), some species are included due to levels of threat or endangerment that make them 'special', but this group is also inconsistent: some species that are covered by the Convention on International Trade in Endangered Species (CITES), such as the medicinal tree *Prunus africana*, are found on the list, but others, such as the timber species *Pericopsis elata*, are not.

Article 12 of the 1994 law establishes national sovereignty over genetic resources and describes requirements for prior informed consent and benefit sharing with the government; articles 64 and 65 of the Environmental Framework Law of 1996 likewise lay down requirements for genetic resources. But here, too, definitional problems arise: distinctions between genetic resources supplied for bioprospecting and medicinal plants traded in bulk as commodities remain poorly elaborated, and these articles only add to regulatory confusion.

The 1994 law also includes fish and fauna in its scope, and provides three classes of protection that regulate the hunting and exploitation of different species through a system of permits and controls (article 78). These species are not included in the Special Forest Products list, which is focused on botanical resources. This is appropriate given the enormous role of bushmeat and fish in the country's economy and livelihoods, the different regulatory issues raised by mobile species, and the need for a distinct legal and policy framework for these resources.

A wide range of diverse and complex forest uses are covered by the 1994 law, but most are poorly elaborated. The NTFP elements are particularly inconsistent, in part

---

Box 2.2 Special Forest Products regulated by the 1994 Forestry Law

The list of permits for Special Forest Products is revised annually by the Cameroon Ministry of Forestry and Wildlife (MINFOF, 2009). The 2006 Decree listed 13 products as Special Forest Products:

- ébène (*Diospyros crassiflora*)
- eru (*Gnetum africanum, G. buchholzianum*)
- pygeum (*Prunus africana*)
- yohimbé (*Pausinystalia johimbe*)
- wild rubber (*Funtumia elastica*)
- rauvolfia (*Rauvolfia macrophylla*)
- rattan (*Eremospatha* spp., *Laccosperma* spp.)
- gomme arabique (*Acacia senegal, A. seyal*)
- tooth sticks *Massularia* (syn. *Randia*) *acuminata, Garcinia mannii*
- candle stick (*Canarium schweinfurthii*)
- charbon de bois (23 species identified in Cameroon: *Albizia zygia, A. adianthifolia, Alstonia boonei, Bridelia micrantha, Dacryodes macrophylla, Entandrophragma utile, Ficus thonningii, Lannea welwitschii, Macaranga asas, Maesopsis eminii, Mangifera indica, Milicia excelsa, Morinda lucida, Piptadeniastrum africanum, Phyllanthus discoideus, Persea americana, Rauvolfia vomitoria, Theobroma cacao, Tetrapleura tetraptera, Voacanga africana, Xylopia aethiopica* (FAO, 1999))
- aniegré (*Aningeria robusta*)
- poteaux d'eucaltyptus (*Eucalyptus* spp. especially *E. robusta, E. globulus, E. grandis*)

The mix of French, English, local and scientific names found in the Special Forest Products lists contributes to confusion about this group of products. This leads to problems on many levels, beginning with uncertainty about which species fall under the law. It is also difficult to monitor and control trade when several species are known locally under the same name, or – as is often the case – a single species has multiple local names.

---

due to a lack of information and understanding of this category of products within government, and thus confusion about which products to regulate and why. The limited formal value of NTFPs compared with timber also means that few resources are allocated to understanding and monitoring the sector and building capacity, and even fewer to developing, drafting and implementing effective measures (Njamnshi et al, 2008). The result is that, in the end, NTFPs are regulated much as they always have been under statutory law – through a system of quotas, permits and taxes, allocated by the most powerful in government to the most powerful exploiters or brokers.

Quotas and permits

The NTFP quota and permitting system is bureaucratic and expensive (with both 'informal' and formal taxation), involving a number of different governmental bodies. It often takes many months, or more than a year, to receive a permit, and one needs 'connections' in government to get this result. The system places enormous burdens on traders and exporters in ways that increase costs and discourage both trade and compliance with laws.

A positive feature of this system is that it regulates only species in trade, most with significant value, and does not focus on the majority of species traded locally or consumed for subsistence. More than 20 species are traded in high volumes nationally and close to 200 locally (Ndoye 1995; Ndoye et al, 1997/1998; Sunderland et al, 1999; Nkuinkeu, 2000; Awono et al, 2002b; Pérez et al, 2003; Sunderland and Ndoye, 2004). However, only 13 Special Forest Products were defined in 2006 (Box 2.2).

One set of permits – those for Special Forest Products – originate in the 1994 Forestry Law. The list of Special Forest Products changes annually, which creates confusion since products may move on and off the list. Quotas for Special Forest Products are granted for a year, from defined areas and for a set amount of material. Annual quotas are set by an interministerial committee headed by MINFOF, and in theory are based on surveys of species populations. In practice, however, quotas are determined by demand from exploiting companies, and quantities harvested regularly exceed the official quotas (Awono et al, 2009). Quotas are allocated primarily to private individuals who are rarely harvesters or exploiters themselves, but have political power of some kind, and are able to assemble the necessary paperwork to receive permits. This parallels allocation of permits in the timber sector, where political patronage is an art form (Assembe, 2009). Most holders of NTFP quotas act as brokers and sell them on to harvesters in the form of the waybills (*lettres de voiture*) used to monitor the transportation of Special Forest Products.

High-value NTFPs *not* included on the list of Special Forest Products are also regulated by the government through a system of quotas and permits, but in this case one which pre-dated the 1994 law. These permits are granted by the Minister of Forests through *gré à gré* (mutual agreement), while permits are issued for Special Forest Products after review by the interministerial committee. Examples of products granted exploitation permits by the government in the past four years include those on the 2006 Special Forest Product list such as rattans, charcoal and eru (*Gnetum* spp.), as well as others such as bush mango (*Irvingia* spp.) that are not included in most years' lists.

Special Forest Products destined for export require an additional permit issued by the Minister of Forests. In 2008, species receiving such permits included *Prunus africana*, *Diospyros egrettarum*, *Cinchona pubescens*, *Voacanga africana* and *Pausinystalia johimbe*. The myriad of bureaucratic and financial obligations associated with permitting for NTFPs traded as commodities has presented significant challenges to the economic viability of this sector. The requirement of annual permits for commercially traded NTFPs makes it impossible for businesses to plan a few years in advance, and the uncertainty associated with permitting means export companies cannot respond to overseas customers in a timely manner. Combined with the generally unsupportive business climate, these factors have discouraged a number of international investors from working in Cameroon (Transparency International, 2008; World Bank, 2009; Laird et al, in press).

Community forests

The 1994 Forestry Law also created 'community forests' (article 37), which provide new opportunities for the local control and management of resources, and enable local communities to manage forest areas of less than 5000ha for commercial exploitation, as well as conservation and subsistence use. Introduced in 1997, the number of community forests peaked in 2004. Just over 400 are now at some stage in the attribution process, although only 43 per cent have approved management plans (Oyono, 2004; RIGC, 2008). These are situated in diverse ecological, political, economic and institutional landscapes, with the vast majority found in the lowland forest zone (Adeleke, 2006).

NTFPs are often included in community forest management plans, but most attention to date has focused on commercially valuable timber (Vabi et al, 2002; Akoa, 2007; Ngum, 2009). Community forests appear to offer little advantage when it comes to NTFPs, and can create an added layer of bureaucracy and cost. Overall, the impact of community forests on NTFPs is modest, with most species continuing to be harvested according to customary law and on an individual basis, rather than through a community forest management plan and on a communal basis. Even when NTFPs are included in management plans, this does not appear to ensure sustainable harvesting practices. In a few cases – notably *Prunus africana* in the North-West Province – the institutional capacity built through community forests has, in fact, contributed to the overexploitation of the resource (WHINCONET, 2005; Nsom et al, 2007).

Community forests have helped some communities achieve greater control over forest areas and more significant benefits from timber production, which are real gains. In other cases, however, they have led to conflict across and within communities and have created competition between traditional and newly established community forestry institutions. This is further aggravated by the absence of a definition for what constitutes a 'community' in the 1994 law (Egbe, 1997; Nuesiri, 2008). Concerns have also arisen over the ways benefits from timber exploitation are dispersed within communities (Ngum, 2009).

Community forests are a well-intentioned initiative but, promoted largely by the donor community, the concept was poorly adapted to local conditions (MINEF, 1998; Vabi et al, 2000; CFDP, 2002; Etoungou, 2003; Awasom, 2005; Adeleke, 2006; CIFOR, 2008; Assembe, 2009). In the case of NTFPs, the additional layer of regulation provided by community forests has proven largely unnecessary or ineffective. Customary law generally works to regulate products in local trade or consumed for subsistence, and it is not clear that community forests can solve sustainability and equity problems resulting from commercial demand. In the absence of a sustainability crisis associated with NTFPs – and unlike many other countries, there have been few in Cameroon[1] – government involvement at the community level is likely to backfire, making communities' lives more difficult and contributing little to species conservation.

FORMAL AND 'INFORMAL' TAXATION

Finance and taxation measures directly impact on the use, management and trade of NTFPs, and the broader equity and sustainability of the sector. This aspect of the NTFP trade is regulated by the Ministry of Finance, the Ministry of Small and Medium Enterprises, and the Ministry of Employment, with limited coordination between them. Taxes levied on NTFPs include those on businesses, 'regeneration' taxes linked to quotas, export taxes, taxes levied in markets and a range of 'informal taxes' (or bribes) extracted throughout the trade network.

Most NTFP traders and organizations are small-scale and informal (Ndoye et al, 1997/98; Erasmus et al, 2006; Tchatat and Ndoye, 2006; Awono et al, 2008; Njomaha, 2008). However, since 1996 traders are required to pay a flat business tax or *impôt libéra-toire* of CFA12,000 (about US$26) per year. In addition, taxes are imposed in markets on traders by municipal authorities. The total tax burden for small-scale traders can be significant.

Larger traders and companies are also subject to significant taxation, including regeneration taxes set at CFA10 (US$0.02) per kg of Special Forest Products exploited. The export of raw, unprocessed Special Forest Products requires payment of another, progressive and volume-based tax. In the mid-1990s, a poorly conceived export tax of 15 per cent was instituted on all NTFPs, but this was reduced over the following years as it became apparent that the tax pushed the trade underground, promoted tax-avoidance and forced many companies to close (Laird et al, in press).

In addition to formal taxes, NTFP harvesters, traders and companies must pay 'informal taxes'. Between supply zones and markets, payments to gendarmes, police, forest guards, customs agents and others can consume up to 20 per cent of traders' gross sales (Ndoye and Awono, 2005). For example, between Sa'a and Idenau, a distance of 400km, traders have reported paying US$530 in informal taxes per truck of *Gnetum* spp., even when they possessed the necessary permits (Case Study B). In part this situation results from a broader deterioration in government institutions and a rise in corruption over the past 20 years (Transparency International, 2007; Tieguhong and Betti, 2008; Assembe, 2009). But it is also due to ignorance of the legal require-ments associated with NTFP harvest and trade on the part of producers, traders and government authorities, which creates openings for abuse. In the case of the extensive

Source: Abdon Awono

Figure 2.2 *Eru* (Gnetum spp.) *loaded onto taxi in transition from Cameroon to Nigeria*

cross-border trade of NTFPs between Cameroon and its neighbours, this ignorance extends to free trade agreements (such as the Economic and Monetary Community of Central Africa, or CEMAC), and also results in informal taxation. Multiple levels of formal and informal taxation have created significant burdens on the NTFP sector, making it difficult for producers and traders to profit, and creating incentives for illegal and unsustainable activities.

GOVERNMENT INSTITUTIONAL FRAMEWORK

A range of government institutions are involved in the regulation of NTFPs. In the 1990s, MINFOF, at that time the Ministry of Environment and Forests, created a subdirectorate for NTFPs. This was located in the newly established Directorate of Promotion and Trans-formation of Forest Products (DPT) that elaborates and executes government policy relating to the commercialization, transformation and development of forest products. The DPT was also tasked with centralizing data collection for these products. However, the DPT is forced to compete with more powerful directorates in MINFOF for influence and resources and accomplished little as a result.[2] The institutional arrangements within MINFOF have been streamlined since the late 1990s, but the same problems continue, and the DPT has limited influence compared with the directorates concerned with timber, in particular the Directorate of Forests.

Since its inception, the NTFP subdirectorate has depended on foreign donors for its operating budget on foreign donors, many of whom are also influential in setting priorities. Even so, the subdirectorate has so few resources that it is unable to collect basic statistics on the vast majority of NTFPs (Walter and Mbala, 2006; Betti, 2007; Ingram et al, 2009). The absence of basic data is a major obstacle to drafting, implementing and monitoring effective NTFP regulation, but it has not been overcome in the more than ten years since the subdirectorate was established.[3]

A host of other government ministries and departments are also involved in regulating NTFPs in one way or another. On the ground, work intended to 'regenerate' and reforest land falls under the auspices of ANAFOR (the National Forest Development Agency), which is developing a forest plantation programme. This programme integrates the regeneration of forest resources (including NTFPs), the protection of water catchments, fuelwood production, climate change and efforts to combat desertification. Other ministries intersect with NTFPs in more narrow, but still significant, ways. For example, honey – an important NTFP in many regions – is regulated by the Ministry of Livestock, Fisheries and Animal Husbandry (MINEPIA). Since honey is classed as an animal product, its processing and trade falls under the Veterinary Sanitary Inspection Law of 2000. This is the case whatever the scale of activities or source of the honey – whether wild bees, forest hives or farm hives. In practice, this law is little known, either by harvesters and traders of forest honey, or by the forest, agricultural and MINEPIA authorities themselves. This means that honey is often seized, and 'informal taxes' are regularly levied by government officials.

Coordination within and among ministries (e.g. the Ministries for Forests, Environment, Livestock, Finance, Customs, Territorial Planning, Small and Medium Sized

Enterprises, and Social Economics and Crafts) on NTFP policy is clearly necessary, but does not happen on a regular or planned basis. The lack of collaboration and coordination is exacerbated by a constant turnover in government staff. NTFPs have been the subject of numerous donor-funded research projects and meetings over the past 15 years, and a number of these have addressed the legal and policy issues surrounding the management, harvest and trade of these products. However, little concrete in the way of policy development and implementation has resulted from these processes.

Outside of government, communities must register as institutions when they wish to harvest timber or Special Forest Products. If communities do not have a Community Forest and want to exploit timber on a communal basis, they must form a legal organization – a common initiative group (GIC) or company. Communities wishing to exploit Special Forest Products must apply to MINFOF for a Special Forest Products permit. To do this they must be registered as companies or approved and accredited as 'forest resource harvesters', something few communities have yet to manage and on which there are no statistics. As a result, the vast majority of Special Forest Products are harvested without official permits (Awono et al, 2008; Tajoacha, 2008; Ewane et al, 2009; Ndumbe et al, 2009).

CONCLUSION AND RECOMMENDATIONS

Revision of the forestry and environment laws in Cameroon over the past 15 years has opened discussion around a range of forest values beyond timber, including NTFPs. Steps have been taken to develop a legal, policy and institutional framework for NTFPs, but largely under pressure from outside agencies and with little internal political will. The legal and policy framework today remains inconsistent and incomplete, and the government's institutional capacity limited. Conflicts between texts are compounded by the absence of implementing decrees and regulations that could address broader concepts in practical terms. For the laws and the decisions that do exist, a very low level of awareness is found in the harvester and trader communities, as well as among government authorities, particularly the local and regional delegations that interact with rural communities.

As a result, most features of the NTFP regulatory framework undermine this sector. For example, the products regulated are not well-defined, and so uncertainty dominates; NTFPs are taxed in formal and informal ways that are inconsistent and often heavy-handed; the long-term management of species populations is not considered when granting quotas, nor are there controls or monitoring that might limit overharvesting; and bureaucracy and costs eat away at profits and limit the groups that might legally participate in the sector.

At the same time, the regulatory framework undermines the livelihoods of small producers and traders, in favour of the politically powerful few and 'feeding the belly'. For example, community land and resources are under ambiguous legal title, and community groups must jump bureaucratic hurdles to become legal entities in order to manage, harvest and trade their own resources and forests; informal and formal taxes are levied at multiple levels and consume the bulk of profits; communities cannot file the necessary paperwork, and do not have the requisite political power, to acquire quotas of Special Forest Products and so must buy waybills from quota holders, or enter into semi-legal or illegal activities in order to trade NTFPs.

In the absence of a functioning and legitimate statutory legal framework, most NTFP activities are regulated through customary law. However, for species that are under strong commercial pressure, statutory law is an important and often necessary complement to customary law. Dramatic changes are clearly needed on a number of fronts in order to develop and implement a legal and policy framework for NTFPs that supports harvesters, traders and rural communities, encourages a vibrant commercial NTFP sector, and promotes sustainable and equitable practices. Following are some – perhaps ambitious – recommendations for those changes.[4]

The range of values NTFPs hold for local communities – economic, environmental, cultural and social – should be acknowledged. Subsistence use of NTFPs should be recognized as central to rural livelihoods and cultures, and made exempt from taxation and direct government oversight and intervention, as should small-scale local trade of NTFPs.

Land tenure and resource rights for local communities should be rationalized. All trees growing on lands used and managed by communities should be their property.

Customary law regulating NTFPs should be respected and seen as an important complement to statutory law.

The regulatory framework for NTFPs should be streamlined and made clear. This will improve its effectiveness, minimize opportunities for corruption that thrive on confusion and ambiguity, reduce the bureaucracy and cost associated with following the law, and encourage harvesters, traders and companies to participate legally in what might widely be viewed as a legitimate and helpful legal framework.

Comprehensive and ongoing consultations with the wide range of affected stakeholders – such as harvesters, traders and companies – should inform any revision of the NTFP legal and policy framework.

Forestry and environment laws should strengthen the clarity and consistency of their NTFP elements. The nature and scope of the products regulated under a revised forestry law should be better elaborated and defined. Objectives for regulating NTFPs (e.g. to promote sustainability, improve local livelihoods, strengthen the NTFP sector and raise government revenues) should be explicit, and the trade-offs between objectives made clear (e.g. that raising government revenues might depress local livelihoods). NTFPs should be integrated into management plans for timber and other land uses.

Taxation and trade levies should be rational, legitimate and just, and the law communicated to the many levels of government that are involved in these activities, as well as to producer and trader groups. Informal taxation should be actively prohibited.

Government institutional capacity to regulate these products should be improved. Staff should be trained and their capacity built, and resources provided to relevant institutions (e.g. the NTFP subdirectorate). The government's understanding of the vast range of NTFP uses, values and roles in local livelihoods, and their relationships with each other, should be strengthened. The collection of information and statistics on NTFPs in trade should be expanded and systematic, with resources allocated for this purpose. Cooperation and coordination within and between ministries around NTFPs should be improved.

Government and other groups should undertake outreach with traders, harvesters and others, informing them about the laws and policies regulating NTFPs, and learning from their experiences.

NOTES

1   The sustainability crisis around *Prunus africana* sourcing in the past few decades is an exception (Cunningham and Mbenkum, 1993; Cunningham et al, 1997; Ndam and Tonye, 2004; Ingram et al, 2009).
2   The directorates within MINFOF that compete with the DPT for resources and power include the cabinet; finance; inventory and forest management; protected areas; wildlife conservation; the valorization and exploitation of wildlife; wood promotion; wood processing; and community forests.
3   The only government sources of data on NTFPs in trade are the government's SIGIF (Information System for the Management of Forestry Parameters) system of data collection and the COMCAM (Cameroon Timber Marketing) database of forest product exports from the Port of Douala, which includes Special Forest Products. Waybills recorded for Special Forest Products checked at MINFOF checkpoints also yield some data, as does reporting on the export of the two CITES species *Prunus africana* and *Pausinystalia johimbe*. The customs centre in Douala also documents exports from that site. Reports on harvests of Special Forest Products from MINFOF regional delegates are often unavailable, and are not summarized annually at a national level. International agencies (e.g. the Center for International Forestry Research and the Food and Agriculture Organization), academics, and conservation and development organizations have undertaken research that fills gaps in understanding of the trade, but this should complement, rather than substitute for, government records.
4   See FAO (2008 and 2009) for additional recommendations.

REFERENCES

Adeleke, W. (2006) *Analysis of Community Forest Processes and Implementation in Cameroon*, WWF-CARPO, Yaoundé
Akoa, R. J. A. (2007) 'Economic analysis of community forest projects in Cameroon', MSc thesis, Faculty of Forest Science and Wood Ecology, Georg-August University of Göttingen, Göttingen

Assembe, S. M. (2009) 'State failure and governance in vulnerable states: An assessment of forest law compliance and enforcement in Cameroon', *Africa Today*, vol 55, no 3, pp85–102

Awasom, B. N. (2005) *Review of Community Forest Management through Decentralisation in Cameroon*, University of Yaoundé, Yaoundé

Awono, A., Manirakiza, D. and Ingram, V. (2008) *Baseline Study in Pilot Sites in the South West and North West Provinces of Cameroon on* Prunus africana, CIFOR, Yaoundé

Awono, A., Manirakiza, D. and Ingram, V. (2009) *Etude de base du Ndo'o (*Irvingia *spp.) dans les provinces du Centre, Sud et Littoral Cameroun*, edited by CIFOR, Yaoundé

Awono, A., Ndoye, O., Schreckenberg, K., Tabuna, H., Isseri, F. and Temple, F. (2002a) 'Production and marketing of safou (*Dacryodes edulis*) in Cameroon and internationally: Market development issues', *Forests, Trees And Livelihoods*, vol 12, pp125–147

Awono, A., Ngono, D. L., Ndoye, O., Tieguhong, J., Eyebe, A. and Mahop, M. T. (2002b) *Etude sur la commercialisation de quatre produits forestiers non-ligneux dans la zone forestière du Cameroun:* Gnetum *spp.*, Ricinodendron heudelotii, Irvingia *spp.*, Prunus africana, FAO, Yaoundé

Barume, A. K. (2004) 'La Loi Forestière de 1994 et les droits des habitants des forêts au Cameroun [The 1994 Forestry Law and the rights of forests peoples in Cameroon]', *Terroirs*, vol 1–2, pp107–119

Berg, J. v. d., Wiersum, K. F. and Dijk, H. v. (2007) 'The role and dynamics of community institutions in the management of NTFP resources'. *Forest, Trees and Livelihoods* 17, pp183–197

Betti, J. L. (2007) Stratégie/Plan D'action pour une meilleure collecte des données statistiques sur les produits forestiers non ligneux au Cameroun et recommandations pour les pays de la Comifac. Ministère Fédéral d'Allemagne pour l'Alimentation, l'Agriculture et la Protection des Consommateurs, COMIFAC, FAO

Burnham, P. (2000) 'Whose forest? whose myth? – Conceptualisations of community forests in Cameroon', in A. Abramson and D. Theodossopoulis (eds) *Land, Law and Environment: Mythical Land, Legal Boundaries*, Pluto Press, London

Burnham, P. and Sharpe, B. (1997) *Social, Economic and Political Dimensions in Cameroon's Forestry Sector*, Department of Anthropology, University College London

Cameroon Forestry Law (1994) Law No 94/01 of 20 January 1994, laying down Forest, Wildlife and Fishery regulations. Yaoundé

CFDP (2002) *Forêts communautaire au Cameroun: Mieux comprendre et apprécier le plan simple de gestion (PSG)* [Community forests in Cameroon: Improving our understanding and adequately assessing simple management plans], DFID, Yaoundé

CIFOR (2008) *The Community Forestry Program in Cameroon: A Participatory Management Option?* Center for International Forestry Research (CIFOR), Yaoundé

Cunningham, A. B. and Mbenkum, F. T. (1993) Sustainability of harvesting *Prunus africana* bark in Cameroon: A medicinal plant in international trade, WWF/UNESCO/Kew People and Plants Working Paper No. 2

Cunningham, M. (2006) Etude CITES du commerce important: *Prunus africana*, PC16 Doc. 10.2/, Etude du commerce important de spécimens d'espèces de l'Annexe II. Convention sur le commerce international des especes de faune et de flore sauvages menacees d'extinction

Cunningham, M., Cunningham, A. B. and Schippmann, U. (1997) *Trade in P. africana and the implementation of CITES*, German Federal Agency for Nature Conservation, Bonn

Cuny, P., Ango, A. A. and Ondoa, Z. A. (2007) 'An experience of local and decentralised forest management in Cameroon: The case of the Kongo Community Forest', *SNV* 15

Dijk, J. F. W. v. (1999). *Non-Timber Forest Products in the Bipindi-Akom II Region, Cameroon: A Socio-economic and Ecological Assessment*, Tropenbos-Cameroon Series 1 Programme, T.-C. Kribi

Egbe, S. (1997) 'Forest Tenure and Access to Forest Resources in Cameroon: An overview', Forestry Participation Series 6, IIED, London

Ekoko, F. (1999) 'Environmental Adjustment in Cameroon: Challenges and Opportunities for Policy Reform in the Forestry Sector'. Paper presented at The Workshop on Environmental Adjustment: Opportunities for Progressive Policy Reform in the Forest Sector, 6 April. World Resources Institute, Washington, DC

Erasmus, T., Hamaljoulde, D., Samaki, J., Njikeu, M. T. and Nyat, G. M. (2006) *Highlands Honey and Bee Products Market Study*, Netherlands Development Organisation (SNV), Bamenda, Cameroon

Etoungou, P. (2003) *Decentralization Viewed from Inside: The Implementation of Community Forests in East Cameroon*, working paper, World Resources Institute (WRI), Yaoundé

Ewane, M., Awono, A. and Ingram, V. (2009) *Baseline Study on* Irvingia *Spp. in the South-West and East Regions of Cameroon*, FAO-CIFOR-SNV-World Agroforestry Center – COMIFAC, Yaoundé

FAO (1999) *Données statistiques des produits forestiers non-ligneux du Cameroun*, Programme de partenariat CE-FAO (1998–2001) Ligne budgétaire forêt tropicale B7-6201/97-15/VIII/FOR PROJET GCP/INT/679/EC, FAO/European Commission Directorate-General VIII Development: 36

FAO (2008) Directives sous-régionales relatives à la gestion durable des produits forestiers non-ligneux (PFNL) d'origine végétale en Afrique Centrale, FAO/GTZ/COMIFAC, Yaoundé, Cameroon

FAO (2009) *Cadre légal et réglementaire régissant l'exploitation et la commercialisation des produits forestiers non ligneux (PFNL) au Cameroun*, Mobilisation et renforcement des capacités des petites et moyennes entreprises impliquées dans les filières des produits forestiers non ligneux (PFNL) en Afrique Centrale, Yaoundé, GCP/RAF/408/EC

Geschiere, P. (1993) 'Chiefs and colonial rule in Cameroon: Inventing chieftancy, French and British style', *Africa* vol 63, no 2, pp151–175

Ingram, V. (2008) *Lobby for sustainable harvesting of Prunus in Cameroon*. Working Paper. Netherlands Development Organization (SNV). Bamenda

Ingram, V., Ndam, N., Awono, A. and Schure, J. (2009) *Draft Outline of a Management Plan for Prunus africana in Cameroon*, FAO-CIFOR-SNV-World Agroforestry Center – COMIFAC, Yaoundé

Konings, P. (1999) 'The "anglophone problem" and chieftancy in anglophone Cameroon', in E. A. B. van Rouveroy van Nleuwaal and R. van Dijk (eds) *African Chieftancy in a New Socio-political Landscape*, LIT Verlag, Hamburg, pp181–206

Laird, S., Nkuinkeu, R., Lisinge Fotabong, E. E. and Tanjong, T. (in press) *Medicinal Plants in the Export Trade in Cameroon: The Impact of Laws and Policy on Sustainability, Equity, and the Viability of a Sector*, People and Plants International, Vermont USA

MINEF (1998) 'Current developments, opportunities, and issues for community forestry in Cameroon', in *Community Forestry Development Project*, Ministry of Environment and Forests (MINEF), Yaoundé

MINFOF (2009) *Décision No 0020/D/MINFOF/SG/DF/SDAF du 06 Jan 2009 Portant octroi des quotas d'exploitation des produits forestiers spéciaux*, D. d. F. Secrétariat Général, Ministère des Fôrets et de la Faune, Yaoundé

Ndam, N. and Tonye, M. M. (2004) 'Chop, but no broke pot: the case of *Prunus africana* on Mount Cameroon', in T. Sunderland and O. Ndoye (eds) *Forest Products, Livelihoods and*

Conservation: Case Studies of Non-Timber Forest Product Systems, Volume 2: Africa. Center for International Forestry Research, Bogor, Indonesia

Ndoye, O. (1995) *The Markets for Non-Timber Forest Products in the Humid Forest Zone of Cameroon and Its Borders: Structure, Conduct, Performance and Policy Implications,* Center for International Forestry Research (CIFOR), Yaoundé

Ndoye, O., M. R. Pérez and A. Eyebe (1997/98) 'The Markets of Non-timber Forest Products in the Humid Forest Zone of Cameroon.' *Rural Development Forestry Network* (Network Paper 22c)

Ndoye, O. (2005) 'Commercial issues related to non-timber forest products', in J. L. Pfund and P. Robinson (eds) *Non-Timber Forest Products between Poverty Alleviation and Market Forces,* Intercooperation, Berne, Switzerland, pp14–19

Ndoye, O. and Awono, A. (2005) 'The markets of non-timber forest products in the provinces of Equateur and Bandundu, Democratic Republic of Congo', in *Congo Livelihood Improvement and Food Security Project, Supported by the US Agency for International Development,* Center for International Forestry Research (CIFOR), Central Africa Regional Office, Yaoundé

Ndoye, O. and Kaimowitz, D. (2000) 'Macro-economics, markets and the humid forests of Cameroon, 1967–1997', *Journal of Modern African Studies,* vol 38, no 2, pp225–253

Ndumbe, L., Ingram, V. and Awono, A. (2009) *Baseline Study on* Gnetum *Spp. in the South-West and Littoral Regions of Cameroon,* FAO-CIFOR-SNV-World Agroforestry Center – COMIFAC, Yaoundé

Ngum, B. (2009) 'Assessing community forest management: Case study of CO.VI.MOF and C.C.I. community forest', dissertation presented in partial fulfilment of requirements for an award of higher Specialized Diploma (D.E.S.S.) in Forest Sciences, University of Yaoundé I

Ngwasiri, C. N., Djeukam, R. and Vabi, M. B. (2002) *Legislative and Institutional Instruments for the Sustainable Management of Non-Timber Forest Products (NTFP) in Cameroon: Past, Present and Unresolved Issues,* Community Forestry Development Project (CFDP), Yaoundé

Njamnshi, A., Nchunu, J. S., Galega, P. T. and Chili, P. C. (2008) *Environmental Democracy in Cameroon: An Assessment of Access to Information, Participation in Decision Making, and Access to Justice in Environmental Matters,* The Access Initiative Cameroon (TAI), Yaoundé

Njomaha, C. (2008) Etude Socio-economique de la Filiere Gomme Aarabique dans le Nord et L'Extreme-Nord Cameroun. IRAD/CEDC., FAO-CIFOR-SNV-World Agroforestry Center-COMIFAC, Maroua, Cameroun, p133

Nkuinkeu, R. (2000) 'Les Oléagineux non-conventionnels du Cameroun: Inventaire succinct et potentialités de développement', in J. Kengue, C. Kapseu, G. J. Kayem (eds) *Actes du 3e séminaire international sur la valorisation du safoutier et autres oléagineux non-conventionnels,* Presses Universitaires d'Afrique, Yaoundé, Cameroun, pp555–561

Nsom, A., Tah, K. and Ingram, V. (2007) 'Current situation of *Prunus africana* in the Kilum forest, Emfveh-Mii and Ijim Community Forests', paper read at workshop with stakeholders on conflict resolution in *Prunus africana* management, Oku, June

Nuesiri, E. O. (2008) 'Forest Governance Challenges on Mount Cameroon', *IHDP Update,* 2. 2008

Oyono, P. R. (2004) 'One step forward, two steps back? Paradoxes of natural resources management decentralisation in Cameroon', *Journal of Modern African Studies* vol 42, no 1, pp91–111

Oyono, P. R. (2005) 'Profiling local-level outcomes of environmental decentralizations: The case of Cameroon's forests in the Congo Basin', *Journal of Environment & Development* vol 14, no 2, pp1–21

Pérez, M. R., Ndoye, O., Eyebe, A. and Puntodewo, A. (2000) 'Spatial characterization of non-timber forest products markets in the humid forest zone of Cameroon', *International Forestry Review* vol 2, no 2

Pérez, M. R., O. Ndoye, A. Eyebe and D. L. Ngono (2003) 'A Gender Analysis of Forest Product Markets in Cameroon,' *Africa Today*, 96–126

RIGC (2008) Projet Renforcement des Initiatives de Gestion Communautaire des ressources forestières et fauniques, Database of Community Forests. Yaoundé, Cameroon

Sharpe, B. (1998) '"First the forest": Conservation, "community" and "participation" in South-West Cameroon', *Africa: Journal of the International African Institute*, vol 68, no 1, pp25–45

Sunderland, T. and Ndoye, O. (eds) (2004) *Forest Products, Livelihoods and Conservation: Case Studies of Non-timber Forest Product Systems*, Volume 2 – Africa, Center for International Forestry Research, Bogor, Indonesia

Sunderland, T. C. H., Clark, L. E. and Vantomme, P. (eds) (1999) *Non-Wood Forest Products of Central Africa: Current Research Issues and Prospects for Conservation and Development*, FAO, Rome

Tajoacha, A. (2008) *Market Chain Analysis of the Main NTFPs in the Takamanda/Mone Forest Reserves, South West of Cameroon and the Cross River State of Nigeria*, University of Dschang, Dschang, Cameroon

Tchatat, M. and Ndoye, (2006) 'Étude des produits forestiers non ligneux de Afrique centrale: réalités et perspectives', *Bois Et Forêts Des Tropiques* vol 288, no 2, pp27–39

Tieguhong, J. C. and Betti, J. L. (2008) 'Forest and Protected Area Management in Cameroon', *ITTO Tropical Forest Update*, vol 18, no 1, Yokohama, Japan

Tonye, J., Meke-Me-Ze, C. and Titi-Nwel, P. (1993) 'Implications of national land legislation and customary land and tree tenure on the adoption of alley farming', *AgroforestrySystems* no 22, pp153–160

Transparency International (2007). Enquete Nationale 2006 aupres des Entreprises sur la Corruption au Cameroun. T. I. C. Transparency International. Yaounde, Cameroon, Transparency International, Centre de Recherche Et d'Etudes en Economie Et Sondage: 156.

Transparency International. (2008) «Corruption Perceptions Index 2008.» Retrieved 20/05/2009, from www.transparency.org/policy_research/surveys_indices/cpi/2009

Vabi, M. B., Ngwasiri, C. N., Galega, P. T. and Oyono, R. P. (2000) *Devolution des responsabilités liées à la gestion forestière aux communautés locales: contexte et obstacles de mise en oeuvre au Cameroun* [The devolution of forest management responsibilities to local communities: Context and implementation hurdles in Cameroon], World Wide Fund for Nature (WWF), Yaoundé

Vabi, M. B., Njankoua, D. W., Muluh, G. A., Lumumba, T. M., Tchakoa, J. and Tene, A. (2002) *The Costs and Benefits of Community Forests in Selected Agro-Ecological Regions of Cameroon*, DFID-Funded Community Forestry Development Project (CFDP)/DFID, Yaoundé

Walter, S. and Mbala, S. M. (2006) 'Etat des lieux du secteur produits forestiers non-ligneux en Afrique centrale et analyse des priorités politiques', in *Note d'Information*, F. COMIFAC, European Commission (EC) and Food and Agricultural Organization of the United Nations (FAO), Malabo, Equatorial Guinea

WHINCONET (2005) *Report on the Illegal Harvesting of* Prunus africana *in the Kilum-Ijim Forests of Oku and Fundong*, B. a. t. C. B. C. S. C. in collaboration with SNV Highlands, Bamenda, Cameroon; Western Highlands Nature Conservation Network (WHINCONET), Yaoundé

World Bank (2009) 'Cameroon at a glance', in *At a Glance* series, World Bank, Washington, DC

Case Study B: Policies for *Gnetum* spp. Trade in Cameroon: Overcoming Constraints that Reduce Benefits and Discourage Sustainability

Ousseynou Ndoye and Abdon Awono

INTRODUCTION

The Congo Basin is the second-largest tropical forest in the world, and home to many NTFPs. In the past, NTFPs were used mainly for subsistence by local communities, but they are now widely traded in local, national, regional and international markets, and are important parts of most urban dwellers' lives as well. The economic crisis of the 1980s, the decline of cocoa and coffee prices and the devaluation of the CFA franc in 1994 provoked a dramatic increase in the trade of and demand for NTFPs in the central African region, including Cameroon. Many urban and rural dwellers turned to the forest to satisfy their nutritional and medicinal needs and to diversify their sources of income.

Among the most highly exploited NTFPs in the Congo Basin are the two leafy vegetables *Gnetum africanum* and *G. buchholzianum* (Shiembo et al, 1996). In Cameroon *Gnetum* species are commonly referred to as 'eru' by the anglophones and 'okok' by natives of the Centre Province. They are highly sought after for their taste and high nutritional and medicinal value. Eru is one of the few leafy vegetables available throughout the year, and is such a favoured food that it is also dried and exported to Europe, North America and elsewhere (Tabuna, 2000).

Gnetum leaves contribute significantly to local livelihoods along the value chain in Central Africa, from harvesters to traders to sellers in markets. For example, in an indication of its value to stakeholders, in Cameroon producers earn a monthly average of $98 (Mungo division) and $111 (Manyu division). Meanwhile, in the Équateur province of the Democratic Republic of Congo, traders of *Gnetum* spp. earn a monthly average of US$131, which is higher than the average wage of secondary school teachers (US$50–70). Traders in other regions shipping *Gnetum* spp. to Kinshasa have an average monthly income higher than that of a medical doctor (US$190–250) (Ndoye and Awono, 2007; Ndoye et al, 2007).

However, this intense commercial pressure and extremely high value mean that *Gnetum* species are increasingly scarce due to overharvesting, and in some cases the impact of other land uses. In the Lékié Division of Cameroon, for example, intensive agriculture has degraded forests and made eru difficult to find, requiring long travelling distances for collection. In many rural areas, the scarcity of eru, combined with intense commercial pressure, has led to conflicts (Awono et al, 2002a).

Figure B.1 *Selling* Gnetum *in the market, Cameroon*

THE MAIN *GNETUM* SPECIES

The *Gnetum* genus consists of 30 species, most of these lianas, distributed in the humid tropics of Asia, America and Africa (Martens, 1971). In Africa, and particularly in Cameroon and the Democratic Republic of Congo, two species are commonly known and used, the lianas *Gnetum africanum* and *G. buchholzianum*. These are distributed in the forest zone of Nigeria, Cameroon, the Central African Republic, Gabon, the Republic of Congo (Brazzaville), the Democratic Republic of Congo and Angola (Mialoundama, 1996). *Gnetum* spp. grow best in fallows, secondary forests and gallery forests, and are even found in abandoned or neglected cocoa, palm and banana plantations (Eyebe, 2002; Awono et al, 2002a; Awono et al, 2002b; Asaha et al, 2000; Duguma et al, 2001). *Gnetum* spp. are also found in some domesticated field sites, and a number of local institutions – such as the World Agroforestry Centre, Cameroon's Institute of Agricultural Research for Development, the Limbe Botanic Garden and the Center for International Forestry Research – are carrying out trials on its potential for widespread domestication (Tchatat et al, 2002). *Gnetum* is harvested by either cutting the shoots or uprooting the plants and tying these into bundles for transport to markets (Shiembo, 2000).

TRADE AND REGULATORY ENVIRONMENT CONSTRAINTS

The trade in *Gnetum* in Cameroon is officially restricted to holders of quotas. An inter-ministerial committee, led by the Ministry of Forestry and Wildlife, allocates quotas to traders. Many traders involved in the *Gnetum* spp. value chain are not harvesters (or exploiters), and must purchase part of the quotas of others in the form of waybills (*lettres de voiture*), paying prices above the normal cost of the quotas (which are paid in the form of 'regeneration taxes' intended to support the resource base). Added to this, traders must pay significant sums at roadblocks – 'informal taxation' or bribes – and these increased transaction costs and difficulties associated with transporting *Gnetum* reduce their profit and in turn result in lower prices paid to harvesters and producers, and higher prices paid by consumers. To achieve the dual objective of improving the livelihoods of forest dwellers and preserving this valuable resource, these constraints need to be tackled.

A summary of key blockages and constraints in the trade and regulatory environment that result in reduced benefits for harvesters and traders follows.

QUOTA SYSTEM

- Quotas are allocated by the government without a solid understanding of the resource base because a national inventory of *Gnetum* species has never been carried out.
- Quotas are allocated in an inequitable way to a few individuals with political power, which does not allow traders and rural communities to officially participate in *Gnetum* export and pushes many into operating at the margins of the law.
- Companies or individuals receiving the quotas are rarely directly involved in the chain of custody of the product, and instead act as brokers, selling portions of quotas to traders as waybills or *lettres de voiture* at a much higher price than they pay for their quotas in the form of 'regeneration taxes' paid to the government.
- There is no established procedure for monitoring if quotas are respected, which implies that there are no procedures to control abuse.

Table B.1 *Amount paid in informal taxes by traders (for a car of about 1.5 tonnes) from Lékié Division to Idenau market, Cameroon*

Controlling agents	Amount paid per truck (US$)
Police	$140
Gendarmes	$220
Forest services	$108
Other services	$62
Total	$530

Source: Ndoye and Awono, 2007.

'CHECKPOINTS' (ROADBLOCKS) AND 'INFORMAL TAXATION'

- There are a large number of checkpoints from the supply zones to the different main markets, and at each traders are required to pay 'informal taxes' or bribes (see Table B.1). For example, between Sa'a and Idenau (a border market which supplies Nigeria), a distance of 400km, traders have reportedly paid US$530 per truck in informal taxes (Table B.1). This payment is made despite the fact that traders and transporters follow all the required procedures and have all the required legal papers.
- Informal taxes can absorb up to 20 per cent of the gross sales of traders (Ndoye, 2005). The burden of the taxes is transferred to village communities and consumers in the form of lower prices for producers and higher prices for consumers.
- A range of government agencies operate checkpoints throughout the trade route – gendarmes, police, forest guards, custom agents, etc. – which makes the roles of each difficult to determine, and creates layers of 'informal tax' collection or bribes in which each takes a cut.

RECOMMENDATIONS

To improve the regulatory framework for *Gnetum* spp., the government should consider the following:

- To improve benefits for producers, the government should promote policies that encourage: domestication in rural areas, allowing farmers to plant *Gnetum* spp. in home gardens and increase production; group marketing of the raw materials; and local processing (drying and packaging) for international trade. The government should also invest in local infrastructure such as roads to support trade in rural areas.

- Regulatory policies should accommodate domesticated as well as wild harvested *Gnetum*. At present they are directed only at wild raw material and discourage cultivation.
- Cost-effective inventories need to be developed so that the distribution of species can be better understood and yearly quotas more effectively allocated.
- Allocations of quotas should be made more transparent through a system of bidding. This would allow more competition among market participants and increase tax revenues for the government.
- The government should prevent quota holders from selling their quotas through waybills.
- Eliminating unnecessary road controls, or checkpoints, would improve the efficiency of the marketing system and the welfare of farmers, traders and consumers.

REFERENCES

Asaha, S., Tonye, M. M., Ndam, N. and Blackmore, P. (2000) *State of Knowledge Study on Gnetum africanum Welw. and Gnetum Buchholzianum Engl.*, report for the Central African Regional Program for the Environment, Cameroon, Limbe Botanic Garden Library, 11–5 March

Awono, A., Lema, N., Ndoye, O., Tieguhong, J., Eyebe, A. and Tonye, M. (2002a) *Etude sur la commercialisation de quatre produits forestiers non-ligneux dans la zone forestière du Cameroun: Gnetum spp., Ricinodendron Heudelotii, Irvingia spp., Prunus africana*, unpublished study, CIFOR, Cameroon

Awono, A., Ndoye, O. and Eyebe, A. (2002b) *Produits forestiers non-ligneux et génération des revenus: Le cas des plantes médicinales, nutritives et de service*, papier présenté au séminaire organisé par le MINEF sur l'état des lieux des produits forestiers non-ligneux au Cameroun, Janvier

Duguma, B., Gockowski, J. and Bakala, J. (2001) 'Smallholder cacao (*Theobroma cacao* Linn.) cultivation in agroforestry systems of West and Central Africa: Challenges and opportunities', *Agroforestry Systems*, vol 51, no 3, pp177–188

Eyebe, J. P. (2002) *Les niches écologiques du* Gnetum *spp. et l'urgence de l'accélération du processus de sa domestication dans les villages de production*, rapport rédigé en vue de l'obtention du diplôme de techniciens supérieurs en agroforesterie, Université de Dschang, Septembre

Martens, P. (1971) *Traité d'anatomie végétale: Les Gnetophytes*, vol 12(2), Gebrüder Bomtraeger, Berlin

Mialoundama, F. (1996) 'Intérêt nutritionnel et socio-économique du genre *Gnetum* en Afrique Centrale', in C. M. Hladick, A. Hladick, O. F. Linares, H. Pagezy, A. Semple et M. Hadley (éds) *Alimentation en forêt tropicale: Interactions bioculturelles*, UNESCO, Paris

Ndoye, O. (2005) 'Commercial issues related to non-timber forest products', in J.-P. Pfund and P. Robinson (eds) *Non-Timber Forest Products Between Poverty Alleviation and Market Forces*, Intercooperation, Bern, Switzerland, pp14–19

Ndoye, O. and Awono, A. (2007) 'Regulatory policies and *Gnetum* spp. trade in Cameroon', *Forest Livelihood Briefs*, no 6, Center for International Forestry Research, April

Ndoye, O., Awono, A., Preece, L. and Toirambe, B. (2007) 'Marchés des produits forestiers non-ligneux dans les provinces de l'Equateur et de Bandundu: Présentation d'une enquête de terrain', in C. Croizer et T. Trefon (éds) *Quel avenir pour les forêts de la République Démocratique du Congo? Instruments et mécanismes innovents pour une gestion durable des forêts*, Coopération Technique Belge, Bruxelles, www.btcctb.org/doc/UPL_2007080616534312165.pdf

Shiembo, N. P. (2000) *Pour une gestion durable des Okok (*Gnetum africanum *et* Gnetum Buchholzianum*): Des produits forestiers non ligneux surexploités dans les forêts d'Afrique Centrale*, FAO, Rome

Shiembo, N. P., Newton, A. C. and Leakey, R. R. B. (1996) 'Vegetative propagation of *Gnetum africanum* Welw., a leafy vegetative legume from West Africa', *Journal of Horticultural Science*, vol 71, no 1, pp149–155

Tabuna, H. (2000) *Evaluation des échanges des produits forestiers non ligneux entre l'Afrique subsaharienne et l'Europe*, FAO, Rome

Tchatat, M., Vabi, M. and Bidja, R. (2002) *PFNL au Cameroun: Etat du secteur et stratégies nationales de gestion durable*, MINEF, Cameroon

Case Study C: Regulatory Issues for Bush Mango (*Irvingia* spp.) Trade in South-west Cameroon and South-east Nigeria

Terry Sunderland, Stella Asaha, Michael Balinga and Okon Isoni

THE IMPORTANCE OF NTFPS IN SOUTH-WEST CAMEROON AND SOUTH-EAST NIGERIA

The inhabitants of the forest region straddling the border between Cameroon and Nigeria depend heavily on the exploitation of the forest resource base for their livelihoods (Malleson Amadi, 1993). NTFPs, in particular, help to stabilize incomes as in some cases they can be harvested during periods of low farm labour demand and at times of peak NTFP production (Arnold and Ruiz Perez, 2001). In the Takamanda region of south-west Cameroon, it is estimated that 70 per cent of the total population of the area (about 16,000 people) collect forest products for consumption and sale, representing an estimated income of US$1 million per annum (based on a 12-month study of trade routes for NTFPs in the Takamanda region by Ayeni and Mdaihli, 2001). In Cross River State, Nigeria, a much larger region with a population of 2.9 million (Cross River State, n.d.), the trade in forest products is highly lucrative, with the total annual trade in major NTFPs estimated to be 321 million naira, or US$2.4 million (Sunderland and Isoni, 2001). While demand for NTFPs in Nigeria is high, the resource base is diminishing, resulting in a large cross-border trade with Cameroon (Malleson Amadi, 1993). The highly porous border between the two countries offers considerable trade and livelihood opportunities to those in the region. The most valuable NTFP in this cross-border trade is bush mango, *Irvingia gabonensis* and *I. wombolu*, the subject of recent surveys on both sides of the border, the results of which are reported in part below (Sunderland and Isoni, 2001; Sunderland et al, 2003; Asaha et al, 2006).

THE RESOURCE: *IRVINGIA GABONENSIS* AND *I. WOMBOLU* (IRVINGIACEAE)

Vernacular names: bush mango; bojep (Boki); ogbono (Igbo); uyo (Efik); uyo (Ibibio); eloweh (Ovande); kelua (Basho); gluea (Anyang).

The two botanical species that comprise the resource known as bush mango are large forest trees up to 35m tall. *I. gabonensis* is restricted to the forested region from eastern Nigeria to the northern Congo Basin, while *I. wombolu* has a wider range through West Africa, reaching as far as Senegal (Harris, 1996). The cotyledons of both species are used as a soup thickener and as a condiment (Ainge and Brown, 2004). However, the period of production varies: *I. gabonensis* is the rainy season bush mango and *I. wombolu* the dry season type. On both sides of the border, *I. gaboneis* is by far the more common species, and *I. wombolu* is not particularly well distributed. However, *I. wombolu* is more common in secondary forest and on farmland, and it is the preferred species for planting due to the slightly higher revenues it earns from the sale of the cotyledons during the dry season.

With two main production periods, June–September (*I. gabonensis*) and February–April (*I. wombolu*), there are peaks and troughs in production, and corresponding fluctuations in prices, but very little attempt is made at the community level to dry and store bush mango to ensure a more consistent supply and hence more steady household incomes (Sunderland and Isoni, 2001).

LIVELIHOOD IMPORTANCE

From harvest to final consumption, the trade in NTFPs is generally part of the 'hidden economy' of the forested regions of Nigeria and Cameroon, despite the considerable interstate and international trade in some products. Most harvesters and traders are self-employed, have little access to capital and earn their incomes in labour-intensive ways.

The harvest and sale of bush mango is a major source of income for rural communities in both Cameroon and Nigeria. A survey undertaken in five communities in Nigeria found that 90 per cent of households were involved in the collection and sale of bush mango and that sales of bush mango accounted for as much as 85 per cent of annual income for some households (Asaha et al, 2006). A similar study undertaken on the Cameroon side of the border showed that over 90 per cent of households were involved in the harvest and sale of bush mango, which contributed 60 per cent of the total annual household income (Sunderland et al, 2003). On both sides of the border both 'poor' and 'wealthy' households were involved in bush mango collection – unlike most NTFPs, which are harvested primarily by poorer members of the community – and some households made significant gains in wealth through the harvesting and trade of the product. Although women are the primary harvesters of NTFPs, particularly those used for subsistence, men assist with the bush mango harvest and are often responsible for its sale (Asaha et al, 2006).

SUSTAINABILITY AND MANAGEMENT

Despite its value and widespread use, bush mango is not threatened by overharvesting. Stocking estimates suggest that there is good regeneration in both Nigeria (6.31 stems per ha, or an estimated 2.5 million trees in the high forest alone in Cross River State

(Otu et al, 1994)) and Cameroon (8.8 stems per ha in Takamanda (Sunderland et al, 2003)). It is likely that faunal predators disperse many seeds prior to human collection, and that collection and transport by people also play an important role in seed dispersal. Bush mango regeneration takes place along paths where seeds have probably fallen from the basins used to carry them from the forest. In one area near Takamanda in Cameroon these areas are called 'bush mango groves' (Malleson, pers. comm.). Trees planted on farmed plots have also increased the supply of bush mango. A reliable market, coupled with regular increases in value, makes the cultivation of bush mango a viable activity for many communities. It is also compatible with other forms of land use (e.g. as a shade crop for cocoa).

CUSTOMARY AND STATUTORY REGULATIONS REGARDING BUSH MANGO HARVEST

The majority of communities in Takamanda, Cameroon, and in the forested areas of Cross River State, Nigeria, have clear customary regulations governing the harvest of high-value NTFPs from their forests. Access rules for key resources are well-developed and vigorously enforced for locally important products such as bush mango. At the same time, somewhat confusing and inconsistent statutory regulations are in place in both countries.

Who can collect bush mango?

Under customary law in both areas only native residents are allowed to harvest economically important resources such as bush mango. Communities often impose large fines on outside parties who enter their forest to collect NTFPs without explicit permission.

Source: Terry Sunderland.

Figure C.1 *Bush mango storage by sticking fruits to the outer walls of the house for later use. This image was taken in the village of Ekuri, Cross River State, Nigeria*

Harvesters can rarely afford to pay the fines, and the produce is commonly confiscated and sold. Funds from the sale of confiscated fruit are sometimes placed in a community fund but are more commonly the preserve of the chief and his council, who determine how such proceeds should be spent. The enclaved community of Ekong-Anaku within the Cross River National Park in Nigeria goes further by charging a tariff for the transport of bush mango through their customary land, even though the seeds may not have been collected in the vicinity of the village (Sunderland and Isoni, 2001). In recent years encroachment from neighbouring Nigerian communities to harvest bush mango in the Takamanda region of Cameroon has been a cause of considerable resource conflict.

In some cases outsiders who are not indigenous to an area are granted an exception and can harvest bush mango. This includes migrants who have settled and become assimilated into a particular community (Sunderland and Isoni, 2001). Some communities might also allow collections by outsiders ('strangers') after they have paid a fee to community leaders. However, the 'sale' of access rights to valuable community resources, such as bush mango, by the community leaders can cause intracommunity conflict if the leaders do not use the revenues in ways that benefit the community as a whole.

Government regulations require permits for both the collection and the transportation of bush mango. To collect bush mango from within a forest reserve in Nigeria, harvesters must obtain a permit from a local forestry 'charge office'. This permit also allows the holder, including traders who purchase from communities in Cameroon as well as from those in Cross River, to transport bush mango. The permit does not specify the quantities that can be collected or transported. Charge offices can check collectors and transporters to see if they have the necessary documentation and issue on-the-spot fines of US$150–900 if they do not. If the transporter is unable to pay, which is usually the case, the produce is then impounded and later auctioned. Permits are not necessary for the collection of bush mango by local communities in their communal forest lands. Traders are also subject to 'informal taxation' along their trade routes by forest officers, police and others who encounter NTFPs as they are being transported. There is a need for less ambiguous measures for NTFPs, since this ambiguity creates additional opportunities for corruption along the harvest, transport and market chain.

Who 'owns' the trees?

Within the community, bush mango trees found in the forest are not owned by individuals or families, and access to the fruits is on a first-come, first-served basis. Trees planted or nurtured on farmland are owned by the landowner, and no one else is allowed access to the bush mango without permission. As bush mango has increased in value, some people have begun to clear land around bush mango trees in the forest, so as to establish long-term collecting rights to those trees (Morakinyo and Ekpe, 2000).

Felling of trees and harvesting of fruit

Customary laws include a prohibition on the felling of individual bush mango trees, whether on farmland or in the wild. Collectors are also not allowed to climb bush mango trees to harvest the fruit; the fruit may be harvested only after it has ripened and fallen to the ground. In Nigeria, statutory laws in Cross River State support customary law by including *I. gabonensis* on the list of strictly protected species under the 1999 Forest Law, and specifying that trees must not be felled. In Cameroon there is no such prohibition on felling bush mango trees.

Who can buy and trade in bush mango?

The sale of harvested bush mango usually takes place in the village, where outside traders come to purchase the product from local harvesters. Most of the bush mango in both Cross River State and Takamanda is bought by Igbo and Ibibio traders who transport large quantities to warehouse facilities outside Cross River State. In many communities on both sides of the border, Igbo bush mango buyers will meet individuals prior to the fruiting season to 'book' their bush mango for purchase (Asaha et al, 2006). Most communities on both sides of the border require outside buyers of bush mango to register in the village before they are permitted to purchase any product. Revenues raised this way are sometimes contributed to a village community fund, but not always, as mentioned above.

Local governance structures

Village chiefs and councils control access to NTFPs, particularly by outside parties. Their decisions are commonly criticized by community members who do not benefit from these types of revenues; such institutional arrangements are a cause of internal conflict and strife within many forest-edge communities where expropriation by 'elites' is a common characteristic of forest resource exploitation. Bush mango unions, more common in Nigeria, provide support to traders in times of need, rather than setting prices for communities (Asaha et al, 2006). However, most institutional structures in the bush mango production sector are weak (Omuluabi and Abang, 1994).

Government institutions are all but absent in the NTFP sector on both sides of the border, apart from the various bribes or 'settlements' insisted upon by government officials along the trade routes. Improved systems of accountability and transparency are clearly needed. In addition, lack of adequately trained staff, basic infrastructure and logistical support hinders the implementation of much of the formal forestry legislation in both Cameroon and Nigeria. The result is demotivated and poorly organized staff, and an environment where the culture of bribes, or 'private settlement', has become standard practice. At the same time, confusion exists as to which institutions of government are in charge of regulating NTFPs and how, even in the case of Cameroon, whose Ministry of Forestry and Fauna has established a department responsible for NTFP development, promotion and revenue generation. It remains unclear what mandate or instruments this department has in relation to the production and promotion of non-timber resources.

CONCLUSION

A range of overlapping measures combine to regulate the harvest of bush mango in the cross-border region of Nigeria and Cameroon. Customary laws serve as the best-recognized and most effective measures when it comes to harvesting practices, resource rights and the allocation of benefits within the community, although corruption at the village council level often siphons off benefits that accrue from granting permission to outsiders to collect. Trade is overseen primarily by statutory laws, although these tend to be vague and ambiguous, creating opportunities for corruption. Unlike some commercially traded NTFPs in the region (e.g. *Prunus africana*), bush mango is not threatened, and the status of wild and planted populations appears strong; at the same time, it provides an important and significant livelihood benefit to a wide range of communities. Additional regulation is therefore probably not called for in this case, since a lack of transparency and accountability in government means that new laws are not likely to improve either the conservation or livelihood profile of this species.

REFERENCES

Ainge, L. and Brown, N. (2004) 'Bush mango (*I. gabonensis* and *I. wombolu*)', in L. E. Clark and T. C. H. Sunderland (eds) *The Key Non-Timber Forest Products of Central Africa: State of the Knowledge*, USAID Technical Paper no 122, pp15–36

Arnold, J. E. M. and Ruiz Perez, M. (2001) 'Can non-timber forest products match tropical forest conservation and development objectives?', *Ecological Economics*, vol 39, pp437–447

Asaha, S., Balinga, M. P. B. and Egot, M. (2006) *A Socially Differentiated Overview of the Significance of Non-Timber Forest Products for Rural Livelihoods in Cameroon and Nigeria*, DfID Forestry Research Programme, final socio-economic report (Project R7636/R7636E)

Ayeni, J. and Mdaihli, M. (2001) *An Evaluation of the Exit Routes for Non-Timber Forest Products within the Takamanda Forest Reserve Area*, report for GTZ/PROFA

Cross River State (n.d.) 'Preview of the Cross River State Economy', in *Destination Cross River: The Nation's Paradise*, Cross River State government website, www.crossriverstate.gov.ng/portal/modules/mastop_publish/?tac=economy, accessed 21 June 2009

Harris, D. J. (1996) 'A revision of the Irvingiaceae in Africa', *Bulletin du Jardin Botanique National de Belgique*, vol 65, no 1–2, pp143–196

Malleson Amadi, R. (1993) *Harmony and Conflict Between NTFP Use and Conservation in Korup National Park*, Rural Development Forestry Network, paper 15c, pp16–22

Morakinyo, T. and Ekpe, S. (2000) *Feasibility of Establishing Community-Run Sawmill Enterprises in Abontakon (Biakwan) and Danare Ogar Ndey Villages, Cross River State*, Living Earth Foundation, Nigeria

Omuluabi, A. C. and Abang, S. O. (1994) *Marketing Margins in Non-Timber Forest Products Trade in Cross River State, Nigeria*, Cross River State Forestry Project, Calabar, Nigeria

Otu, D., Dunn, R. and Wong, J. (1994) *A Reconnaissance Inventory of High Forest and Swamp Forest Areas in Cross River State, Nigeria*, Cross River State Forestry Project, Calabar, Nigeria

Sunderland, T. C. H. and Isoni, O. (2001) *Cross River State Community Forest Project: Non-Timber Forest Products Advisory Report*, Department for International Development, Environmental Resources Management, Scott Wilson Kirkpatrick & Co Ltd

Sunderland, T. C. H., Besong, S. and Ayeni, J. S. O. (2003) 'Distribution, utilization, and sustainability of non-timber forest products from Takamanda Forest Reserve, Cameroon', in J. A. Comiskey, T. C. H. Sunderland and J. L. Sunderland-Groves (eds) *Takamanda: The Biodiversity of an African Rainforest*, Smithsonian Institution's Monitoring and Assessment of Biodiversity Program, Washington, DC, vol 8, pp155–72 [SI/MAB Series no 8]

Chapter 3

NTFPs in India: Rhetoric and Reality

Sharachchandra Lele, Manoj Pattanaik and Nitin D. Rai

INTRODUCTION

India is a vast, ecologically and culturally diverse, and densely populated country. While the percentage of forest cover is not very high (16–19 per cent) and many parts of these forests are in various stages of degradation, forests are a vital resource for many communities. State policies on forests, other common land and NTFPs, and the implementation of these policies, therefore have a significant impact on rural livelihoods as well as forest condition. A changing agrarian and industrial context further influences the role that NTFP collection can play in rural livelihoods. Analysing current policies and practices provides important insights about the possible role for NTFPs in rural development. This chapter seeks to do so by using macro-level analyses across several states and detailed case studies.

The term 'NTFP' can mean different things to different people.[1] In south Asia, it is useful to distinguish between two broad categories: high-bulk, low-value products such as firewood, grass and leafy matter that are important as inputs to the domestic, livestock and agricultural sectors; and relatively high-value, low-volume products such as specific fruits, nuts, leaves and herbs that are important in food products, medicines, cosmetics or other applications.[2] Although, in ubiquity and ecological impact, the former are more important, most discussions on 'NTFP policy' tend to focus on the latter. The reason could be that high-value, low-volume products can provide significant direct incomes, even to marginal landholders or the landless, and their collection is simultaneously seen as potentially less ecologically 'damaging'. Reconciling livelihoods and conservation through such NTFP-based enterprises has thus elicited much debate. Further focus is given to these high-value income-generating NTFPs by the shorthand 'commercial NTFPs'.

Policies on commercial NTFPs may seek very different objectives (or a balance between them): the generation of revenue for the state, meeting the demands of NTFP-based industries, protecting harvesters from exploitation by middlemen, enhancing the livelihoods of poorer communities, promoting resource sustainability and meeting broader biodiversity conservation goals. The instruments deployed to achieve these objectives include the manner in which harvesting and marketing rights are assigned, the organizational set-up through which harvesting and marketing are carried out, fiscal strategies such as taxation and subsidies, and investments made in harnessing traditional and modern knowledge and generating market information. Our analysis of NTFP policies in India seeks to critically examine the objectives pursued (both stated and implicit), the instruments used and the impacts on collector livelihoods and ecological sustainability.

We begin this chapter with a brief summary of the role of NTFPs in rural livelihoods in India that indicates how important commercial NTFP collection is, for whom and in which regions. After giving a broad history of state intervention in the NTFP sector in India, we focus on two major regions – the central-eastern dry forest region that straddles the states of Orissa, Madhya Pradesh (including Chhattisgarh), Andhra Pradesh, Jharkhand and small parts of Bengal and Maharashtra, and the Western Ghats moist forest region that spreads across the states of Maharashtra, Goa, Karnataka, Kerala and Tamil Nadu (Figure 3.1). For the central-eastern forest region, we cover several states, but focus on Orissa and Madhya Pradesh. For the Western Ghats region, we present

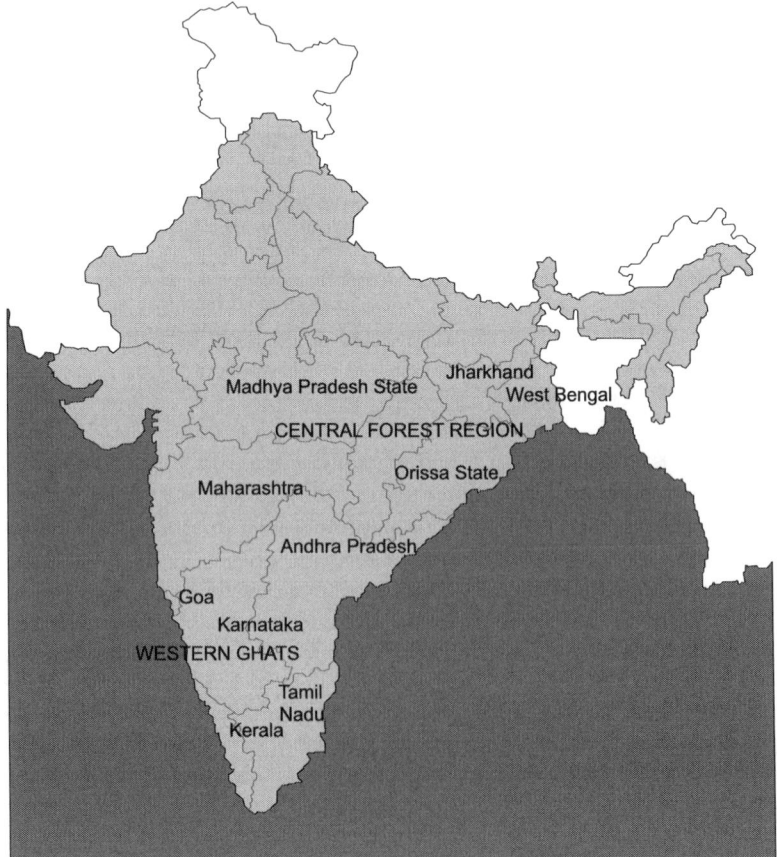

Source: Forest Survey of India, 2003.

Figure 3.1 *Location of central dry forest region and Western Ghats region in India*

case studies from two parts of Karnataka that highlight issues specific to a mixed tribal and non-tribal context, in one case providing insights into the ecological complexities around NTFP extraction.

SIGNIFICANCE OF NTFPS IN RURAL LIVELIHOODS, REGIONS AND THE NATIONAL ECONOMY

Some 3000 wild plant species in India are used for purposes including food, fodder, medicines, spices and condiments, dyes, fibres, gums and resins, essences and oils, plates and furniture (Tewari, 1994). Estimates of the contribution of NTFPs to rural incomes vary widely. Tewari and Campbell (1995) estimate that about 25 per cent of India's rural labour force derives up to 50 per cent of its income from NTFPs, which translates into around 100–150 million persons. Between 55 and 70 per cent of the wage employment in the forestry sector is attributed to NTFPs (Gupta and Guleria, 1982). Other estimates of the rural employment obtained through NTFP collection and processing range from 3.3 million person-years (Mitchell et al, 2003) to 'significant' employment for 50 million persons a year (MOEF, 2001). What accentuates the importance of NTFPs is the fact that their collection often complements agriculture-based livelihoods, as it is largely carried out during the dry season.

NTFPs play an even more important role in the livelihoods of communities near or in forests. These communities, invariably among the poorest, may be broadly called 'forest-dwelling tribal communities', a subset of the ethnic groups designated as Scheduled Tribes in the Indian constitution.[3] They depend on NTFPs not just for income (Malhotra et al., 1991; Prasad, 1999; Hegde et al., 1996), but also for subsistence (MoEF, 2001), and forests are an integral part of their cultures.

The region where commercially valuable NTFPs play the most significant role in rural livelihoods is undoubtedly the central-eastern forest belt, where the dry deciduous forests are rich in various industrially important oil seeds (such as sal), soap nuts (myrobalans) and cigarette leaves (tendu), and where most of the country's tribal population is located.

Finally, it is worth noting that these so-called 'minor' forest products represent a significant source of revenue for certain governments. In 1986, NTFPs accounted for nearly 40 per cent of the revenue of state forest departments and 75 per cent of net export earnings from the forest sector (Mitchell et al, 2003, quoting M. P. Shiva). With several states banning the felling of timber in natural forests, the importance of NTFPs as a source of state revenue may be increasing.

NATIONAL OVERVIEW[4]

Historical shifts in NTFP policy

Systematic state intervention in forestry was initiated by the British colonial government. The objective of British forest policy was primarily the maximization of state

Source: RCDC, Bhubaneswar, India.

Figure 3.2 *Collection of Kalabhalia seed by the villagers of Kalasulia, Boudh*

revenues, meeting the needs of British industries and expanding state control over the country. It was effected by reserving large chunks of forest for exclusive state use and declaring valuable products off-limits to local users. The main focus was on timber (and later softwood) extraction, but where NTFPs had significant commercial value, the objective of revenue maximization was clearly visible, such as in pine resin extraction in the Himalayas (Guha, 1989) or in tree gum or *Acacia catechu* extraction in peninsular India (Gadgil and Chandran, 1988). Following protests, some concessions were made regarding firewood collection and grazing, but commercially valuable NTFPs were kept under state control. In the central forest belt, the objective of suppressing tribal rebellions and establishing state control was also a priority, for which forest control was one instrument.

In the decades immediately following independence, forests continued to be seen as a resource that supported industrialization and nation-building. Forests were thus managed to maximize the production of commercially valuable products to provide the raw materials for industries and urban areas and revenue for the state. Exploitation of the forests for bamboo, resin and other 'minor' products continued and expanded. Revenues did not necessarily increase, because many of these products were given to industries at highly subsidized prices, as in the case of bamboo in Karnataka (Gadgil et al, 1983).

The role of NTFPs in the livelihoods of forest-dwelling communities began to gain attention in the 1950s. Sporadic protests against exploitative or revenue-oriented state policies occurred, such as the representations by 1000 tribal women to the Chief Minister of Orissa in 1953 (Das, 1996). In 1961, the Dhebar Commission urged state governments to provide for intensive collection and local processing of MFPs (GoI, 1961). The Committee on Tribal Economy in Forest Areas also recommended the establishment of forest corporations and tribal development cooperative corporations for the collection, processing and marketing of NTFPs (GoI, 1967) as did the Bawa Committee on Cooperative Structures in Tribal Areas (GoI, 1971).

These pressures resulted in a more proactive state policy on NTFPs. Legal and administrative initiatives were taken in different states to regulate and support NTFP collection and trade. The ostensible goals of this policy, as summarized by Prasad (1999), were to:

- reduce exploitation of NTFP collectors and ensure fair returns to them;
- maximize the collection of produce and ensure supply to industries using them; and
- increase revenues to the state.[5]

There was no explicit commitment to 'sustainable harvest'.

Over the next two decades, these initiatives led to the creation of complex institutional arrangements around the collection and marketing of different NTFPs, including laws and administrative orders, NTFP-related organizations and financial support policies. These are particularly prevalent in the states of the central forest belt. Nationally the main direct intervention was the formation of the Tribal Cooperative Marketing Development Federation of India Ltd (TRIFED) in 1987 for marketing NTFPs and other agricultural produce harvested by tribal communities. These were the core arrangements concerning NTFPs for several decades. As we shall see below, the non-implementation of subsequent legislation in the context of NTFPs means that these are still largely the de facto arrangements.

The new National Forest Policy of 1988 marked a significant change in forest policy rhetoric. The goals of maintaining 'ecological balance' and meeting the needs of villagers, especially tribal communities, were given top priority. The involvement of local communities in forest management was also considered, leading to the central circular in 1990 that initiated 'joint forest management' (JFM) in the country.

The implications of the new policy for the structures and forms of NTFP collection have been mixed but limited. Under JFM programmes, some attention has been given to increasing incomes from NTFPs. JFM orders issued by many states[6] have increased the villagers' shares in NTFPs from both regenerated and standing forests. The practical implementation of these orders is still highly contingent upon the overall set of

NTFP policies in each state (see e.g. Lélé et al, 2005). At the same time, restrictions on the collection of NTFPs from national parks and wildlife sanctuaries have been tightened, especially since 2004, as a result of the Supreme Court's strict (overly so, in our view) interpretation of the Wildlife Protection Act.

Other changes in the wider governance system since the 1990s could also influence NTFP rights and management. The landmark 73rd Constitutional Amendment in 1992 prescribed that states should enact legislation creating three additional tiers of government at the district, subdistrict and village level. Ownership of several resources, including fuelwood and fodder, social forestry plantations and NTFPs, should vest in the lowest tier. Although most states have passed the necessary laws, the transfer of control has not happened (Mathew, 1995; Mathew, 2000; Mathew, 2004).

A more radical law passed in 1996 – the Panchayats (Extension to Scheduled Areas) Act, 1996, or PESA – seeks to give wide-ranging powers to the village general body in Scheduled Areas (i.e. tribal majority districts). PESA makes the radical provision of granting 'ownership of minor forest produce' (and several other natural resources) to the Gram Sabha (village general body). But again, the provision has been rendered ineffective by state governments leaving ambiguity about which forests the rights are to be exercised in, making the provision subservient to JFM rules and other MFP-related rules and laws, or completely ignoring the provision, as in Madhya Pradesh (Upadhyay, 2004). Even at the central level, the Ministry of Environment and Forests has undermined the provisions by excluding bamboo and cane from the definition of 'MFP' and by recommending against any change in forest rights (MoEF, 1998). Perhaps the only state in which PESA has had some impact is in Orissa, where the government recently framed a new set of rules transferring rights over some NTFPs to Gram Panchayats (village councils).

A very recent law – the Scheduled Tribes and Other Traditional Forest Dwellers (Recognition of Forest Rights) Act, 2006 – confers NTFP rights on the hamlet-level bodies of forest-dwelling communities. This act is yet to be implemented and its implications still need to be considered. Due to the similarities in the MFP-related rights conferred under this act vis-à-vis the PESA, the impacts are likely to be similar unless larger issues are addressed.

Thus, in most cases, the main policies and structures that shape NTFP use continue to be those set up in the 1970s and 1980s.

Basic elements of current policies

According to existing forest law, the state is the 'owner' of all NTFPs.[7] The state may grant 'lease rights' or 'usufructory rights' of collection and possibly of transport and sale to certain individuals, organizations or state agencies. Depending largely on the commercial value of the NTFPs, the state varies the extent of its direct involvement in and control of the collection, procurement and sale of the NTFPs as follows.

The most valuable NTFPs are 'nationalized',[8] that is, the resource is treated as entirely state property and its harvesting, transport, storage and sale are carried out entirely by state agencies or are very strictly regulated. In practice, state agencies such as forest development corporations declare a state-wide procurement price, and all collectors are required to sell the produce to the state agency or its appointed agents at this price.

Other commercially valuable NTFPs are 'controlled' or 'specified' – that is, less stringently regulated – with extraction rights being granted to agencies or individuals in different locations or years, with no restrictions on storage, but some monitoring of transport. Typically, the state 'auctions' the rights of extraction for a particular forest area. The one who wins the auction gets the sole rights of extraction for two years in the form of a lease and pays a royalty to the state forest department. This 'contractor' then announces a procurement price. Actual collection may be done by local households, but they must sell the produce only to the contractor at the procurement price specified. The contractor may bring in outside wage labour to carry out the collection.

Other less valuable NTFPs are completely unregulated, and may be freely extracted and consumed or sold by any individuals. The sale may be in the open market or to traders. There may or may not be any tax on the sale (typically not). Finally, some NTFP species might be declared as 'lease-barred', which means that their extraction is not permitted due to fears of their extinction. In all cases where extraction takes place, the forest department, as the custodian of the forests, is supposed to enforce sustainability norms.

The immediate implication of these differences in state control is a difference in the capacity of the state to extract the surplus produce. For unregulated produce, this capacity is virtually zero. For controlled produce, the state can extract some of the surplus as royalties, with the rest going to the contractor. For nationalized produce, the state can extract the entire surplus, since it is, in effect, also the contractor.

Greater control need not necessarily translate into greater surplus extraction by the state itself. Much depends upon how prices are set and what structures are set up for improving prices obtained by collectors. In many cases, the state created primary collector cooperatives – often called 'Large Area Multi-Purpose Societies' or 'Large-Scale Adivasi Multi-Purpose Societies' (LAMPS)[9] – and gave them exclusive harvesting rights in specific forest areas and the mandate to carry out collective marketing of the products. In some cases, state-level federations of these primary cooperatives were also created, and they had to market their produce through the federations, ostensibly to obtain economies of scale and thereby ensure higher returns to collectors. Andhra Pradesh went the furthest in this direction, creating the Girijan Cooperative Corporation, a state-level and state-supported tribal cooperative with no lower-level primary cooperatives. In some cases it was made mandatory to sell the produce through the national-level TRIFED. In other words, a coercive 'cooperative' was pursued by several states in parallel with state control. The policy of leases to private companies, state forest corporations or other bodies continued in some pockets.

How these policy shifts and variations in structures have worked in practice is what we will now examine using three case studies. The first case explores 'nationalized' and 'controlled' NTFPs in the central Indian forest belt, specifically Orissa and Madhya Pradesh. Second, we examine a cooperative NTFP collection in Karnataka. Third, we investigate a 'non-nationalized' and 'non-cooperative' NTFP in the Western Ghats, for which detailed ecological studies are also available. In all cases, we seek to understand what the ostensible policy goals are, what has been done to achieve them, and their impacts, particularly in terms of collector livelihoods and (where possible) the sustainability of use.

CASE 1: NTFP POLICIES IN THE CENTRAL FOREST REGION: LIP-SERVICE TO TRIBAL INTERESTS?

The context

The central forest region is perhaps the most important region in the country in terms of the availability of commercially valuable NTFPs and also the existence of a large, forest-dwelling, largely tribal population that has been historically engaged in collecting these NTFPs. It is estimated that 70 per cent of NTFP collection for sale takes place in this region. About 77 million people reside in villages that have forest area within their boundary. These villages constitute 20 per cent (Andhra) to 40 per cent (Orissa) of the total villages in the state.

The most important NTFPs across these states are tendu or kendu leaf (KL) (*Diospyros melanoxylon*), sal seeds and leaf (*Shorea robusta*), mahua flowers and seed (*Madhuca indica*) and bamboo (including *Dendrocalamus strictus*, *Bambusa arundinacea*, *Bambusa nutans* and *Bambusa tulda*). Tendu leaf, sal seed, mahua and bamboo were 'nationalized' in most of these states in the 1960s and 1970s. Other products were brought into the 'controlled' category and, in a few cases, the 'lease-barred' category. The exact list of NTFPs in each category, the dates when they were so categorized and the pertinent legislation are given in Table 3.1.[10]

As mentioned above, many policies and institutions have been set up and modified over the past few decades to collect, sell and otherwise regulate NTFPs in each category. The impacts of these policies and institutions are described below using a detailed history of some NTFPs in each of the 'nationalized' and 'controlled' categories in Orissa and Madhya Pradesh. The case of tendu leaf shows how much the state continues to covet the revenues from this very valuable resource, but also the variety of approaches adopted. The history of other resources highlights the complex interplay in the institutional arrangements, but also the halting progress on the ground with long-term changes in the status of the NTFP collectors.

Tendu leaf: A coveted resource

The leaves of *Diospyros melanoxylon* (family Ebenaceae), called 'tendu' in Madhya Pradesh and 'kendu' in Orissa, are used as wrappers for bidis (Indian cigarettes). They are the most valuable NTFP in the central forest region, in terms of total revenue generated.

Orissa pioneered the state control of tendu leaves. Some control was introduced in 1949 through the Kendu Leaf (control and distribution) Order under the Essential Commodities Act. Partial nationalization took place in 1961 with the Orissa Kendu Leaf Act (see Table 3.1), and full nationalization in 1973. Under nationalization, the procedure has now been set as follows. The procurement price is fixed every year by the government. The procurement of KL from the collectors or growers, and preliminary processing such as drying, binding and storage, are done by the KL department, which falls under the state forest department. Many KL procurement ('collection') centres have been set up in the KL producing districts for this purpose. Seasonal staff are engaged as agents of the KL department to carry out the procurement. KL movements

Table 3.1 *NTFP categories and relevant legislation*

	Orissa	MP/ Chhattisgarh	Andhra Pradesh	Bihar/ Jharkhand	Maharastra
Nationalized NTFPs (year of nationalization)	Tendu leaf (1961, strengthened in 1973); sal seed (1983, denationalized in 2006, but not clear how this will work); bamboo (1988)	Tendu leaf (1964), harra, gums and sal seed (1975)	Bamboo and tendu leaf. (1970)	Bamboo (1984), tendu leaf (1972–3), sal seed (1977), mahua seed, mahulan leaf and harra	Tendu leaf (1969) and mahulan leaf
'Controlled' NTFPs	69 MFPs	No controlled NTFPs	24 NTFPs	Sabai; all others are completely unregulated	33 MFPs given to gram panchayats and 88 NTFPs auctioned at deputy conservator level
Lease-barred NTFPs (if any)	9 NTFPs: sal leaf (but lease has been given!), sal resin, gums, khair, barks, *Rauwolfia serpentina*, tassar cocoons, cane, sandalwood	No lease- barred NTFPs	No lease- barred NTFPs	No lease- barred NTFPs	No lease- barred NTFPs
Relevant Acts	Orissa Kendu Leaf (Control of Trade) Act, 1961; further modified in 1973; Orissa Forest Produce (Control of Trade) Act, 1981	MP Tendu Patta (Vyapar Viniyaman) Adhiniyam, 1964; MP Van Upaj (Vyapar Viniyaman) Adhiniyam, 1969	AP Abnus Leaves Act 1956; AP NTFP (Regulation of Trade in Abnus Leaves) Act & Rules, 1970	Bihar Kendu Leaf (Control of Trade) Act, 1972	Maharashtra MFP (Regulation of Trade) Act, 1969 and its 1997 amendment

are strictly monitored. Marketing of the procured KL is done by the Orissa Forest Development Corporation Ltd (OFDC) through bulk auction. OFDC now gets 5 per cent commission on this work, which covers costs and profits. Around 1 million people are engaged in plucking KL for about 20–45 days and some 6 million person-days of employment are created for the processing of KL in a season. From the mid-1990s to the early 2000s, state earnings from the sale of KL generally ranged from Rs400–700 million (US$10–17.5 million) per year.[11] KL is estimated to have contributed around 74 per cent of the state's total earnings from forests during this period (Government of Orissa, 2005).

Collectors were supposed to benefit from nationalization in at least two ways: high prices and guaranteed prices regardless of the amount collected. However, neither benefit accrued in practice. The state (or its agents) had the authority to reject the KL offered by a collector if not satisfied with the quality. More importantly, the prices paid to the collectors were dramatically lower than the price at which OFDC finally auctioned the produce. Table 3.2 gives the price received by the collectors as

Table 3.2 *Percentage share of collectors in final KL auction price*

Year	Orissa	Madhya Pradesh	Andhra Pradesh	Bihar
1989–90	7	16		
1990–91	15	45		
1991–92	19	32		
1992–93	21			69
1993–94	19		44	71
1995–96		37	49	

Source: Vasundhara and Vikalpa, 1998.

a percentage of the price obtained by OFDC at its auction. Over the ten-year period (1984–1994), the average share of the final auction price received by the collectors was an appalling 16 per cent. Even setting aside transportation, storage and handling costs and losses (estimated at around 28 per cent), it turns out that the state got a hefty 56 per cent of the final auction price over the same period.

The state was aware early on that retaining most of the profits for itself would not be a popular policy. It passed orders in 1986 under the Kendu Leaf Act that 50 per cent of the profits from the KL trade would be shared with local government bodies (Panchayat Samitis and Gram Panchayats). In practice, however, the government has persistently claimed that it cannot calculate the profits from KL trade and hence it has not released these 'KL grants' systematically. Only Rs100 million (US$2.5 million) has been released annually in the form of ad hoc KL grants to the Panchayats, whereas the actual amounts that should have been released were of the order of Rs160 million (US$4 million) to Rs290 million (US$7.25 million) a year during the period 1992/93–1995/96.[12] As Vasundhara and Vikalpa (1998) say, it is indeed 'appalling that for the last over 15 years the Government has ... used [not being able to work out the profits] as an excuse to forfeit its legal commitment to share KL profits with local people'.

There are further problems with the very concept of KL grants. First, the grants made to the Panchayats are not proportional to the collection of KL from those areas. For example, although Bolangir district contributed 25 per cent of the state's total KL collection from 1993 to 1996, its share of ad hoc KL grants given to Panchayats during that period was 14 per cent. Second, and more important, even if KL grants had been proportional to the KL contribution of each Panchayat, the Gram Panchayats represent all the residents within the Panchayat boundary, both collectors and non-collectors. Transferring profits to the Panchayats instead of paying high prices to the collectors amounts to transferring income from the collectors to the non-collectors in a Gram Panchayat. Typically, the latter are the better off households and elite in the villages. Even the recent order to increase the KL grants to 90 per cent of the profits does not address this unfair transfer of income.

Finally, the implementation of the KL Act even as it exists has several shortcomings in practice. Delayed payment to the collectors is common, leading to the collectors borrowing funds from the KL department agents. Also common is these agents' practice of underpaying the collectors by demanding more leaves in a bundle than the official measure.

In Madhya Pradesh, the structures have evolved somewhat differently. Madhya Pradesh was quick to follow Orissa in nationalizing tendu leaves in 1964. The initial approach retained many elements of the contractor system and hence failed to yield the desired results. Payments to collectors were delayed and collection undervalued. In 1984, the Madhya Pradesh government set up the State Minor Forest Produce (Trading & Development) Co-operative Federation (SMFPCF) as an apex body that would pool the individual collections of the LAMPS and primary cooperative societies (PCSs) that were set up in a few districts. In 1988, the government further rationalized the arrangements by setting up primary forest produce cooperative societies (PFPCSs) at the bottom, district forest produce cooperative unions (DFPCUs) in the middle and the SMFPCF at the top. At present, Madhya Pradesh has 1066 PFPCSs and 58 district unions, with nearly 1.5 million members in the PFPCSs. After an initial period of nominated office-bearers, the first elections for a president and other office-bearers were conducted in 1995. The forest officer at the territorial forest division acts as the managing director of the DFPCU. Similarly, the chairmanship and all other top executive positions in the state federation are held by state officials.

Since 1989 the PFPCSs have been engaged in the procurement of tendu leaf. Each PFPCS covers approximately 10–12 phads (collection centres). The collection of leaves is done with the involvement of local forest officials. The transportation and storage of the leaves is done by district unions. The funds for various operations are made available to the district unions by the state-level MFP federation. The district unions provide funds for procurement to the PFPCSs. The phad munshi of primary society and phad abhirakshak, who is a government employee, purchase the leaves. The manager of the primary cooperative society and the nodal officer, who is a government servant (mostly deputy Ranger or Forester), withdraws cash from the society's bank account and the nodal officer carries the cash to the collection centres for payments by the phad munshi or phad abhirakshak. Each family is given a collector's card. The phad munshi enters the collector's daily collection on the card. Payment for the collection of leaves is made weekly and the payment made is entered on the card. The sale of the leaves is done by the MFP federation, generally through either a nationwide tender or an auction.

The primary cooperative societies initially received commission at Rs10 (US$0.25) per standard bag. The district unions were paid at Rs3 (US$ 0.07) per standard bag. The SMFPCF received a commission of 2 per cent on the amount received from the sale of leaves in the whole state. The government later reduced the commission of the SMFPCF and the DFPCUs to Rs1 per annum (a token amount of less than 3 US cents). The idea was that the primary collector cooperatives and their upper-level federations would have a strong interest in ensuring timely and appropriate payments to their collector members.

The ultimate impact of these structures on the returns received by the collectors seems to have been slightly, but perhaps not dramatically, different from that in Orissa, at least until the mid-1990s. The data in Table 3.3 show that although the collectors' price increased substantially in absolute terms, their share remained in the range of 16–45 per cent of the final sale price received by the state-level federation. Even allowing for handling costs, the state made a significant profit. For example, in 1995–96, the state made a profit of Rs340 (US$8.50) per standard bag (42 per cent of the final price), while the collectors got Rs310 (US$7.75, or 38 per cent of the final price).

Table 3.3 *Collector share in final sale price of tendu leaf in Madhya Pradesh*

Year	Collector price	Final sale price	Collector share
1989–90	150	932	16%
1990–91	250	554	45%
1991–92	250	758	32%
1995–96	310	810	38%

As the state's profits soared, pressure to share this profit with the collectors mounted. The state put in place a system of distributing some of the profits back to the collectors in the form of 'incentives'. In 1989, following a bumper profit, the state distributed Rs1500 million (US$37.5 million) back to the collectors. Subsequent years saw much lower incentive distribution. This payment was discontinued in 1990 but restarted in 1995. For the 1995–1997 seasons, nearly 20 per cent of net income was paid as 'incentive wages'.

As a consequence of the 73rd Constitutional Amendment of 1992 and PESA, the Madhya Pradesh government decided, in 1998, to pass on all the net profit from the trade of tendu leaf to the primary collector cooperatives. The cooperatives, in turn, have to distribute 60 per cent of this to the tendu leaf collectors as incentive wages and spend 20 per cent on NTFP development and 20 per cent on infrastructure development.[13] The Madhya Pradesh Forest Department continues to report revenues received from tendu leaves, indicating that perhaps not all the profits are being passed on to the cooperatives. Moreover, as in the case of the KL grants, the sharing of profit is fixed at the top, not by the owners of the produce (which are now supposed to be the Gram Sabhas). This results in a sense of paternalism and insecurity about the process and a general delay of one or two years between collection and payment of incentive wages. Furthermore, the functioning of the primary cooperatives is not particularly democratic, with elections not having been held for a long time, the forest department remaining in control in practice, and the improved prices being offset by the manipulation of quantities (Ranu Bhogal, personal communication, based on unpublished study conducted for CIFOR).

A potentially positive element in the Madhya Pradesh government's policy has been the setting up of a group insurance scheme for the KL collectors since 1991. This scheme covers around 2.4 million collectors. Collectors do not pay fees for this insurance, and get different levels of compensation for death, disability, etc. For the period 1991–2005, the federation reported that 150,820 claims were settled and Rs561 million (US$14 million) was paid to the nominees of deceased collectors.[14] There is, however, at least anecdotal evidence of some amount of mismanagement of this system also (Ranu Bhogal, personal communication).

On the whole, the tendu leaf policy in Madhya Pradesh is somewhat more supportive of the NTFP collectors than that in Orissa. Another way of looking at it is that Orissa is slowly moving along a trajectory that Madhya Pradesh has already traversed. Madhya Pradesh state also started out extracting significant fractions of the profit from tendu leaves. But Madhya Pradesh has moved more quickly to a somewhat better sharing of the profits in the post-PESA period, while Orissa tried a conceptually faulty KL-grants approach. The common feature in both states is that control is entirely top-down and even administrative functions at the lower levels are manned by government servants, with significant involvement of the forest department. This raises serious questions about the level of democracy and autonomy in the functioning of the so-called collectors' cooperatives. This issue is further highlighted in the analogous case of LAMPS in Karnataka.

Controlled products: Fuzzy organizational arrangements

The approach in Orissa to the management of non-nationalized NTFPs has been characterized by the presence of multiple organizations and shifts and variations in arrangements. LAMPS were set up in Orissa and elsewhere in the mid-1970s. By the 1980s, there were 222 LAMPS. There are a few other cooperatives also involved in NTFP collection and marketing, such as the agency marketing cooperative societies (AMCSs) and the Orissa Rural Marketing Society (an autonomous agency under the Department of Panchayati Raj, involved in the formation of self-help groups for micro-enterprise development). The government also set up the Tribal Development Cooperative Corporation (TDCC) in 1973 as an apex cooperative, of which 202 LAMPS, 35 other cooperatives, 47 panchayat samitis and the state government are members.

Orissa government policy towards these cooperatives and their federations has been wavering, paternalistic and not well thought out. For instance, the TDCC was given rights to sal seed procurement in 1984 following the 'nationalization' of sal seeds. However, this right was taken away in 1991 and handed over to the OFDC. In 1990, the following complicated allocations were made:

- The TDCC was given the exclusive right to four MFP items: tamarind, hill broom, honey and mahua in all 27 forest divisions of the state.
- Utkal Forest Products Ltd (UFPL), a joint sector company, was given the exclusive right to collect 29 other NTFP items in all the forest divisions of the state.
- AMCSs were given leases to operate in three divisions for all products except the ones given to TDCC and UFPL.
- TDCC was additionally given rights over all products except those given to UFPL in 19 divisions.
- The OFDC was given rights over all products in five divisions not allocated to AMCSs, the TDCC or UFPL.

This policy ensured that there was only one buyer per product in a division. It was assumed that since these were also state or state-controlled agencies, they would not misuse their monopsonist position. But neither efficiency nor support to collectors could be achieved. The agencies either did not buy all the produce or set unrealistically high procurement prices, thereby incurring losses. For example, the TDCC was given sole rights to mahua flower procurement in 1991. The procurement price was set at Rs3 (US$0.07) per kg plus overheads. Traders in Orissa purchased mahua flowers from neighbouring Bihar state at Rs1 (less than US$0.03) per kg and sold them to the TDCC at Rs3, thereby making a large profit and leaving the TDCC with a huge loss. The government de-specified mahua flowers in 1992 (Vasundhara and Vikalpa, 1998).

Not surprisingly, the TDCC, which was created to protect tribal collectors from exploitation, has itself turned out to be a liability to the government, with huge losses. At the end of March 2000, the accumulated loss of the TDCC was Rs410 million (US$10.25 million). According to the balance sheet of 2004/05 the accumulated loss was Rs611 million (US$15.275 million). As a result of these losses, the TDCC seems to have become largely defunct and is now not involved in the procurement and trade of NTFPs in Orissa.

In Madhya Pradesh, the arrangements are different and the ultimate impact seems even less favourable to NTFP collectors. The three-tier organization for tendu leaves does not seem to be involved in the collection and marketing of other produce. 'Specified' products such as sal seeds are regulated by putting quantitative restrictions on transport and sale. Licences are required for the growing, transport or sale of quantities beyond the specified limits. A price-spread analysis (Vasundhara and Vikalpa, 1998) indicated that the collectors got significantly lower prices than those at which sal seeds sold in local towns, although such analysis ignores the transaction costs incurred for transport and marketing.

As mentioned above, the post-PESA period saw changes in policies regarding the sharing of profits for nationalized produce in both Orissa and Madhya Pradesh. In the case of non-nationalized NTFPs, the most significant post-PESA change occurred in Orissa. The changes were announced in 2000 but were legally implemented in 2002, with the new Orissa Gram Panchayats (Minor Forest Produce Administration) Rules being framed under the state's Panchayati Raj Act. The key features of this policy change were:

- The royalty system was abolished, as was the system of assigning monopoly buying rights to individual contractors or organizations, and also the transit permit system.
- The Gram Panchayats were to be given the power to regulate the procurement and trading of MFP, whether produced in government lands and forest areas within the limits of the village, or collected from the Reserved Forests and brought into the village. This regulation would take the following form:
  - All traders would have to register themselves with the Gram Panchayat. Unregistered traders would not be able to procure NTFPs.
  - The traders would have to pay at least the minimum price specified by the Gram Panchayat.
  - The ecological aspect of collection would be regulated by the Forest Department, which could impose temporary bans on collection if collection was found to be unsustainable.

(For more details, see RCDC, 2007).

These changes were encouraging, as they established greater control by the local communities over the product and resource and liberalized the trade and movement of the produce. But many details and nuances are still being worked out and the final outcome of these changes is yet to be understood.

CASE 2: LAMPS IN KARNATAKA: WHOSE COOPERATIVES AND WHOSE PRODUCE?

The state of Karnataka in south-west India contains the largest portion of the Western Ghats – a hilly, forested region considered to be a global biodiversity hotspot. Although the population in this region is predominantly non-tribal, there are many pockets in the Karnataka portion of the Western Ghats with a significant tribal population.

Following the adoption by the government of India in 1971 of the recommendations of the Bawa Committee mentioned earlier, the first LAMPS (in this state, the expansion being Large-scale Adivasi Multi-Purpose Society) was set up in Karnataka in Hunsur taluka (subdistrict). Five more were set up in other parts of Mysore district during 1982 and 1983 (Kamath, 1988). There are now 20 active LAMPS in Karnataka, covering more than 100,000 adult tribals across four districts. Each LAMPS typically covers one taluka, and its membership is supposed to be open to all adult tribals in that area. The general body elects a tribal as president and five to ten other tribals to the board of directors. Several other government officials are ex-officio members of the board and, more important, the secretary of the LAMPS is provided by the Department of Cooperatives.

NTFP collection and marketing is supposed to be the major activity of the LAMPS and the only income-generating activity undertaken by it. Each LAMPS applies to the Karnataka State Forest Department (KFD) for grant of a lease to collect NTFP from forests in that taluka. The KFD grants the lease for some designated areas in return for royalties. The LAMPS auction the produce to the highest bidder. Other activities include acting as a channel for the public distribution system, selling subsidized agricultural inputs and channelling government soft loans to members. Until a few years ago, there was no state-level NTFP marketing federation. Although such a federation now exists, the LAMPS are not required to sell to it alone. Karnataka thus represents a situation in which NTFPs were not 'nationalized', but rather rights of NTFP collection and sale were given to primary cooperatives, with no compulsion to sell the NTFPs to higher level federations or state corporations.

Performance of Karnataka LAMPS[15]

How well have the LAMPS functioned as NTFP marketing cooperatives, which was their primary goal? This question needs to be answered from an economic, organizational and ecological perspective. Economically speaking, the performance is generally poor. First, as Table 3.4 shows, the LAMPS pay their collector members only 40–70 per cent of the final sale price. Collectors acknowledge that the presence of the LAMPS has made private traders offer higher prices than they would have otherwise, but they are bitter at such a large proportion of what would be their legitimate income being lost to the LAMPS. Second, even after retaining such high margins, the majority of the LAMPS show annual operating losses in most years and almost all show long-term accumulated losses (which have been periodically written off by the government) (Table 3.5). Third, from time-series data it is apparent that the range of products sold and the total revenues from MFP sale fluctuate wildly and are declining in several

Table 3.4 *Prices offered to collectors for select NTFPs and LAMPS margins*

Name of NTFP		Use	Price paid to collector	Final sale price	LAMPS margin
Common	Scientific				
Honey		Medicine, food	27.0	37.0	27%
Aralekai	(*Terminalia chebula*)	Leather softening, medicine	1.5	2.5	40%
Amla (fresh)	(*Phyllanthus emblica*)	Pickles, medicine	2.3	3.3	30%
Amla (dried)		Medicine	6.9	8.9	22%
Gum		Bookbinding, silk reeling, starching	28.8	38.8	26%
Lichen		Paint, condiment	20	25	20%
Tamarind	(*Tamarindus indica*)	Condiment	2.0	3.0	33%
Dhoopa	(*Vateria indica*)	Cooking fat	0.8	1.8	56%
Sheekakai	(*Acacia concinna*)	Soap, shampoo, medicine	4.0	5.0	20%
Ramapathri	(*Myristica malabarica*)	Paint	35	50	30%
Almaddi	(*Ailanthus malabarica*)	Agarbatti	30	80	63%

Note: All figures are in Rs per kg. In case of HD Kote LAMPS, margin includes commission paid to their agents.
Source: Interviews with traders and LAMPS records.

LAMPS, while increasing in others. One might thus conclude that while the formation of LAMPS has benefited the tribal collectors to some extent in some locations, the cooperatives are not financially sustainable. Moreover, the gains are far below the potential gains, are not consistent from year to year, are possibly accompanied by a shrinking product base and have come at enormous public cost (as the state has to intermittently write off the losses).

An organizational analysis of the LAMPS showed that they hardly functioned as proper cooperatives. Membership rolls had not been revised for years, and so current members constituted only around half the total number of tribal collectors in the locality. Decision-making was not transparent or democratic, with the Presidents often being unelected non-tribals (government officials). Non-tribal secretaries are in control in almost all LAMPS.

The ecological performance of the LAMPS is difficult to judge in the absence of any quantitative data on NTFP availability. Qualitative discussions with collectors suggest that factors other than harvest, such as shrinkage in forest area, destruction of certain habitats and invasion by weeds may be more significant factors affecting availability in most cases. Some cases of extraction-driven extinction may also have occurred, such as a *Cinnamomum* species used for making essence-sticks.

Explaining the performance

This poor economic performance is linked to the organization of the LAMPS. Although a cooperative is meant to be owned and operated by its members, the paternalistic

Table 3.5 *Revenues, margins and profit/loss for all LAMPS in Karnataka, 1994–95*

Sl. no.	Name of LAMPS	NTFP revenue		Non-NTFP revenue		Gross margins		Profit or loss?	
		Gross	Per member	Gross	Per member	NTFP	Non-NTFP	Current year	Accumulated
		['000 Rs]	[Rs]	['000 Rs]	[Rs]			*[1994–95]*	
1	Yalandur	169	195	161	186	32%	26%	loss	loss
2	Hunasur	104	16	10	2	45%	3%	profit	loss
3	HD Kote	10	2	15	4	17%	6%	profit	loss
4	Chamarajanagar	535	557	5	5	44%	2%	profit	profit
5	Kollegal	39	15	0	0	32%	0%	profit	loss
6	Gundlupet	13	8	5	3	5%	2%	profit	loss
7	Somavarpet	127	43	89	30	43%	5%	profit	loss
8	Virajpet	80	16	83	17	26%	8%	profit	loss
9	Madikere	124	43	3	1	29%	21%	profit	loss
10	Koppa	11	5	204	93	5%	4%	profit	loss
11	Moodigere	43	2	197	55	NA	3%	profit	profit
12	Puttur	86	28	62	20	36%	4%	loss	loss
13	Udupi	0	0	228	51	NA	7%	profit	profit
14	Sulya	6	2	90	39	7%	4%	loss	profit
15	Belthangadi	75	20	95	26	5%	15%	loss	loss
16	Mangalore	0	0	27	43	NA	3%	loss	loss
17	Karkala	74	17	193	44	6%	7%	profit	loss
18	Bantwal	0	0	88	36	NA	7%	loss	loss
19	Kundapur	75	31	122	51	29%	9%	profit	loss

Notes: Actual magnitudes of profits/losses are not given because the numbers were found to be inconsistent.
NA = no business reported in 1994–95 in that category.
Source: Returns filed by LAMPS secretaries with Registrar of Cooperative Societies, Bangalore.

stipulation that the secretary should be an official assigned by the Department of Cooperatives completely undermines this concept. The non-tribal, educated government official who comes as a secretary to the society wields all the power and almost invariably mismanages the LAMPS for personal gain, co-opting some of the elected directors or the president in the process. There have been many cases of obvious swindling of funds, but, in this paternalistic arrangement by which the Department of Cooperatives still controls the LAMPS (and annually subsidizes them), the tribal members have no recourse beyond requesting that a particular secretary be replaced.

Furthermore, the LAMPS have actually internalized some of the exploitative practices of the trader world. A LAMPS will appoint a few tribals as 'commission agents', both for procuring the forest produce from members and for advancing seasonal credit. Their functioning is not transparent or accountable, and in some cases they have become the new money-lenders. Finally, in several places, the forest department has also moved into the game: in return for their 'cooperation', forest officers have insisted on becoming the presidents of these cooperatives.

The last point relates to larger institutional questions and also ones of ecological sustainability. In theory, the LAMPS are required to ensure that harvesting is within sustainability norms, but these norms are never prescribed or debated openly. Moreover, a necessary condition for the sustainable harvest of any product is that the harvesters have secure and clear tenure over the resource. This is completely missing in the LAMPS arrangement. The KFD controls the forest land and the NTFP resources in it. The LAMPS do not have a statutorily assigned right to this resource; this is granted by the KFD on a two-year lease. The renewal of the lease almost invariably requires major efforts on the part of the tribal community, making it clear that this is not a right but a 'privilege' that can be discontinued any time. Indeed, it has been discontinued in many LAMPS, off and on.

The KFD also decides how much forest area to lease to the LAMPS, and in several cases only parts of the forest have been so assigned, while other parts have been allocated to private traders. The assigned areas have also shrunk over time, largely in the name of wildlife conservation measures, but without any proof that NTFP collection is harmful to wildlife. Eight out of the 20 LAMPS do not have any forest area assigned to them and hence do not have significant NTFP collection going on. The KFD also interferes in the day-to-day process of NTFP extraction, by dictating areas of extraction and entry permits for individual collectors.

In other words, the individual tribal collectors neither have secure rights over the resource they harvest nor adequate control over 'their' cooperative that has been ostensibly created to facilitate marketing. Lack of secure rights over the resource creates a lack of incentive to get involved in resource management and ensure sustainable harvesting. Lack of control over their cooperative means they are often subject to almost the same level of exploitation as in the pre-LAMPS period. The de-linking of the resource from the cooperative in some cases means that the main function of NTFP marketing is impossible, and the members either move to non-forest activities or let the LAMPS lie defunct.

CASE 3: UPPAGE (*GARCINIA GUMMI-GUTTA*) IN THE KARNATAKA WESTERN GHATS: ECOLOGICAL COMPLEXITY

The Western Ghats forests harbour a great diversity of plant and animal species. Several of these are collected and sold for medicinal, culinary or other purposes. The case of *Garcinia gummi-gutta* extraction from a certain part of the Western Ghats highlights the complexities that may confront makers of NTFP policies in tropical developing regions, particularly regions with a long history of forest use that are looking towards expanding markets for NTFPs. Ecological factors, forest rights and markets have together shaped the manner in which the economic gains from this NTFP have been distributed over time and across different players, and resulted in ecological impacts across the landscape.

The species and the product

Garcinia gummi-gutta (L.) Robson (family Guttiferae), locally known as uppage, is endemic to the Western Ghats of India and Sri Lanka. It is found in evergreen and lower 'shola' forests up to a height of 1000m. Uttara Kannada district in Karnataka state is at the northern end of its range and seems to have the highest density of uppage trees. Before the commercialization of the product, uppage seeds, which are rich in fat, were used by some local households for making a kind of margarine. The main consumption of uppage is in the state of Kerala, where the dried rind is used extensively as a souring agent in fish curries.

The commercial collection of uppage rind in Uttara Kannada commenced in the late 1970s with the realization that a market for the rind existed in Kerala. The price of the dried rind started at around Rs3 (US$0.07) per kg and increased slowly to Rs12–16 (US$3–4) per kg in the early 1990s. At these prices collection hovered around an esti-mated 50,000kg for Sirsi forest division, one of the three forest divisions in Uttara Kannada district that report a significant uppage harvest (Shivannagowda and Gaonkar, 1998).

Uppage economics, local livelihoods and markets

As with other commercially valuable NTFPs in Karnataka and elsewhere, once uppage became valuable, its collection was controlled by the state forest department, which wanted a share in the profits. Since the late 1980s, rights to uppage harvest in different administrative units (typically forest ranges) have been auctioned for two-year periods by the forest department. Those who win such auctions (the leaseholders) can either send in their own labourers to collect the product or insist that all local collectors sell what they collect to them at prices they set. The forest department, having auctioned the rights, plays a mostly passive role, not identifying or enforcing sustainability norms, and only occasionally ensuring that 'leakages' (villagers selling produce to persons other than the leaseholder) do not occur.

Uppage is traded through complex private channels, which further changed during the boom period (see below). The final sale price of uppage is therefore not easy to determine. However, it is clear that the state is extracting a substantial royalty. The royalty paid by contractors to the forest department for Sirsi forest division alone increased from Rs388,300 (US$9707) in 1989 to Rs3,545,600 (US$88,640) in 1995 (Rai, 2003, and Saxena et al, 1997, quote a somewhat higher figure). Unfortunately, the state does not really utilize these funds to ensure resource sustainability or other conservation measures. On the other hand, the contractors are clearly making a hefty profit, around Rs20 (US$0.50) per kg (Saxena et al, 1997), which is a margin of 25–30 per cent. It may be noted that the state practice of auctioning NTFP rights to the highest bidder has not changed, not even since the introduction of JFM in this forest division in 1993 and the explicit statement in 2000 that NTFP rights in JFM areas would belong to the village-level committees.

Boom and bust

Many NTFPs show a boom–bust cycle. Typically, the boom is because of some unique application and the bust is a result of domestication, or substitution with alternatives.

In the case of uppage, the story is slightly different, but no less dramatic. In the late 1980s, some studies (Sergio, 1988) showed that hydroxycitric acid (HCA), a secondary compound present in the rind of uppage fruit, might be effective in weight loss and therefore a natural solution to obesity (Majeed et al, 1994). Over-the-counter drugs derived from uppage, such as Citrin and Citrimax, were aggressively marketed. As a result, the price of dried uppage rind received by the collectors increased rapidly, reaching Rs75–90 (US$1.87–2.25) per kg at its peak in 1998. The annual extraction of uppage in Sirsi Division shot up to 1,600,000 kg in 1999 (Rai, 2004). From being a specialized activity carried out by a few households in each village located in the evergreen forests, uppage collection and drying became a booming industry in which people from across the socioeconomic spectrum and far-off villages participated, scouring deep in the forests, harvesting fruit before it was fully ripe, cutting branches and sometimes even felling entire trees to harvest the fruit.

The shaky claims regarding the effectiveness of HCA did not stand up to scrutiny. More research showed that HCA did not provide the claimed weight loss benefits (Heymsfield et al, 1998). The price of HCA in the international market dropped from US$30–35 in 1994 to US$9–11 in 2000. The price of rind paid to collectors in Uttara Kannada dropped dramatically from about Rs60 (US$1.50) per kg in 1999 to Rs28 (US$0.70) per kg in 2000. Processors of uppage also point to two additional reasons for the drop in prices: the low quality of rind due to the harvesting of unripe fruit, and the importing of fruit from Sri Lanka at cheaper rates.

What was the government's response to the boom and bust? The state forest department did little more than cash in on the boom – royalties from auctioning the licence to collect uppage went up tenfold from 1989 to 1999. The energies of the department were devoted to policing the movement of uppage – not to keep it sustainable, but rather to ensure that the contractors who had won the auction for a particular area then got all the produce from that area. Even quality control was missing, resulting in a fall in prices for uppage from Uttara Kannada (as compared to that from Sri Lanka). Funds generated from the royalties simply went into the state treasury, with no additional allocation for forest protection or conservation. After the bust, the response was equally ineffectual – the collectors were left to fend for themselves, while some forest officers were relieved that the bust had reduced the harvest.

Uppage ecology, harvesting practices and harvest impact

Uppage is harvested whole by humans, so the seeds are removed from the forest. In principle, high levels of such seed removal might result in inadequate regeneration, which should be visible in lower seedling numbers. Rai's detailed study of uppage regeneration (Rai, 2003), however, shows that this impact is not discernible. The size–class distribution of individual plants showed the 'reverse J' pattern typical of stable plant populations (Figure 3.3). The seedling density was high at all sites, with even sites that experienced high harvest intensity showing high seedling numbers. This might be due to harvesters not collecting fruit from inaccessible parts of trees, or from trees that are difficult to climb or have not produced enough fruit to justify the effort. The fruits thus left behind are eaten by animals, which disperse the seeds.

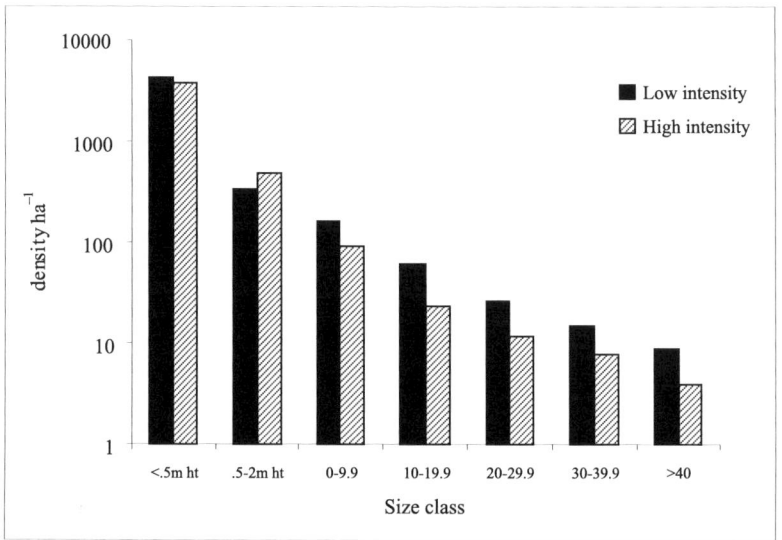

Source: Rai, 2003.

Figure 3.3 *Population structure of uppage individuals in low-intensity and high-intensity harvest sites*

A feature of the harvesting process that might, however, result in negative impacts on future uppage availability and population growth itself is destructive harvesting practices. While the impact of light pruning may be ambiguous, that of cutting off major branches and felling whole adult trees (which harvesters do when in a hurry to extract the fruit) is deleterious to the availability of the resource in subsequent years and to long-term uppage population growth (Rai, 2007). Whether such destructive harvesting takes place or not is a function of the tenurial arrangements (see below).

Forest tenure and harvest practices

What is the pattern of uppage harvest today and why? Our observations suggest that the pattern varies significantly and is clearly the combined result of the extent of competition among collectors and the nature of forest tenure. The semi-evergreen forests of Uttara Kannada are typically under one of three regimes. The majority is reserve forest, where rights of local communities are very limited, although enforcement varies. Other parts have been declared minor forests, which are meant for local use and are, in effect, open-access. There are, however, pockets of forest where individual farmers or groups of farmers have been given exclusive rights to the harvest of firewood, leaf manure, fodder and other products. These patches, called soppinabettas, are generally adjacent to cultivated land and are often fenced off by the farmers. Whereas the reserved forests and minor forests are de facto open-access for harvest, the soppinabetta holders can prevent anyone else from extracting NTFPs from those lands, and they are thus de facto the sole NTFP collectors in those patches.

Not surprisingly, it has been observed that harvesting practices vary significantly between soppinabettas and other areas, especially in times of high demand. In many soppinabettas, collectors actually wait for the fruit to ripen and fall to the ground or for the rind to be discarded by frugivores. In such cases, it is often the women members of the household who pick up the rind, obviating the need to climb the trees or to beat or cut the branches.

When uppage prices were low, collection methods in open-access areas were also somewhat similar. When, however, the price of uppage increased dramatically during the boom in the mid to late 1990s, local collectors began scouring deeper and deeper in the forest. People from villages far away came to these forests to harvest uppage. Contractors also began sending in their own 'gangs' of labourers. Whatever caution 'traditional' harvesters may have exercised was thrown to the wind, as collectors grabbed whatever they could as fast as they could. Collectors routinely climbed trees and beat branches, cut the branches and even occasionally cut down whole trees. Parikh et al (1999) reported that the percentage of undamaged trees dropped from 97 per cent in soppinabettas to 86 per cent in open-access areas. Rai and Uhl (2004) reported an even higher percentage of trees (up to 50 per cent) experiencing branch cutting or felling in open-access patches.

Furthermore, Parikh et al (1999) also report more 'impatient' behaviour in the open-access forests (93 per cent of collectors reported unripe fruit harvests) than in the private-access forests (only 11 per cent reported unripe fruit harvest). As cited above, harvesting unripe fruit is one of the factors contributing to the fall of prices for uppage from Uttara Kannada. Moreover, harvesting fruit (rather than collecting empty rind) means that the fruit pulp and seeds are transported out of the forest and are unavailable for regeneration or for animal consumption.

Conclusions

The case of uppage is both typical and unique. Typical are the state's totally revenue-oriented and sustainability-neglecting NTFP approach, its lack of attention to what might constitute fair returns for collectors and to quality control issues, and its refusal to transfer harvesting rights to local communities even when overarching policies have ostensibly changed. Also typical are the thin markets that are susceptible to boom–bust and the presence of state-backed monopoly purchasing systems.

But the case is unique in its ecology, which offers the possibility of almost 'totally sustainable' rind extraction while leaving fruit and seed for predators and for the future regeneration of the resource. It is also unique in the unusual existence of exclusive private-access regimes in this region, which demonstrates how exclusive and secure tenurial arrangements can result in sustainable extraction, although the current inequitable distribution of such secure tenure results in an inequitable distribution of the gains from uppage. The detailed ecological studies carried out on uppage, the like of which are not available for most other NTFP species, also highlight the complexity of the life cycles of NTFP species and the possibility that such species can survive high levels of extraction, but also the possibility of negative side effects that NTFP managers and policy-makers need to be aware of.

SUMMING UP: THE GAP BETWEEN
RHETORIC AND REALITY

NTFPs in central and peninsular India are clearly important for the livelihoods of several million people. The diversity of the NTFPs available also speaks to the diversity of the forests from which they are collected. State policy towards NTFPs has, however, combined indifference and the favouring of state interests (revenue maximization or support to industries) for a long time, starting with the British period but extending several decades into the post-independence period.

In response to pressure from tribal development groups, various arrangements were introduced in the 1970s to improve the returns to tribal forest dwellers from NTFP collection and sale. But even then, the major changes in NTFP policy appear driven by a desire to appropriate the maximum possible surplus for the state (especially for high-value produce), while paying lip service to the interests of the NTFP collectors.

For medium-value NTFPs, where collector livelihoods were perhaps given greater priority than state revenues, several arrangements have been initiated. Cooperatives and cooperative federations have been the forms of organization promoted by the state. Even here, the top-down and paternalistic approach of the bureaucracy has kept cooperatives from achieving income enhancement, let alone empowerment and broader tribal development. Ham-handed monopsony powers given to so-called cooperative federations have often worked to the detriment of NTFP collectors and their primary cooperatives, while also constituting a big drain on the exchequer. Lack of secure rights to NTFPs in a particular forest for a particular group makes unsustainable harvesting highly possible.

Recently, due largely to changes in political devolution, some states in central India have initiated steps to transfer NTFP rights to local communities. One approach is to transfer more income to collectors within the elaborate framework already set up, without modifying the rights on the ground. Another approach is to try to devolve NTFP regulation rights to local bodies. Both of these are overlaid on changes that have been introduced in JFM areas. All this, however, pertains largely to the medium- or low-value products, not 'nationalized products'. Similarly, in peninsular India, contracting out of collection rights to valuable NTFPs remains the norm, even in JFM areas.

Little is known about the ecological sustainability of NTFP harvests in India (Shahabuddin and Prasad, 2004). While the open-access nature of most harvests and the lack of monitoring and incorporation of local knowledge into their management suggest the likelihood of unsustainable harvesting, the complex ecology of the products makes impacts unpredictable. In some cases, such as uppage, the impacts may become visible only at very high levels of extraction, and may be manageable with some innovative changes in tenurial arrangements.

Strengthening NTFP-based livelihoods of forest-dwelling communities in an ecologically sustainable and economically viable manner thus continues to be a major policy challenge. While there are encouraging signs of the state shifting towards a more responsive mindset, there is a long way to go.

ACKNOWLEDGEMENTS

The analyses presented in this article are based on several years of work in different parts of India that has benefited from interactions with researchers, activist groups and local communities, and from the support provided by the authors' organizations. Lele and Pattanaik specifically wish to acknowledge the institutional support provided by the Ford Foundation. Rai would like to acknowledge Christopher Uhl for ideas; Parameshwar Hegde, Arun Hegde and Purushotam Gowda for field assistance; and the Pennsylvania State University and the Conservation, Food and Health Foundation for support.

NOTES

1   Another term prevalent in India is minor forest produce (MFP), sometimes used synonymously with NTFPs and sometimes excluding firewood, fodder, cane and bamboo.

2   Bamboo and cane are bulky but also high value, and tend to be treated like the other 'commercial' NTFPs. Strictly speaking, animal products – including meat – are also 'non-timber forest products'. Given the ban on hunting, however, only a few animal products are included in the common understanding of NTFPs, the main ones being wild honey and deer antlers.

3   There are more than 250 distinct tribal communities in India, constituting about 8 per cent of the population.

4   This overview is limited by lack of information about NTFP policies and laws in the north-eastern states.

5   For example, the act passed by the Madhya Pradesh government to regulate tendu leaves states its goals to be stopping pilferage in government forest and other lands, providing definite value for tendu leaves to growers, increasing revenue to the state, providing adequate wages to labour, improving the quality and quantity of leaves by regular pruning and ensuring the supply of leaves to small and medium manufacturers of bidis (Indian cigarettes).

6   Since 'forests' are part of the concurrent list, i.e. under the dual control of the central and state governments, the states actually control and manage the forests and implement programmes within an overall national forest policy.

7   This is supposed to change under PESA, but has not yet happened.

8   Although the term suggests that the product has somehow been appropriated by the nation as a whole, the central government has actually no role to play in the decision of a state government to 'nationalize' any product.

9   The concept of LAMPS was mooted by the Bawa Committee in 1971 as cooperative societies for integrated tribal development through the marketing of MFPs and the provision of credit, agricultural inputs and rationed goods. By 1989, 2912 LAMPS had been established across the country, more than 80 per cent of them in the five states of Madhya Pradesh, Bihar, Maharashtra, Rajasthan and Orissa that have large tribal populations (Mahalingam, 1992).

10 Note that the 'completely unregulated' NTFPs are not listed because the products are many, varying from state to state, and add up to a very small fraction of the commercial NTFP trade.

11 One rupee is currently worth 2.5 cents (US), but ranged in value from 2 to 13.3 cents during the period under discussion.

12 Profits as per the finalized proforma accounts were Rs495 million (US$12.375 million) in 1992–93, Rs587 million (US$14.675 million) in 1993–94, Rs451 million (US$11.275 million) in 1994–95 and Rs313 million (US$7.825 million) in 1995–96.

13 See www.mfpfederation.com/content/about_us.html.

14 See www.mfpfederation.com/.

15 This section is based upon Lélé and Rao (1996).

REFERENCES

Das, V. (1996) 'Minor forest produce and rights of the tribals', *Economic and Political Weekly*, vol 31, no 50, 14 December, pp3227–3229

Gadgil, M. and Chandran, M. D. S. (1988) 'On the history of Uttara Kannada forests', in J. Dargavel, K. Dixon and N. Semple (eds) *Changing Tropical Forests*, Centre for Resource and Environmental Studies, Canberra

Gadgil, M., Prasad, S. N. and Ali, R. (1983) 'Forest management and forest policy in India: A critical review', *Forest, Environment and People*, vol 33, no 2, pp127–155

GoI (1961) *Report of the Commission on Scheduled Areas and Scheduled Tribes 1960–61 (Dhebar Commission)*, Ministry of Home Affairs, Government of India, New Delhi

GoI (1967) *Report of the Committee on Tribal Economy in Forest Areas (Hari Singh Committee)*, Department of Social Welfare, Government of India, New Delhi

GoI (1971) *Report of the Committee on Cooperative Structures in Tribal Areas (K S Bawa Committee)*, Department of Agriculture, Government of India, New Delhi

Government of Orissa (2005) *Economic Survey 2005–2006*, Department of Planning and Co-ordination, Bhubaneshwar

Guha, R. (1989) *The Unquiet Woods: Ecological Change and Peasant Resistance in the Himalaya*, Oxford University Press, Delhi

Gupta, T. and Guleria, A. (1982) *Non-wood Forest Products in India: Economic Potentials*, Oxford & IBH, New Delhi

Hegde, R., Suryaprakash, S., Achot, L. and Bawa, K. S. (1996) 'Extraction of non-timber forest products in the forests of Biligiri Rangan Hills, India: 1 Contribution to rural income', *Economic Botany*, vol 50, no 3, pp243–251

Heymsfield, S. B., Allison, D. B., Vasselli, J. R., Pietrobelli, A., Greenfield, D. and Nunez, C. (1998) 'Garcinia cambogia (hydroxycitric acid) as a potential antiobesity agent: A randomised controlled trial', *Journal of the American Medical Association*, vol 280, pp1596–1600

Kamath, S. U. (ed) (1988) *Karnataka State Gazetteer: Mysore District*, Government of Karnataka, Bangalore

Lele, S. and Rao, R. J. (1996) 'Whose cooperatives and whose produce? The case of LAMPS in Karnataka', in R. Rajagopalan (ed) *Rediscovering Cooperation*, Institute of Rural Management, Anand, Gujarat

Lele, S., Kiran Kumar, A. K. and Shivashankar, P. (2005) *Joint Forest Planning and Management in the Eastern Plains Region of Karnataka: A Rapid Assessment*, Centre for Interdisciplinary Studies in Environment and Development, Bangalore

Mahalingam, S. (1992) 'Institutional support for marketing of minor forest produce in India', *Indian Journal of Agricultural Marketing*, vol 6, no 2, pp76–83

Majeed, M., Rosen, R., McCarty, M., Conte, A., Patil, D. and Butrym, E. (1994) *Citrin: A revolutionary herbal approach to weight management*, New Editions Publishing, Burlingame, California

Malhotra, K. C., Deb, D., Dutta, M., Vasulu, T. S., Yadav, T. S. and Adhikari, M. (1991) *Role of NTFP in Village Economy*', IBRAD, Calcutta

Mathew, G. (1995) *Panchayat Raj: From Legislation to Movement*', Institute of Social Sciences, New Delhi

Mathew, G. (ed) (2000) *Status of Panchayati Raj in the States and Union Territories of India*, Institute of Social Sciences and Concept Publishing Company, New Delhi

Mathew, G. (2004) *Ownership, Control and Management of MFP in Schedule V Areas: The Role of Gram Sabha/Village Panchayat (Covering States of Andhra Pradesh, Rajasthan and Maharashtra)*, Institute of Social Sciences, New Delhi

Mitchell, C. P., Corbridge, S., Jewitt, S., Mahapatra, A. K. and Kumar, S. (2003) *Non-Timber Forest Products: Availability, Production, Consumption, Management and Marketing in Eastern India*, Universities of Aberdeen, Cambridge and London

MoEF (1998) *Report of the Expert Committee on Conferring Ownership Rights of MFPs on Panchayats/ Gram Sabhas*, Ministry of Environment and Forests, Government of India, New Delhi

MoEF (2001) *National Forestry Action-Programme India*, Ministry of Environment and Forests, Government of India, New Delhi

Parikh, A., Lele, S., Shrinidhi, A. S. and Rao, R. J. (1999) 'Influence of forest rights regimes on sustainability and returns from NTFP harvest: The case of Garcinia gummi-gutta in the Western Ghats of India', in *National Workshop on JFM*, VIKSAT, Gujarat Forest Department and Aga Khan Foundation, Ahmedabad

Prasad, R. (1999) 'Joint forest management in India and the impact of state control over non-wood forest products', *Unasylva*, vol 50, no 198, pp58–62

Prasad, R. and Bhatnagar, P. (1991) *Socio-Economic Potential of Minor Forest Produce in MP*, State Forest Research Institute, Jabalpur

Rai, N. D. (2003) 'Human use, reproductive ecology, and life history of *Garcinia gummi-gutta*, a non-timber forest product, in the Western Ghats, India', PhD thesis, Pennsylvania State University, University Park, PA

Rai, N. D. (2004) 'The socio-economic and ecological impact of Garcinia gummi-gutta fruit harvest in the Western Ghats, India', in K. Klusters and B. Belcher (eds) *A Global Comparison of Non Timber Forest Products, Volume 1: Asia*, Center for International Forestry Research, Bogor

Rai, N. D. (2007) 'The ecology of income: Can we have both fruit and forest?', in G. Shahabuddin and M. Rangarajan (eds) *Making Conservation Work*, Permanent Black, New Delhi

Rai, N. D. and Uhl, C. F. (2004) 'Forest product use, conservation and livelihoods: The case of uppage fruit harvest in the Western Ghats, India', *Conservation and Society*, vol 2, no 2, pp289–313

RCDC (2007) 'Policies governing NTFP: Orissa', *Banajata* website, Regional Centre for Development Cooperation, Bhubaneshwar, www.banajata.org/l/a25.htm, accessed 10 November 2008

Saxena, N. C., Sarin, M., Singh, R. V. and Shah, T. (1997) *Independent Study of Implementation Experience in Kanara Circle*, Karnataka Forest Department, Bangalore

Sergio, W. (1988) 'A natural food, the Malabar tamarind, may be effective in the treatment of obesity', *Medical Hypotheses*, vol 27, no 1, pp39–40

Shahabuddin, G. and Prasad, S. (2004) 'Assessing ecological sustainability of non-timber forest produce extraction: The Indian scenario', *Conservation and Society*, vol 2, no 2, pp235–250

Shivannagowda, B. and Gaonkar, D. S. (1998) *Kanara Circle-Sirsi: NTFP Management Strategies*, Karnataka Forest Department, Bangalore

Tewari, D. D. (1994) 'Developing and sustaining non-timber forest products: Policy issues and concerns with special reference to India', *Journal of World Forest Resource Management*, vol 7, no 2, pp151–178

Tewari, D. D. and Campbell, J. Y. (1995) 'Developing and sustaining non-timber forest products: Some policy issues and concerns with special reference to India', *Journal of Sustainable Forestry*, vol 3, no 1, pp53–77

Upadhyay, S. (2004) 'Tribal self-rule law and common property resources in Scheduled Areas of India: A new paradigm shift or another ineffective sop?' presented at *The Commons in an Age of Global Transition: Challenges, Risks and Opportunities*, Tenth Conference of the International Association for the Study of Common Property, Oaxaca, Mexico, 9–13 August

VASUNDHARA and VIKALPA (1998) *NTFP Policy in Orissa and a Comparative Analysis of NTFP Policy and Prices with Neighbouring States*, VASUNDHARA, Bhubaneshwar

Chapter 4

Policy Gaps and Invisible Elbows: NTFPs in British Columbia

Darcy A. Mitchell, Sinclair Tedder, Tim Brigham, Wendy Cocksedge and Tom Hobby

INTRODUCTION

Timber and non-timber values in Canada

In 2005–06, forest products – principally softwood lumber, pulp and paper – contributed US$35.7 billion to the Canadian gross domestic product (GDP) (Natural Resources Canada, 2006). In the same fiscal year, the three NTFPs documented by Natural Resources Canada (NRC) – Christmas trees, maple syrup and wild pelts other than sealskins – contributed some US$238 million to the GDP (NRC, 2006). The Canadian Forest Service (CFS) estimates, however, that the actual contribution of all NTFPs is between US$689 million and $1.26 billion annually (NRC, 2005) – still less than 3 per cent of the documented value of timber and pulp products.

The gulf between the reported economic value of timber and that of non-timber products provides one of the more obvious reasons for the virtual absence of NTFPs from policy agendas everywhere in Canada.

However, the wide range of services provided by forest ecosystems also falls outside most official documentation of the economic value of Canada's forests. Despite this omission, the production, maintenance and protection of air, water, soils, climate and biological diversity, as well as the provision of social, cultural, spiritual, aesthetic and educational benefits associated with forests, are generating a significant – and growing – public policy discussion (Daily et al, 1997). As a result government, NGOs and industry are expending significant resources to support research, advocacy and action on these services and values. In contrast, NTFPs, and those who use them for commercial, subsistence, recreational or cultural uses, rarely feature on the policy agenda in Canada.

The case of British Columbia

This chapter explores a number of issues connected with NTFP policy in Canada as a whole through a review of the sector in British Columbia (BC), Canada's westernmost province. We chose BC as our focus because the NTFP sector there is relatively well developed and researched. The province also leads the country in the degree to which the sector has been studied, although most dimensions of both the production and the use of NTFPs remain poorly understood.

We first describe the range of products harvested in BC, for both commercial and non-commercial uses, and discuss resource users, volumes and values. We then review the history and implications of the NTFP policy debate in BC and analyse major factors that we believe contribute to its current status, which we characterize as consisting of a mixture of 'policy gaps' and 'invisible elbows'.

Policy gaps and invisible elbows

The title of this chapter refers to two kinds of policy circumstances that arise with respect to NTFPs. The first, 'policy gaps', is self-explanatory – the policy required has

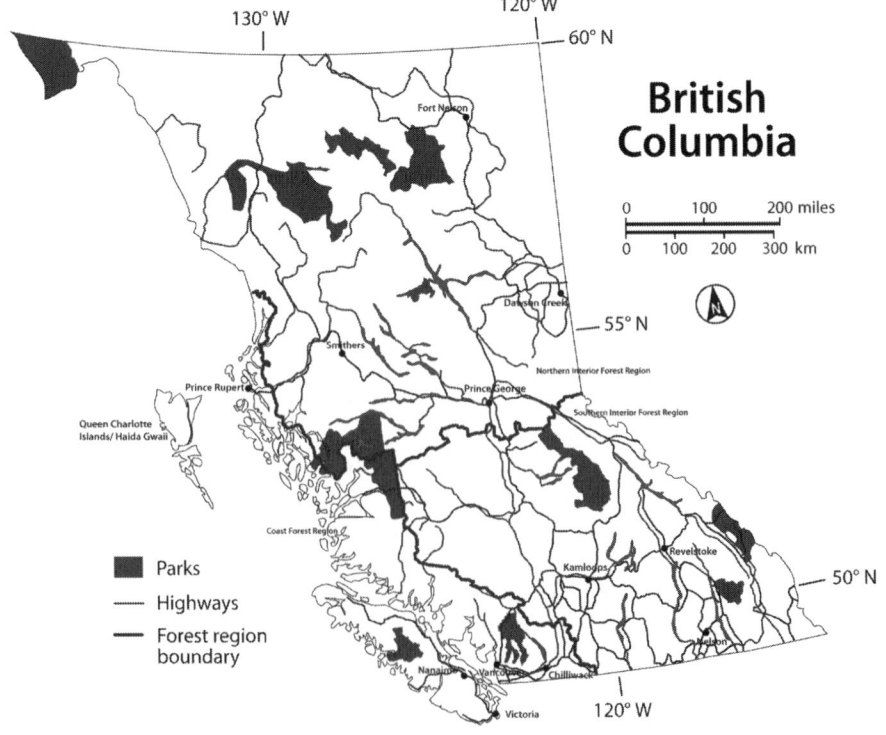

Source: Used with permission, British Columbia Ministry of Forests and Range.

Figure 4.1 *Map of British Columbia, Canada*

not yet been developed. The second term, 'invisible elbows', is drawn from Jacobs (1993). Jacobs contrasts the 'invisible elbow' with Adam Smith's 'invisible hand'. In both cases, market forces operate through the combined results of many private decisions. Where the result of the 'invisible hand' is to bring about general prosperity, Jacobs argues that the 'invisible elbow' can bring about general ruin.

> *Elbows are sometimes used to push people aside in the desire to get ahead. But more often elbows are not used deliberately at all; they knock things over inadvertently... Outcomes occur overall, because small individual decisions add up inexorably to large collective ones, and no one is counting. (Jacobs, 1993, p25)*

In the context of this paper, we use 'invisible elbow' to refer to general problems of collective action, where individual actions consistently fail to support the interests of the group (Bickers and Williams, 2001, p62). More specifically, the phrase describes policy decisions made by a wide array of actors that, while not intended to create problems or impede development in the NTFP sector, tend to have those effects.

Implications and possibilities

Following a discussion of the current state of NTFP policy in BC, we consider the implications of the policy environment and explore if, and how, policy might be developed to enhance stewardship of NTFP resources and increase investment in the NTFP sector. In this policy arena, as in any other, problem definition lies at the heart of how, and whether, policy will be developed and implemented. It seems likely that NTFPs have not emerged in BC – or any Canadian jurisdiction – as a pressing policy issue because the sector does not:

- present itself as a significant policy problem;
- clearly offer itself as a solution to other already recognized problems; or
- compete in significance with problems already on the policy agenda.

There is evidence, however, that NTFPs are attracting more attention in forestry, rural development and environmental circles as conventional resource opportunities continue to decline and as public sentiment shifts more strongly toward the 'non-timber' values of forests. In this context, our tentative conclusions suggest that there may be greater opportunities now than in the past for adaptive and collaborative approaches to policy development in partnership with First Nations, rural communities, non-governmental organizations and the private sector.

GEOGRAPHY AND POPULATION

BC is Canada's westernmost province, with a land area of about 95 million hectares, of which 95 per cent is provincial Crown (public) land and the remainder is privately owned or federal land, including Indian reserves. About 63 per cent (or 60 million hectares) is forested (BC Ministry of Forests, 2010). BC is geographically and biologically highly diverse, with ecosystems ranging from temperate rainforest on the coast and boreal forest in the north to Canada's only true desert.

The population of BC is predominantly urban (85 per cent in 2001) (Statistics Canada, 2006). Most British Columbians live in the south-west corner of the province, particularly in Vancouver and the lower mainland. As in other parts of the world, rural–urban migration accelerated after World War II, producing a shift from 54 per cent urban in 1941 to 78 per cent urban by 1981, with a continued upward trend in urbanization since that time. The decline of resource extraction industries in the province – notably timber harvesting and processing, and fishing – has drastically reduced the employment opportunities available in rural and remote communities. The trend to urbanization is markedly lower among BC's Aboriginal[1] (Statistics Canada, 2006) population. Almost 50 per cent of First Nations people live in rural areas, many on the large number of small Indian reserves. As the rate of population growth among First Nations is higher than the general BC population, it is likely that Aboriginal peoples will form a larger percentage of the rural population in years to come.

PRODUCTS, USERS, VOLUMES AND VALUES

While the NTFP sector in BC is relatively well developed, many questions remain about the volume and values of resources being harvested, and about the number and characteristics of harvesters and others working in the industry.

NTFPs in BC may be categorized as:

- floral greenery;
- wild edibles;
- medicinals and nutraceuticals (also known as functional foods);
- landscaping and restoration products;
- crafts and art;
- miscellaneous products (essential oils, smoke woods, soaps, etc.); and
- forest-based cultural tourism or ecotourism with a NTFP component.

Commercial harvesting

While there is no definitive list of all the NTFPs harvested in BC, de Geus (1995) estimates that over 200 products have been commercially harvested in the province. Wills and Lipsey (1999) estimate direct revenues at approximately US$266 million (including ecotourism-related activities). These figures provide an indication of the economic importance of the sector, especially when its impacts on rural BC are considered.

Wild mushrooms and floral greenery dominate commercial trade in NTFPs in BC. Pine mushrooms (*Tricholoma magnivelare*), chanterelles (principally *Cantharellus formosus*) and morels (*Morchella* spp.) are the most commonly marketed wild mushrooms, while salal (*Gaultheria shallon*) and boughs of various coniferous species account for over 90 per cent of the floral greens output. Work by the Centre for Non-Timber Resources (Hobby and Cocksedge, 2006) estimates the value of the trade in wild mushrooms in BC at US$9.5–40 million per year over the past decade, with an annual average of US$27.5 million. The export value of the floral greens sector is estimated at US$25.5–62 million per year over the past five years, with an annual average value of approximately US$38 million. The significant variations in values are attributable to changing environmental conditions and the impact of global production and prices, although the relative contribution of these factors is not well understood.

Data for the many other NTFPs are not available, nor is there a consistently collected set of data for NTFPs in aggregate, or for either domestic use or export. In part this is because there is no generally accepted definition of the term 'NTFP'. Also, distinct NTFPs 'hide' in undifferentiated product categories. In addition, the harvesting and sale of most NTFPs are not licensed, monitored or otherwise regulated and there is no systematic collection of data by government entities.

Harvesting NTFPs for sale is a small-scale economic activity in many parts of the province, although tens of thousands of people engage in NTFP harvesting as an occasional, part-time and sometimes full-time occupation across the province (Wills

Source: Wendy Cocksedge.

Figure 4.2 *Salal (Gaultheria shallon) harvest*

and Lipsey, 1999). Most NTFP products are harvested and sold on a piecework basis by individuals working alone or in small groups. Generally, these individuals do not have worker's compensation coverage or other benefits and are not established as formal business operations. The potential income from NTFP collecting is fairly modest, perhaps in the area of US$38,500 per year (Hobby et al, 2006), but may compare favourably with other opportunities available to individuals who wish to remain in rural communities, who lack education or formal job skills, and who may also face literacy challenges in English.

Buyers, distributors and wholesalers of some products (particularly floral greens and mushrooms), however, are well established as businesses in BC. In recent years, businesses engaged in the floral greens trade based in the Pacific Northwest of the United States have expanded into coastal BC, apparently in response to an increasing regulatory burden and perhaps to declines in product quality and quantity in Washington and Oregon (Lynch and McLain, 2003).

Other uses of NTFPs

Subsistence, recreational and cultural benefits of NTFPs are even less well-documented than commercial uses. In BC, these products have traditionally played an essential role as sources of food, clothing and medicines for Aboriginal peoples and featured in their cultural and spiritual practices. Research has shown the extent of the use of these products and also the range of resource management strategies and ownership patterns First Nations employed to control, maintain and enhance these resources (Turner and Cocksedge, 2001). Early European settlers failed to recognize many of these activities, probably because they did not correspond to European views of 'management' or 'ownership'.

Although traditional knowledge held by First Nations of forest plants and fungi has diminished as a result of acculturation and other factors, the use of non-timber forest resources by First Nations remains widespread and their knowledge continues to be a rich source of information for NTFP management and use. In some communities, non-timber forest products and services are seen as tools for the revitalization of Aboriginal culture. People from youth to middle age are seeking to reverse the loss of cultural knowledge, including the knowledge of plants and their uses. Some are exploring the potential for traditional and non-traditional NTFPs to form the basis for new community-owned businesses that can help address at least some of the challenges faced by Aboriginal communities (Mitchell, 1997, 2004).

Picking berries, mushrooms and other wild foods is a popular activity in rural communities throughout BC. As noted by Pouta et al (2006), harvesting wild berries and other wild foods is strongly associated with rural residency and the practice of a rural 'producer' lifestyle, as opposed to an urban 'consumer' lifestyle. As British Columbia has become more urbanized, hunting and gathering activities generally have declined as important contributions to household incomes. In 1981, for example, about 6 per cent of BC residents purchased hunting licences; this percentage declined to 2 per cent in 2003 (BC Stats, 2005).

At the same time, urban residents are increasingly interested in nature-based tourism and recreational activities. For example, wildlife viewing contributed over

US$665 million (measured in direct expenditures) to the BC economy in 2005 (BC Ministry of Water, Land and Air Protection, 2005). This compares to US$45.5 million from hunting in 2003. Wild foods are developing a considerable cachet, similar to that of organic foods a few decades ago. Several businesses in BC and across Canada cater to this trend, while other organizations and businesses offer guided mushroom forays and sponsor mushroom identification workshops (Cowper, 2007). BC's resource guide to buyers and sellers of NTFPs, *Buy BCwild,* includes listings for over 170 enterprises offering more than 300 products and services (CNTR, 2007). While no province-wide statistics are available, a 2006 survey of residents in the East Kootenay region of BC found that 35 per cent of the region's total population harvested NTFPs for mainly recreational, but also commercial, purposes (Hobby and Cocksedge, 2006). These results indicate the importance of the non-commercial values for NTFPs as key resources for traditional and recreational purposes.

NTFP LAW AND POLICY – HISTORY AND CURRENT STATUS

Even though policy discussions about the NTFP resource began to figure on the political agenda in BC in the mid-1980s, there is very little law and regulation specifically governing NTFPs (described as 'botanicals' in government legislation)[2] (BC Forest Practices Board, 2004). Initial policy efforts concerning NTFPs in BC reflected a response to immediate utilization pressures on single species rather than a formal strategic approach to resource management or to economic development in forest-dependent communities. The following discussion describes the evolution from this ad hoc approach to recognition of the need to address non-timber issues from a broader policy perspective. ('Policy' is used here to refer to formal legislation and regulation as well as to operational level policy.)

Management of single species

Between the mid-1950s and the mid-1990s, regulations were developed to manage cascara bark, yew bark, conifer foliage and pine mushrooms in response to external market demand pressures and increasing values. The policy response to these products varied considerably.

The harvesting of cascara bark (used in the development of laxative products) led to the first regulation of a forest product not specifically related to the production of wood products or pulp. In 1958, the provincial government introduced the Cascara Bark Regulation, which required harvesters and buyers to obtain either a permit or a licence. Buyers had to maintain records and purchase bark only from licensed harvesters. In 1981, declining product demand led to the repeal of this regulation. In 1991, the high demand for yew bark (from *Taxus brevifolia,* whose derivative paclitaxel is used as an anti-tumour drug) led to the development of an information package that defined interim provincial procedures requiring harvesters to obtain a letter of authorization from the land tenure holder or a free-use permit from the local forest district manager. Shipments of yew were also to be accompanied by a notice outlining where the bark had been harvested, its weight and the buyer. These measures were only 'recommended', however, and no legislation or regulation was implemented. Demand for yew bark from BC has virtually disappeared, with the shift in purchaser preference to Canada yew (*Taxus canadensis*) from central and eastern Canada.

Some forest districts responded to the increased demand for cedar foliage (used in the production of aromatic oils) by establishing an authorization process for cedar foliage harvested from Crown lands. Harvesters were required to identify the quantity sought and the harvest location, and had to use appropriate pruning methods. This approach was also applied to white pine and other boughs used in Christmas decorations, with the intent of ensuring that the removal of any portion of the tree did not damage the timber resource. The process of permitting and monitoring bough harvest is no longer in effect, although concerns about the effects of bough harvesting on tree growth and survival remain (Natural Resources Canada, 2004).

Pine mushrooms provide another example focusing on a single species, but illustrate a shift in policy focus. In 1979, the BC Ministry of Agriculture produced a pamphlet to promote the harvesting of pine mushrooms (FBM Consulting, 1989) as an economic activity to supply the burgeoning Japanese market. By 1988, reports of mushrooms selling for US$95 per pound (0.454kg) encouraged an intense harvesting effort and a massive influx of people to certain areas of the province. This level of activity led to a variety of local and regional concerns, including damaging collecting practices (particularly raking the forest floor to expose button mushrooms), garbage, the danger of forest fires, conflict between harvesters and buyers and between local and transient harvesters, and increases in drug- and alcohol-related crimes (Gamiet et al, 1998).

After a failed attempt to develop a response to this level of activity in the late 1980s, the Pine Mushroom Task Force was formed in 1993 to 'develop a management approach through which government could achieve and maintain a sustainable pine mushroom industry in the province. The task force considered that such a management approach could be a prototype to manage all commercially harvested agroforestry products' (BC Ministry of Forests, 1994). The process resulted in a set of recommendations and, in 1994, NTFPs (termed 'botanical forest products') were included in the new Forest Practices Code of British Columbia Act. Sections 104 and 216 of the code allowed for the development of a botanical forest products buyer's permit or licensing regime. In 2002, the code was replaced by the results-based Forest and Range Practices Act (FRPA), which opened up the possibility of developing a more comprehensive management regime for NTFPs. None of the sections in either the code or the FRPA have led to implementation at the operational level.

In the past several years, prices for pine mushrooms have declined. With that decline, perceptions of both a resource crisis and an economic gold mine have diminished. Local concerns about various aspects of mushroom harvesting persist, however, and harvesters continue to access known mushroom areas throughout the province.

Broader multi-species management

Property rights

There are few examples in BC of designated rights, such as leases or permits, to harvest or otherwise benefit from NTFPs on Crown lands (Tedder et al, 2002). Only one type of forest tenure, the community forest agreement, provides such rights. Even on private lands, with the exception of southern Vancouver Island, the management of access to NTFPs (such as the collection of fees from harvesters) is uncommon.

This comparative vacuum in specific institutional arrangements for NTFPs exists, however, in a context of active debate about the pre-existing and ongoing rights of First Nations to manage and benefit from these (and all other) resources within their traditional territories. First Nations' rights and title are arguably the most important policy issue affecting NTFPs and other resources in BC.

Aboriginal rights and title

First Nations consider the collection and use of non-timber resources, and of other forest resources, to be part of their aboriginal rights. The BC Treaty Commission (2007) briefly describes the nature of aboriginal title as follows:

> *In 1997, the Supreme Court of Canada ruled in the Delgamuukw case that aboriginal title is a right to the land itself – not just the right to hunt, fish and gather. Crown title refers to the provincial or the federal government's interest in land. Almost all Crown land in BC is held by the province. Delgamuukw confirmed that aboriginal title still exists in BC and that when dealing with Crown land the government must consult with and may have to compensate First Nations whose rights are affected. Aboriginal title is often referred to as a burden on Crown title. (BC Treaty Commission, 2007)*

In most instances, however, the specific nature of property rights in land and resources has not yet been formally defined, as treaties have not been concluded for most of the province. As of June 2007, the Nisga'a were the only nation that had signed a treaty in modern times, although about three-quarters of the Aboriginal population in BC belong to First Nations that are in the process of negotiation. The Nisga'a treaty came into effect in 2000, under which the Nisga'a Lisims government manages NTFPs on their treaty lands. Most management effort is directed to pine mushrooms, for which there are special management zones and requirements that pickers and buyers obtain permits (BC Forest Practices Board, 2004).

Community forest agreements

In 1998, the provincial government introduced legislation to establish community forest agreement (CFA) tenures in BC. The intent of the new tenures was to provide communities with greater control over a portion of their local forest resources. These CFAs are area-based tenures and the only forest tenures in the province to specifically include NTFPs within an agreement. As section 43.3(ii) states, the minister or authorized delegate 'may give to its holder the right to harvest, manage and charge fees for botanical forest products and other prescribed products'. As of 2007, these agreements accounted for about 1 per cent of the timber harvesting land base.

A recent review of CFAs has revealed that many tenure holders would like to manage NTFPs on these lands. However, they are discouraged from doing so by a variety of concerns, including the lack of inventory and market information and the cost barriers associated with developing basic management information resources. In addition, while holders of CFAs may acquire the right to manage and charge fees for NTFPs, they do not have the right to restrict access or regulate use (Meyers Norris Penny LLP and Enfor Consultants Ltd, 2006).

Private lands

As noted above, almost all forest lands in BC are publicly owned, with only about 5 per cent of land in BC under private ownership. Private forest landowners, ranging from single individuals to large forest companies, have property rights to all forest resources on their holdings, including the right to exclude others from benefiting from those forest resources. Most large private forest landholdings are located on Vancouver Island. Some owners, such as TimberWest Forest Corp and Western Forest Products Ltd, have taken steps to manage or benefit from the use of NTFPs. TimberWest, for example, has provided permits to individual salal harvesters for about 15 years. In 2005, however, in response to intense harvesting pressure on its southern Vancouver Island timber lands, the corporation contracted with a single company to manage the issuing of permits for and harvest of salal from some of its lands. Western Forest Products Ltd also contracts out the rights to access salal and other forest products on some of its private landholdings. These permits were the first offered by any landowners in BC, private or public.

Land and resource planning

Formal planning processes

Beginning in the early 1990s, the BC government initiated participatory processes for land and resource management with the initial objective of ending escalating conflict about the province's resource development priorities – an era known as the 'war in the woods' (Jackson and Curry, 2004). Formal planning has continued since that time with varying levels of geographic scope. Few of the plans produced by these processes have addressed NTFPs, although a few, such as the recently completed Central Coast Land and Resource Use Plan, consider NTFPs explicitly. The provincial government is reviewing the resource planning processes with the aim of developing a new government-to-government process with First Nations (A. Lidstone, pers. comm., 2006).

In 1994, the provincial government released the Cariboo-Chilcotin Land Use Plan (BC Forest Practices Board, 2004), which included information related to the management and monitoring of NTFPs. The group negotiating the plan dealt with ensuring access to forested areas. However, the effort to address the management of NTFPs more broadly was hampered by a lack of information about which products were harvested, who harvested them and where, and the productivity of the resource base.

The Kispiox Land and Resource Management Plan (LRMP) (BC Forest Practices Board, 2004), released in 1996, applied to some of the most productive pine mushroom sites in the province. Shortly after the release of the LRMP, the Seven Sisters region of the planning area became a 42,000ha park. A driving force behind protecting the area from high-impact industrial activity was the value of pine mushrooms, the harvesting of which was able to continue. The land-use planning and research efforts within the Kispiox forest district led to the incorporation of a specific management regime for pine mushrooms in the Kispiox timber supply review, and subsequently influenced the setting of the allowable annual cut (BC Forest Practices Board, 2004).

Informal agreements and planning efforts

There are also other examples of provincial efforts to address NTFPs that influence timber management. For instance, small business timber sales in the Nahatlatch water-shed in the Fraser Canyon area of the Chilliwack forest district were modified to take account of high-value pine mushroom habitat. The Blackwater Creek Pine Mushroom Management Area in the Squamish forest district is another example of a so-called 'log-around' (i.e. an area excluded from logging to protect the mushroom area) that has been in place since the mid-1990s[3] (Tedder et al, 2002). On the Queen Charlotte Islands/Haida Gwaii, the forest district withheld approval of logging development plans that included cut blocks in highly productive chanterelle mushroom habitat around Skidegate Lake on Moresby Island (Tedder et al, 2000).

Use of current enabling legislation

Enabling legislation that could be used to develop regulations for the management of NTFPs currently exists in BC. For example, provisions in the BC Land Act currently allow legally binding objectives for NTFPs to be set. The Land Amendment Act, in sections 93.1 and 93.3(1), allows the cabinet to designate areas and establish objectives to conserve or manage natural resources, including NTFPs. The Forest and Range Practices Act (FRPA) and regulations have provisions for establishing legally binding resource management objectives, which do not currently include NTFPs explicitly, but could. Where an objective is specified under the FRPA, agreement-holders must identify measurable and verifiable results and prepare strategies that are consistent with these objectives in their forest stewardship plans (FSPs). The Forest Planning and Practices Regulation establishes objectives for biodiversity, wildlife and the conserva-tion of cultural heritage resources, which can include some aspects of NTFP use by First Nations. A mechanism to do so exists through the establishment of old growth management areas (OGMAs) or wildlife habitat areas (WHA). For example, an OGMA was used to protect pine mushroom habitat in the Kispiox timber supply area, as mentioned above.

In response to public concern about the potential impacts of timber harvesting on NTFPs and the sustainability of NTFP harvesting, the BC Forest Practices Board, which describes itself as 'BC's independent watchdog' and is responsible for promoting the stewardship of forest values and strengthening the regulatory regime, commissioned a special report on non-timber forest products in 2004. The report recommended that the provincial government support research, encourage greater public aware-ness of NTFPs and investigate regulatory options, including the application of existing enabling legislation to establish management objectives.

THE CURRENT POLICY ENVIRONMENT

Research and knowledge transfer

Better information contributes to, although it does not guarantee, the development of good policy, and the expansion of research and knowledge-extension activities in the NTFP field is a promising indication of the growing profile of the sector in BC.

For over 30 years, the Ministry of Forests and Range (formerly the Ministry of Forests) has funded a large body of research on the distribution, ecology and autecology of BC plant species. Beginning in the early 1990s, the ministry undertook research specifically focused on NTFPs. More recently, research effort has been intensified under the leadership of a small group of academic and government researchers associated with Royal Roads University, the Ministry of Forests and Range, the CFS and the BC Ministry of Agriculture and Lands. This group has further explored the ecology and autecology of key NTFP species and undertaken new research on the compatible management of timber and non-timber values, inventory methods and standards, property rights and other institutional aspects of NTFPs, as well as the commercial, recreational and other kinds of contributions made by these resources to the BC economy. BC government-funded programmes have played an increasing role in funding NTFP research, especially in areas related to the compatible production of timber and non-timber products and the production of NTFPs in agroforestry systems.

Collaboration with First Nations and partnership with many of the industry's key stakeholders have been important to the projects carried out in the past ten years. The growth of a network for research and knowledge exchange has contributed significantly to the creation of an identifiable policy network for NTFPs in the province. In 2006, an interagency committee of provincial government ministries, led by the Ministry of Forests and Range and the Ministry of Agriculture and Lands, was established to coordinate and focus provincial policy and other initiatives relative to non-timber resources, particularly NTFPs. It has adopted in principle the recommendation of Tedder et al (2002) that an adaptive 'experimental' approach be taken to policy development. The question of how best to address First Nations' concerns and interests in relation to policy development has been a major focus of the committee's deliberations, as has the epidemic of mountain pine beetle (*Dendroctonus ponderosae*) in the central part of the province.

As noted above, Aboriginal rights and title are among the most significant issues affecting policy development with respect to NTFPs in BC. In April 2005, in accordance with recent court decisions regarding the government's obligation to consult with First Nations when decisions could have an impact on Aboriginal rights and title, the province announced its intent to build a new relationship with First Nations to ensure that Aboriginal people shared in the economic and social development of BC. Policy initiatives of the provincial government, including the 'New Relationship' initiative, which aims to share decision-making about land and resources between First Nations and the Crown, offer new opportunities for First Nations to manage the traditional use of NTFPs as well as to develop NTFP enterprises (BC Ministry of Aboriginal Relations and Reconciliation, 2006).

A second major factor affecting forestry in BC, including NTFPs, is the large-scale mountain pine beetle outbreak in the central interior. By 2013, about 80 per cent of the lodgepole pine volume in central BC is expected to be dead (Eng et al, 2005). There will be significant economic implications for BC communities from the resulting shortfalls in wood supply. The province is investigating alternative economic options for these communities, including possibilities associated with NTFP enterprises (BC Ministry of Forests and Range, 2006a). NTFPs are highlighted in documents published by the BC First Nations Mountain Pine Beetle Working Group (2006), subsequently restructured into the BC First Nations Forestry Council.

POLICY GAPS AND INVISIBLE ELBOWS

As the discussion up to this point has indicated, there has been slow progress in the development of policy that directly addresses the NTFP sector in BC. Property rights in non-timber species remain largely undefined and unallocated, with only a few small islands of management in a sea of unregulated access to forests, whether title to those lands is rightfully public or First Nations. Across Canada, in areas where there is a policy focus, efforts are mainly directed to the production of non-timber products such as maple syrup and wild blueberries on private lands and long-established tenures.

It is not easy to attract the attention of policy-makers and researchers, investors and others to the NTFP sector. First, the term 'non-timber forest products' is itself problematic. The ambiguity of the term and the heterogeneity of products, services and users seriously inhibit clear definition of the 'policy issue' that NTFPs present (Belcher, 2003). 'NTFP issues' are not clearly defined, but are 'squishy' (difficult to measure, politically controversial), 'messy' (deeply entangled with other issues and problems) or both (Pal, 2001, p113). This difficulty in clearly defining the policy issues and then classifying them as problems in need of solution inhibits recognition of the salience of NTFPs and their elevation to the government's agenda. NTFPs can be thought of as problems (or opportunities) in sustainable forest management and forest conservation, rural economic development and Aboriginal relations, at a broad level, and, more specifically, as problems (or opportunities) in food production, natural health products, housing, alternative energy, tourism, international trade, arts and culture, and any of the other myriad groups of products and services of which NTFPs form a part. Because of the heterogeneity of the sector, it is largely ignored by public agencies with more limited and specialized mandates.

Other emerging sectors, such as ecotourism, bioproducts and natural health products, are also ambiguously defined and have far-reaching impacts. Yet each of these is emerging or has emerged as a focus for public and private attention in policy, research and investment. We believe the difference relates in part to the nature of those who use these various products and the opportunities that exist for product and service development. For example, there is a huge worldwide market in nature-based and cultural tourism, and this generic demand supports tourism development in Canada, whether the particular attraction is Aboriginal cultures, whales, bears, mountain climbing or river rafting. Wood-based biofuels are seen as alternatives to declining petroleum resources and, unlike conventional 'low-tech' NTFPs, appeal to the interests of researchers and investors in the development and application of technologies. Natural health products and services (such as organic foods and nature-based tourism) speak to the aging baby boomer demographic, a group with unprecedented levels of disposable income, leisure time and concern with health, appearance and spiritual well-being. NTFPs do not have such obvious broad market appeals, although there is no reason why various NTFPs could not be readily marketed as components of these fast-growing sectors.

Currently those who harvest NTFPs are a heterogeneous group who rely on NTFPs for all or a portion of their income. NTFP harvesting is often a 'lifestyle' choice, and many harvesters have little interest or experience in organizing to pursue common interests or lobbying for policy change. Some may see their potential to influence the policy process as a poor exchange for the loss of anonymity with respect to, for example, the payment of income tax on cash receipts. Others may, understandably, be sceptical of their ability to influence policy and suspect that drawing attention to a viable business opportunity may simply attract unwelcome competition. The NTFP sector often generates bad press about, for instance, trespass, damage or littering by harvesters, conflict among users (e.g. 'mushroom wars') or the existence of an 'underground market' in which tax avoidance or illegal activities figure prominently. Those with a major financial stake in NTFP businesses such as floral greens and mushrooms have shown little interest in policy change, and would probably resist changes that empower harvesters, manage resource use or extract revenues for the Crown or other landowners.

The complexity of the NTFP sector and the lack of readily available information about its potential also make it difficult for rural communities seeking relief from the loss of jobs and investment in fishing, logging and mining to view NTFP enterprises as viable opportunities. In the past, the 'answer' to job loss has been a new mill, mine, or fish plant. The micro- or small-business enterprises that make up the NTFP sector have not attracted much attention yet. In general, NTFPs have not been viewed as a potential source of 'jobs', even though – with appropriate investment in infrastructure and training – they may be very well suited as a component of a strategy to ensure the livelihoods of individuals and families and the resilience of rural communities.

These factors characterize a resource sector dramatically different from the timber product sector that dominates forest policy in BC. The dominance of timber in forest policy is such that other marketable products and services are often overshadowed, even when the timber industry is in decline and new economic opportunities are being sought. Nelson et al (2006) cite the reports of the government-appointed BC Competition Council and the Pulp and Paper and Wood Products Advisory Committees, which describe an industrial sector that is generally in decline and, as far as the coastal sector is concerned, in near term crisis. Only the BC interior lumber industry is flourishing, having gone through a period of consolidation and major investment in new technology and 'super mills'. However this state of affairs is largely attributable to the enormous supply of inexpensive wood resulting from the mountain pine beetle epidemic. Once the epidemic has run its course, the interior 'fall-down' in wood supply will probably be massive.

The historical 'path dependence' of the BC forest policy and industry seriously inhibits their adaptability not only to the needs of non-timber sectors, but to the timber sector as well. The provincial timber industry is now well advanced in the pattern of decline described by Clapp (1998, p131) as a 'resource cycle' in which 'a pattern of over-expansion [is] followed by ecosystem disruption and economic crisis'.

In BC, recent market-related policy changes have been implemented to allow firms to adapt more quickly to competitive market forces (Nelson et al, 2006). The consequence has been the acceleration of the 'resource cycle' described by Clapp; in other words, an increase in the rate of capital consolidation and intensification. This response is exacerbated by a temporary abundance of resources provided by the mountain pine beetle.

In such circumstances, it might be reasonable to think that policy-makers and forest communities themselves would identify the need for, and aggressively pursue, options for diversification. Diversification is indeed a major theme in government plans and funding programmes to address the crisis in the BC resource sector, both forestry and fisheries (see BC Ministry of Forests and Range, 2006b; BC Ministry of Environment, 2007).

However, as Clapp notes, initial abundance and 'natural' cycles in resource production exacerbate the resource cycle.

> *When abundant, resource rents allow regions to postpone making the hard economic choices necessary for the development of a diversified industrial base. Once rents begin to decline, they are likely to be insufficient to support regional diversification, just when it becomes vitally necessary. (Clapp, 1998, p130)*

These factors, taken together, have largely relegated NTFP policy development to small corners of the desks of individuals who, for one reason or another, take an interest in the subject. Tedder (2006), applying Kingdon's (1984) 'policy streams and windows' model to the history of NTFP policy in BC, concludes that the reason for the failure of the effort to establish a formal policy to guide and oversee the use of NTFPs was not the lack of effort by policy entrepreneurs.[4] Rather, these individuals have largely failed, up to this point, to defend the salience of the issue at the regulatory and operational levels and to identify and exploit appropriate policy windows. In addition, the public agency managing the province's forests has historically maintained a path-dependent focus on the timber industry, perhaps suggesting that some other government agency may be a more appropriate location for the consideration of NTFPs in land and resource management and community development. However, given the pre-eminence of timber concerns in provincial forest policy, an agency without strong links to the timber sector might be even less able to influence policy in directions more hospitable to NTFPs.

Change, however, is often driven from within a particular industry sector. The NTFP industry itself, fearful of policy that would negatively affect its ability to access product, has remained largely silent. Sector participants are perhaps concerned that anticipated government interventions would be harmful to their interests, particularly if these interventions take the form of regulatory 'sticks' without any corresponding 'carrots' such as business support, training, information and the facilitation of collaboration within the sector and between the NTFP and other industries. Meanwhile, at a broader economic level, the mountain pine beetle epidemic is producing a vast short-term abundance of timber and associated timber-related jobs that currently obscure the need for economic diversification by resource-dependent communities.

IMPLICATIONS OF CURRENT POLICY ARRANGEMENTS

While it is impossible to evaluate opportunities that have been forgone in terms of resource management or economic and social policy in the NTFP sector, it is certainly true that Canada's track record with respect to resource development and resource-dependent communities has followed a more or less global pattern. This is a pattern of extraction of certain preferred resources, which, as they become less abundant, are replaced by other resources, with increasing infusions of technology to produce high-quality products with lower-quality inputs. The development of laminated beams and artificial crab are examples.

In the process, rural communities grow to support resource extraction and processing, and many subsequently perish (often after long periods on expensive life support). Some forest and fishing communities – especially First Nations communities whose existence pre-dates resource industries – remain in place, but only with large infusions of public funds and often suffering from serious social problems. Meanwhile, biodiversity declines as natural resource systems are manipulated through selective extraction, perhaps changing irreversibly.

This situation vis-á-vis NTFPs reflects policy gaps, or outright holes in the resource management landscape, that severely limit the contribution of NTFPs to forest conservation, rural economic diversification, and community and social well-being. In the case of NTFPs, no property rights have been established for most species in most areas of BC, or for Canada as a whole. This de facto open-access environment for the harvesting of NTFPs may lead, if resource demand and prices are sufficient, to the overuse and degradation of the resource. In such a scenario, Hardin's (1968) 'tragedy of the commons' thesis may be relevant. At the same time, the very large 'elbow' of current forest policy and the underlying dynamics of the resource cycle effectively move forest resources other than timber to the sidelines, as both too complex and too small to be worthy of consideration.

GETTING BEYOND THE GAPS AND ELBOWS

Our explorations in this chapter lead us to believe that the need for 'NTFP policy' is really a need for more integrated and more imaginative policy in many areas of natural resource and rural development. It may be useful to think of NTFPs as case studies for different approaches to policy development: approaches that actively engage those whose interests span the full range of resource values, economic and social, while identifying and protecting environmental values. From this perspective, the current gaps in NTFP policy may provide an opportunity to try out new approaches.

There are relatively few powerful economic interests established in the sector, which is still considered a small and even trivial collection of minor forest products. As Bromley (1992) notes, this state of affairs is one of the few situations in which significant institutional change is possible, since existing claims to resources have not yet become entrenched in the hands of powerful interests. In this context, we may best discover policy solutions by trying a number of experiments that ask consistent questions about how to manage NTFPs in ways that are environmentally sustainable, socially equitable and economically efficient.

This relatively fluid situation can, however, change rapidly. In the context of treaty negotiations, and particularly since the announcement of the 'New Relationship' and the emergence of the mountain pine beetle crisis, First Nations are paying increasing attention to NTFPs and demanding greater participation in policy development, especially for policies concerning the allocation of property rights. There are two potential outcomes. One possibility is that the NTFP sector could become an arena in which the 'New Relationship', with all its positive implications of partnerships and mutual respect between communities, is actually realized. Alternatively, should the assertion of First Nations' rights and interests come to dominate the NTFP policy debate to the exclusion of other concerns, the possibility of facilitating practical, cooperative partnerships among all interests may be inhibited. At present, there is a compelling need to better understand the manner in which NTFPs contribute to the livelihoods of rural communities, both Aboriginal and non-Aboriginal, and to the preservation of non-commercial values, so that all voices and values are heard in the policy debate.

Where such policy experiments should be located is also important. The availability of interested partners is clearly a primary concern. Also the requirement of a fairly major commitment of research funds should dictate careful consideration of the conditions under which focused effort on sustainable NTFP development would bear most fruit. NTFPs will never 'save' a community, but may well be a contributor to the overall diversification, revitalization and resilience of forest-dependent communities, while also providing opportunities for urban employment and investment in areas such as marketing and technical services. Most processing facilities and most markets for NTFPs are, after all, in cities in Canada and abroad.

Consequently, we may wish to consider first identifying target communities where promising conditions exist for promoting community viability and adaptability. Haynes (2003) proposes a number of factors useful for assessing resilience or adaptability, including population size, existing economic diversity, civic leadership, attitudes toward change, social cohesion, civic and natural amenities, and location. Generally, higher resilience is associated with larger, more diverse communities, with strong leadership and other forms of social capital, located on major trade routes or near service centres or other destinations.

These factors may offer guidance for public and private investment in communities with a high potential for enhanced viability. If investment is targeted in this way, however, what are the prospects for small communities in BC, especially the many First Nations communities that are small and isolated, with few amenities and a limited economic base? In the case of such communities, it may be especially important to look beyond conventional economic indicators to other measures of enhanced well-being that may follow from encouraging productive engagement in NTFP occupations. Given the much lower propensity of First Nations individuals to leave remote communities, even where employment opportunities are very limited, initiatives that enhance health and well-being, even if not commercially viable, may pay significant dividends in other ways, such as a reduction in expenditures for health and social services that may follow from higher self-esteem and social engagement.

CONCLUSIONS

As we review the history and current status of NTFP policy in BC, it appears that the comparative lack of government interest in the NTFP sector reflects the absence of the sector from the list of identifiable, let alone urgent, policy issues or problems. Even at times when NTFPs have increased in salience and found their way onto the government's agenda, the design of a comprehensive, defendable and implementable solution has remained elusive, or unpalatable, to decision-makers. However, as we have outlined, NTFPs do contribute to sustainable, resilient communities.

We have also highlighted the interest of government and non-government agents in gaining a better understanding of the resource characteristics of a variety of non-timber species, such as pine mushrooms and salal. Over the past 20 years, research and policy discussion about NTFPs in BC has evolved from a focus on a single species to include a wider range of ecological, social and economic parameters. First Nations' interests and the mountain pine beetle epidemic provide two potential drivers for the greater prominence of NTFPs in resource management and rural development.

The creation of the interagency committee marks a subtle but important acknowledgement by the provincial government of the need for a more coordinated approach to NTFP issues. At the same time, however, we note a resistance to introducing legislation and regulations governing access to and the appropriate use of the resources, to the introduction of more integrated management of multiple uses and to more integrated and supportive approaches to rural economic diversification and the NTFP business sector.

A lack of formal policy is not necessarily a bad thing where serious problems are absent. Although they generate a recognized level of wealth, NTFPs still do not present themselves as a significant and salient problem in need of a solution, nor as a compelling opportunity with the potential to rescue ailing communities or industries. No organized NTFP voice has emerged from within the sector to demand attention; in fact, many within the sector prefer to remain outside government purview. The intensity of use of NTFPs also varies by region and species. This heterogeneity, as noted, makes it harder to characterize NTFP uses in simple, general ways and hence to define a simple 'NTFP problem' in response to which a uniform, easily implementable and inexpensive solution can be devised. It will continue to be difficult to provide problem definitions that could act as motors for the adoption of clear solutions. As a consequence, the prospect of intervention will probably continue to appear costly and the potential gains small.

In the past, as at present, NTFPs have not figured prominently as a component of solutions to other problems. Attempts at 'coupling' NTFPs with other policy initiatives, in order to introduce policy that would otherwise appear to lack salience (Kingdon, 1984), have had only minimal success. For example, the somewhat inappropriate inclusion of NTFPs in the Forest Practices Code, which attached NTFP access and buyer issues to legislation directing forest operational practices, failed at the implementation stage. This trend may be changing, however, as First Nations become more vocal about NTFPs. The need for economic development opportunities arising from the mountain pine beetle epidemic in the interior of BC and the struggling timber sector on the coast may lead to more integrative approaches, or present further challenges – or probably both.

The potential silver lining in this cloudy policy environment is that the absence of an existing policy framework for NTFPs and other minor or emerging resource sectors may provide an opportunity for adaptive policy development that is not possible in better-established sectors. Such experiments are not without cost, however, in financial and human terms. The cost of establishing a multi-year community-based experiment in adaptive policy-making, for example, is almost certainly greater than the cost of drafting a conventional regulatory instrument, and the results of such policy innovation are, by definition, uncertain.

Nonetheless, recent events in BC and other parts of Canada suggest that policy windows for NTFPs may be opening as part of an evolving approach to the use and allocation of forest resources. If this is true, and if Jane Jacobs (2004, p106) is correct to say that 'opportunity is *actually* the mother of invention', then the time to embark on such experiments is now – not when problems have become large, well defined and costly to correct.

NOTES

1   'Aboriginal' and 'First Nations' are used interchangeably throughout this chapter. Many indigenous people in British Columbia prefer the term 'First Nation'.
2   Many laws affect the harvesting and use of NTFPs, including legislation and regulations governing food handling and processing, and laws prohibiting the removal of plant materials from parks. An extensive discussion of the legal regime affecting NTFPs in BC is available at www.cntr.royalroads.ca/law-policy-compendium. These reports are part of a national compendium of NTFP law and policy prepared by the CNTR at Royal Roads University.
3   In May 2007, local First Nations, environmental groups and others protested against the granting of logging rights in the Blackwater Creek area on the grounds of threats to pine mushrooms, traditional First Nations gathering areas and wildlife species including the spotted owl (CanWest News Service, 2007).
4   Policy entrepreneurs are public entrepreneurs who, from outside the formal positions of government, introduce new ideas and help translate and implement them into public practice (Roberts and King, 1991).

REFERENCES

BC First Nations Mountain Pine Beetle Working Group (2006) *Beetle News*, July/August, /www. fnforestrycouncil.ca/news_res/pinebeetlenews4.pdf, accessed 9 June 2008

BC Forest Practices Board (2004) 'Integrating non-timber forest products into forest planning and practices in British Columbia: Special report', Forest Practices Board, Victoria, BC

BC Ministry of Aboriginal Relations and Reconciliation (2006) *The New Relationship with First Nations and Aboriginal People*, www.gov.bc.ca/arr/newrelationship/trust/board.html, accessed 9 June 2008

BC Ministry of Environment (2007) *Oceans and Marine Fisheries Division*, www.env.gov.bc.ca/ omfd/, accessed 9 June 2008

BC Ministry of Forests (1994) 'Workshop results: Pine mushroom task force', Pine Mushroom Task Force, Integrated Resources Section, Ministry of Forests, Victoria, BC

BC Ministry of Forests (2010) British Columbia's Forests, www.naturallywood.com/sustainable_ forests.aspx?id=285, accessed 26 March 2010

BC Ministry of Forests and Range (2006a) *First Nations receive funding for pine beetle response*, 18 October, www2.news.gov.bc.ca/news_releases_2005–2009/2006FOR0137–001242.htm, accessed 9 June 2008

BC Ministry of Forests and Range (2006b) *Mountain Pine Beetle Action Plan 2006–2011*, www.for. gov.bc.ca/hfp/mountain_pine_beetle/actionplan/2006/Beetle_Action_Plan.pdf, accessed 9 June 2008

BC Ministry of Water, Land and Air Protection (2005) *Hunting and Trapping Regulations Synopsis 2005–2006*, Victoria, BC

BC Stats (2005) *British Columbia's Hunting, Trapping and Wildlife Viewing Sector*, BC Stats, Victoria, BC

BC Treaty Commission (2007) *Aboriginal Rights*, www.bctreaty.net/files/issues_rights.php, accessed 9 June 2008

Belcher, B. (2003) 'What isn't an NTFP?', *International Forestry Review*, vol 5, no 2, pp161–168

Bickers, K. N. and Williams, J. T. (2001) *Public Policy Analysis: A Political Economy Approach*, Houghton Mifflin, Boston, MA

Bromley, D. W. (1992) *Making the Commons Work*, ICS Press, San Francisco, CA

CanWest News Service (2007) 'Logging protest on deck near Whistler', www.friendsofgrassynar-rows.com/item.php?723F, accessed 9 June 2008

Centre for Non-timber Resources (2007) *BuyBC Wild*. Royal Roads University, Victoria, B.C. www.buybcwild.com/buy-bcwild-directory, accessed 9 June 2008.

Clapp, R. A. (1998) 'The resource cycle in forestry and fishing', *The Canadian Geographer*, vol 44, no 2, pp129–144

Cowper, D. (2007) *Adding Value to Tourism Businesses*, Latitude West Consultants, for Centre for Non-Timber Resources, Royal Roads University, Victoria, BC ·

Daily, G. C., Alexander, S., Ehrlich, P. R., Goulder, L., Lubchenco, J., Matson, P. A., Mooney, H. A., Postel, S., Schneider, S. H., Tilman, D. and Woodwell, G. M. (1997) 'Ecosystem services: Benefits supplied to human society by natural ecosystems', *Issues in Ecology*, no 2, Ecological Society of America, Washington, DC

De Geus, N. (1995) *Botanical Forest Products in British Columbia: An Overview*, Integrated Resources Policy Branch, British Columbia Ministry of Forests, Victoria, BC

Eng, M., Fall, A., Hughes, J., Shore, T., Riel, B., Hall, P. and Walton, A. (2005) *Provincial Level Projection of the Current Mountain Pine Beetle Outbreak: An Overview of the Model (BCMPB v2) and Results of Year 2 of the Project*, Working Paper 2005–20, Mountain Pine Beetle Initiative, Natural Resources Canada, Canadian Forest Service, Victoria, BC

FBM Consulting (1989) *The Harvesting of Edible Wild Mushrooms in British Columbia*, Integrated Resources Branch, BC Ministry of Forests, Victoria, BC,www.for.gov.bc.ca/hfd/library/documents/bib92985.pdf, accessed 9 June 2008

Gamiet, S., Ridenour, H. and Philpot, F. (1998) *An Overview of Pine Mushrooms in the Skeena-Bulkley Region*, Northwest Institute for Bioregional Research, Smithers, BC, www.northwestinstitute.ca/downloads/mushroom_report_98.pdf, accessed 9 June 2008

Hardin, G. J. (1968) 'The tragedy of the commons', *Science*, vol 162, no 3859, pp1243–1248

Haynes, R. W. (2003) *Assessing the Viability and Adaptability of Forest-Dependent Communities in the United States*, Gen. Tech. Rep. PNW-GTR-567, Pacific Northwest Research Station, Forest Service, US Department of Agriculture, Portland, OR

Hobby, T. and Cocksedge, W. (2006) 'Critical information for policy development and management of non-timber forest products in British Columbia: Baseline studies on economic value and compatible management', Forest Science Program Proposal Y061065, Centre for Non-Timber Resources, Royal Roads University, executive summary (report unpublished), www.cntr.royalroads.ca/files-cntr/File/Baseline%20Studies%20on%20Economic%20Value%20and%20Compatible%20Management.pdf, accessed 9 June 2008

Hobby, T., Dow, K. and MacKenzie, S. (2006) 'Commercial Development of NTFPs and Forest Bio-Products: Critical Factors for Success; A Salal (Gaultheria shallon) Case Study', unpublished working paper, Centre for Non-Timber Resources, Royal Roads University, Victoria, BC

Jackson, T. and Curry, J. (2004) 'Peace in the woods: Sustainability and the democratization of land use planning and resource management on Crown lands in British Columbia', *International Planning Studies*, vol 9, no 1, pp27–42

Jacobs, J. (2004) *Dark Age Ahead*, Random House Canada, Toronto, ON

Jacobs, M. (1993) *The Green Economy*, UBC Press, Vancouver, BC

Kingdon, J. W. (1984) *Agendas, Alternatives, and Public Policies*, HarperCollins, New York

Lynch, K. A. and McLain, R. J. (2003) *Access, Labor, and Wild Floral Greens Management in Western Washington's Forests*, Gen. Tech. Rep. PNW-GTR-585, Pacific Northwest Research Station, Forest Service, US Department of Agriculture, Portland, OR

Meyers Norris Penny LLP and Enfor Consultants Ltd (2006) *Community Forest Program: Program Review*, submitted to the BC Ministry of Forests and Range, Victoria, BC

Mitchell, D. A. (1997) 'Non-timber forest products in British Columbia: The past meets the future on the forest floor', *The Forestry Chronicle*, vol 74, no 3, pp359–362

Mitchell, D. A. (2004) *Many Voices, Many Values: Community Economic Diversification through Nontimber Forest Products in Coastal British Columbia, Canada*, Gen. Tech. Rep. PNW-GTR-604, Pacific Northwest Research Station, Forest Service, US Department of Agriculture, Portland, OR

Natural Resources Canada (2004) *Perfect Pines are Limbless 'cause of Christmas*, National Forestry Week, Pacific Forestry Centre, www.pfc.forestry.ca/news/national%5Fforest%5Fweek/2004/index_e.html, accessed 5 April 2007

Natural Resources Canada (2005) *The State of Canada's Forests – 2004–2005*, Government of Canada, Ottawa

Natural Resources Canada (2006) *The State of Canada's Forests – 2005–2006,* Government of Canada, Ottawa

Nelson, H., Niquidet, K. and Vertinsky, I. (2006). *Assessing the Socio-Economic Impact of Tenure Change in British Columbia,* Synthesis Paper SP06–021, Vancouver, BC: BC Forum on Forest Economics and Policy

Pal, L. (2001) *Beyond Policy Analysis: Public Issues Management in Turbulent Times,* 3rd edition, Oxford University Press, Toronto

Pouta, E., Sievänen, T. and Neuvonen, M. (2006) 'Recreational wild berry picking in Finland: Reflection of a rural lifestyle', *Society and Natural Resources,* vol 19, no 4, 285–304

Roberts, N. C. and King, P. J. (1991) 'Policy entrepreneurs: Their activity structure and function in the policy process', *Journal of Public Administration Research and Theory,* vol 1, no 2, April, pp147–175

Statistics Canada (2006) *Population Urban and Rural, by Province and Territory (BC),* www40.statcan. ca/l01/ind01/l3_3867_375.htm?hili_demo62, accessed 10 June 2008

Tedder, S. (2006) 'The evolution of NTFP policy in British Columbia: An assessment using Kingdon's agenda setting model', unpublished manuscript

Tedder, S., Mitchell, D. and Farran, R. (2000) *Seeing the Forest Beneath the Trees: The Social and Economic Potential of Non-timber Forest Products and Services in the Queen Charlotte Islands/Haida Gwaii,* South Moresby Forest Replacement Account and BC Ministry of Forests, Victoria, BC

Tedder, S., Mitchell, D. and Hillyer, A. (2002) *Property Rights in the Sustainable Management of Non-timber Forest Products,* BC Ministry of Forests and Forest Renewal, Victoria, BC

Turner, N. and Cocksedge, W. (2001) 'Aboriginal use of non-timber forest products in north-western North America: Applications and issues', in M. R. Emery (ed) *Non-Timber Forest Products in the United States: an Overview of Research and Policy Issues,* special issue of *Journal of Sustainable Forestry,* vol 13, no 3/4, pp31–57

Wills, R. M. and Lipsey, R. G. (1999) *An Economic Strategy to Develop Non-timber Forest Products and Services in British Columbia,* Forest Renewal BC, Project no. PA97538-ORE, Victoria, BC

Chapter 5

NTFPs in Scotland: Changing Attitudes to Access Rights in a Reforesting Land

Alison Dyke and Marla Emery

INTRODUCTION

Nearly one-quarter of the Scottish population gathers non-timber forest products (NTFPs), according to recent surveys (Heggie, 2001; TNS Global, 2003; West and Smith, 2003; Snowley and Daley, 2005). The practice of gathering wild plant materials and fungi crosses age, class, ethnicity and socio-economic status. It provides a suite of benefits that contribute to health and well-being, while connecting people to woodlands and countryside in direct and intimate ways. For a small but important subset of Scottish gatherers, NTFPs also are a source of income. However, the legal status of gathering often is unclear and tensions have arisen around the terms of access to land and NTFPs in Scotland. The legal context of contemporary gathering in Scotland is a function of formal law and customary practice, grounded in the 20th-century history of Scottish forests.

In the UK, including Scotland, levels of forest cover have declined steadily over the course of centuries. By World War I a crisis was reached, with the UK in danger of running out of timber altogether. The Forestry Commission was formed in 1919 to deal with this crisis, acting on a policy of planting fast-growing exotic species such as Sitka spruce and lodgepole pine to produce timber in large volumes, if not of high quality. This policy began to change with conservation concerns in the 1970s and 1980s, and although exotic conifers still form the majority of reforestation efforts, new plantings contain greater levels of native species and are more likely to be sensitive to topography and other local landscape characteristics. Despite recent increases in planting, levels of forest cover in Scotland remain at 16.4 per cent (Forestry Commission, 2001) compared with a European average of 46 per cent (FAO, 2003).

The use of NTFPs has followed a trajectory similar to that of forest cover. A gradual decline was interrupted by a brief resurgence as the shortage of resources during World War II forced a return to reliance on some wild gathered products. Medicinals in particular were used to a significant extent. For example, rose hips were gathered

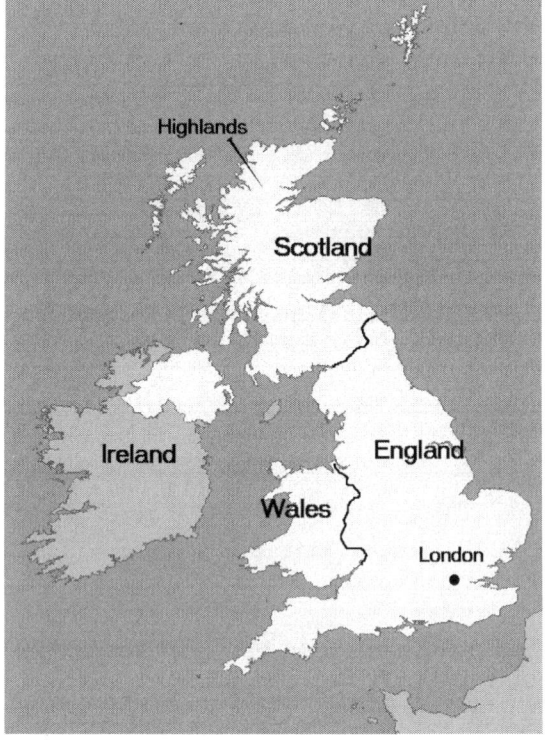

Source: Public domain base map from www.commons.wikimedia.org/wiki/File:United_Kingdom_location_map.svg, modifications by Elizabeth Skinner.

Figure 5.1 *Map of United Kingdom including Scotland*

to make a syrup rich in vitamin C, and sphagnum moss was used to make antibacterial wound dressings. Wild game such as pigeon, rabbit and venison was used to a much greater extent than it is today. Reliance on NTFPs during the war may have served to encourage a decline in gathering afterwards, as it reinforced negative perceptions of wild harvesting as a desperate measure.

Over the past 20 years there have been ebbs and flows in the popular attention focused on NTFPs. In the 1970s there was a resurgence in gathering, epitomized by Richard Mabey's book *Food for Free* (1972), which influenced a generation of environmentally conscious middle-class people to experiment with wild foods. Today gathering is enjoying another wave of popularity. In addition to harvesting for personal use, commercial harvesting is increasing as well, and a wider range of people are involved. Among the drivers of this renewed activity are the development of global markets for wild fungi, the promotion of wild foods by celebrity chefs, a crafts renaissance and the influx of workers from eastern European countries, many of whom arrive with a strong cultural history of gathering.

Wild mushrooms offer a useful illustration of overall trends in NTFP harvesting and use in Scotland. There is little history of fungus use in Scotland. However, after

Source: Alison Dyke.

Figure 5.2 *Cep* (Boletus edulis), *one of Scotland's highest value mushrooms*

World War II, a first wave of Polish and Italian settlers brought with them a culture of harvesting fungi that has endured and expanded beyond their immediate families. As the domestic harvesting of fungi increased in the second half of the 20th century, a commercial industry was established. Initially it supplied international markets and London restaurants. More recently, Scottish markets have also developed. A study conducted in 1998 found that the Scottish mushroom industry was worth around £400,000 per year and was responsible for approximately 20 year-round jobs, with some 350 casual pickers benefiting from seasonal earnings (Dyke and Newton, 1999). By 2006, the industry was worth at least double the 1998 figure.

The adaptation of harvesting rights in the face of these changes is the focus of this chapter. We begin by briefly describing NTFP gatherers and gathering in Scotland at the beginning of the 21st century. Then we discuss current legal terms of access to land and NTFP resources, including specific legislation and common law. We note that perceptions of harvesting rights and attitudes to the products themselves have affected the interpretation of formal legal rights and the development of customary rights. Next, we examine the uneasy fit between actual gathering practices and the legal canons that apply to them before outlining the factors that give gatherers, businesses and landowners differential access to both NTFPs and the legal process. Finally, we suggest that there is a need for change in the management of NTFPs and the legal structures governing them. We propose that gatherers and an understanding of gathering be incorporated in the development of Scottish NTFP policy and suggest steps in that direction.

CONTEMPORARY NTFP GATHERERS AND GATHERING

A recent ethnography of wild harvesting in the Borders and north-east Highlands indicates that today Scottish wild plant materials and fungi are commonly used as food, beverage ingredients, and craft or decorative materials (Emery et al, 2006).[1] Less frequently, they find use as medicinals, cultural or spiritual items, garden implements and toys. Interviews with gatherers provide examples of the diverse socioeconomic backgrounds of contemporary Scottish gatherers. They include a member of the House of Lords, an underemployed gentleman, a retired hotel manager and a young farmer (Emery et al, 2006). Others are a young man scraping together enough cash for a Friday night out, a recent immigrant harvesting moss for a crew boss, and a family on its annual excursion to collect berries for jam-making (Dyke, 2006).

There are great differences from case to case in the contribution that gathering makes to household economies. Nearly all gather for personal use, and most do so primarily or exclusively for this purpose. However, some produce value-added items such as baskets or jams for gift-giving or sale, and a small but important subset sell raw or minimally processed products such as wild fungi to wholesalers or restaurants. Survey results suggest that this pattern of livelihood uses for wild harvested materials is typical of Scottish gatherers (Snowley and Daley, 2005). There may also be considerable variety in the knowledge and skills of these individuals, as well as the quantities harvested.

This diversity makes it clear that there is no particular group or groups that can be seen as being more important, typical or representative of gatherers. Indeed, clearly segmented demographic groups of gatherers are not easily identifiable. Categorizing gatherers by the livelihood function of their activities is equally problematic, as a single individual often engages in gathering for a variety of reasons.

The most invisible form of harvesting is probably also the most commonplace: personal gathering, in which individuals and groups of friends or relatives go out to harvest small amounts of food items once or twice per year. These might include brambles (*Rubus fructosis*) or blaeberries (*Vaccinium myrtillus*) for jam, elderflowers (*Sambucus nigra*) for cordial, or mushrooms (*Agaricus* spp.) for dinner. Although this may be an infrequent activity involving small quantities of plant materials or fungi, it is perhaps the most direct and intimate way in which people interact with the woodlands they visit. This practice often is recognized by local woodland managers but is generally invisible in national policy and law.

Commercial mushroom harvesting is a comparatively new activity and as such it has attracted the attention of the press. Much of the media coverage is negative and often based on second-hand information that reiterates one or two mutually reinforcing stories: 'rape and pillage of the land' or 'fortune to be made on easy pickings'. Given this public image, people who gather for commercial sale are understandably cautious about drawing attention to themselves.

In fact, most gatherers who collect fungi for sale do so in small quantities to earn income for special purposes. Mary, for example, with the knowledge and tacit permission of a local landowner, harvests fungi for her own household use and for sale. She generally uses the income for special purposes such as holiday spending money. When Mary's young son wanted to buy a new bicycle, he helped with the harvesting and was able to raise half the cost. Other special purposes mentioned by gatherers include charitable contributions and savings for children. Only among those with the tightest household incomes do proceeds go into the general household fund.

Additional examples of the role of NTFPs in lives and livelihoods further illustrate the folly of trying to assign even those who sometimes gather for commercial sale to a single category. Aleksy, from Poland, is working in Scotland for a short period. He is amazed by the availability of mushrooms in the Scottish woodlands and harvests commercial species such as chanterelles and boletes to sell. He keeps some of these and harvests a variety of other species to preserve and take home when he returns to Poland. Meanwhile, two permanent Scottish residents exhibit high degrees of livelihood dependence on wild fungi and other NTFPs for both household use and occasional sale or barter, albeit with different motivations. Fiona is an avid member of the community woodland movement in Scotland. Her desire to live in closer harmony with the natural environment leads her to harvest extensively from local woodlands for food and other materials. She sometimes trades wild food or craft materials for other items. Michael has had a patch of bad luck lately and NTFPs provide a critical safety net for him. The income helps him make his rent and put food on the table when money would otherwise run short. It also helps maintain his confidence in his ability to provide for his family.

Regardless of the livelihood use to which wild harvests are put, common to all gatherers is personal pride in the skill that it takes to gather and process these products and derive some income or goods for domestic use when times are hard (Dyke, 1998b). The therapeutic effects of taking time off as a group of family or friends is also cited as a reason to gather year after year, as are the break it offers from ordinary life to observe the impact of the weather and seasons on what is available in the woods and the excuse to take some exercise (Dyke, 1998b; Emery et al, 2006). The aggregate effect of these benefits, regardless of the economic intent, is a considerable positive impact on well-being.

LEGAL TERMS OF ACCESS TO NTFPS

Two political changes in the late 20th century condition the legal context for access to NTFPs in Scotland today. Scotland voted for devolution within the UK in 1997 and now has its own parliament with legislative and tax-raising powers. One result has been the passage of some distinctively Scottish laws that reflect the nation's unique cultural and legal traditions. Simultaneously, the process of harmonization under the European Union is trending toward increased homogeneity of laws governing selected issues. Legislation on species protection particularly has been affected by harmonization. Within the UK, too, the adoption of nearly uniform laws on species protection suggests an assumption that conservation need not be culturally specific.

In contrast, legislation dealing with property and access to land and resources has been strongly affected by devolution and now reflects the distinct traditions of Scotland. The passage of this legislation on access reflects both changes in attitudes and the formalization of long-held customary rights. In this section, we describe the common law, legislation, by-laws and general restrictions relevant to gathering NTFPs in Scotland in terms of their effects on access to land and resources, provisions related to property theft or damage, and species protection.

Access to land and resources

In order to gather NTFPs, one must first gain access to land. Scottish law provides a universal right of access (with sensible restriction) through the Land Reform (Scotland) Act 2003, which went into force on 14 June 2004 (Scotland, Parliament, 2003). This Act codified the customary 'right to roam', which was previously supported by the absence of an offence of trespass in Scottish law. The Act is silent regarding gathering for non-commercial purposes, leaving the legal status of such activities ambiguous. However, the Act is explicit in excluding commercial gathering from the right of access. Since the majority of commercial activity, particularly in the wild mushroom industry, occurs *without* the permission of the landowner (Dyke and Newton, 1999), this legislation effectively criminalizes most commercial gathering.

During the consultation period for the Act, the landowning lobby speculated that the formalization of the 'right to roam' would lead to large increases in visitor numbers. In fact, the Forestry Commission's biannual *Public Opinion of Forestry* surveys show a gradual decline in the percentage of the population who had visited woodlands in the preceding few years (Gillam, 1999; Heggie 2001; West and Smith, 2003; Snowley and Daly, 2005). Despite this fact, there is a perception among landowners that visitor numbers have risen and harvesting has increased. While harvesting may have experienced some increase, we suspect that landowner impressions are more the result of heightened awareness of the right to public access and greater attention to visitors and their activities following debates on and the enactment of the Land Reform Act.

Theft

Under Scottish common law, plants and fungi are included in the 'parts and pertinents' to land, making the produce of the land the property of the owner by 'accession of fruits' (Reid, 1996, p457). Despite such apparent clarity regarding the ownership of resources, user rights are murky. Loss to the owner must be demonstrated in order to establish compensable wrongdoing, and a gatherer's intent to engage in commercial sales must be demonstrated. The latter is a legally challenging task, given that plants and fungi are widely regarded as lacking commercial value. In a suit brought by the Department for Environment, Food and Rural Affairs in England (Box 5.1), the courts found for the gatherer based on the ancient concept of 'right of user' for an individual who has used a resource unchallenged for more than 20 years. However, this right does not exist in Scottish law.

Almost half of Scotland's forest is under the management of the Forestry Commission and therefore is subject to Forestry Commission by-laws, which forbid taking anything away from the land. A strict interpretation of these by-laws, usually invoked to seek remedy in cases of particularly bad harvesting practices, would be out of step with actual agency practice. The Forestry Commission increasingly depends for continued government funding on its position as a provider of public good, through recreational opportunities, environmental protection, etc. Hence it encourages the public to engage with forests by offering events such as fungi forays and wild food walks. Ongoing legislative action may result in the abandonment of Forestry Commission by-laws as more of their provisions are covered by national laws such as the Land Reform Act.

Box 5.1 Mushroom gathering and the 'right of user'

The case of Mrs Tee's wild mushrooms illustrates the coexistence in the UK of legal and customary rights to gathering. Brigitte Tee Hilman and her employees gather fungi in and around the New Forest in south-east England and import from overseas to supply London restaurants. In November 2002 Brigitte Tee Hilman was arrested in possession of 6kg of winter chanterelles (*Cantharellus lutescens*) gathered in the New Forest. The arrest followed several warnings by Forestry Commission staff after it had responded to increased gathering by developing a policy that prohibited harvest for commercial purposes and established a limit of 1.5kg for domestic use (the time period and geographical area to which this limitation applied were not specified).

Brigitte Tee Hilman was prosecuted under the Theft Act 1968 (applicable to England and Wales) (UK Parliament, 1968), which specifically makes the gathering of fungi, flowers, fruit or foliage for commercial purposes, without the permission of the landowner, theft (section 4(3)). The case was dropped on appeal in May 2006, but Brigitte Tee Hilman still faced civil action. Her defence was based on prescription (Collis, 2005), whereby rights to an easement over land (in this case for picking wild mushrooms) can be established by uninterrupted use over a period of 20 years or more. This use must occur at least once per year, take place without force or secrecy and without permission, and be of a type that could lawfully have been granted. Having gathered regularly and unchallenged in the New Forest for over 20 years, Brigitte Tee Hilman felt that she had established a customary right of use. She won in civil court, thereby establishing that she had a personal right (which did not apply to her employees) to gather mushrooms for her own use and for commercial purposes in the locations where she had previously collected on Forestry Commission-managed land.

This opens up the possibility of additional challenges to the Forestry Commission policy by other long-standing users. Given the considerable sum of public money dedicated to the case, it also suggests that there should be consultation with local gatherers prior to restrictions being imposed.

Damage

Under common law, deliberate or negligent damage to wild plants is an offence against the landowner. This was recently formalized through amendments to the Wildlife and Countryside Act 1981 (UK Parliament, 1981) under the Nature Conservation (Scotland) Act 2004 (Scotland, Parliament, 2004). In theory, these amendments could make gathering a product or trampling vegetation a civil offence, although the activity would have to be proven deliberate or negligent. Such activities might also provide grounds for prosecution for vandalism under the Criminal Law (Consolidation) (Scotland) Act 1995 (Scotland, Parliament, 1995). In fact, cases that do reach the courts are often tried under the vandalism clauses of the Criminal Law Act rather than legislation formulated for the protection of wildlife, in part because the former is more commonly used and better understood by the police and Crown Prosecution Service, and because it carries stiffer sentences (Nurse, 2003).

However, successful prosecution requires the demonstration of substantial loss. The case illustrated in Box 5.1 was first brought under the Theft Act and then as a civil action for theft. In both instances, prosecution for the theft of mushrooms with such a small commercial value was widely regarded as a waste of public money and the court's time. Public reaction to the plaintiff's case might have been different had the case been brought under conservation legislation.

Species protection

The Wildlife and Countryside Act also makes it an offence to uproot or destroy any wild plant without the permission of the landowner (section 13(1)(b)),[2] and thus could serve as the basis for legal action to protect localized plant populations and/or the interests of landowners. Some species have complete protection from gathering (including seeds or spores), disturbance and sale or possession with or without landowner consent under Schedule 8 (sections 13(1)(a) and (2)(a)) of the Wildlife and Countryside Act. Bluebells illustrate the application of this provision (Box 5.2).

In addition to national laws, special restrictions apply to land under some types of ownership and designation. For example, plants may not be picked in some nature reserves or on the properties of the Ministry of Defence and the National Trust, which together encompass approximately 1 per cent of the national land area. Picking is also prohibited on sites of special scientific interest (commonly known as SSSIs and covering almost 13 per cent of Scotland's land area) without permission from the appropriate government agency. In addition, nature reserve managers often ask visitors not to gather wild products, arguing that it diminishes the experience for future visitors and reduces habitat quality for wildlife. Aesthetic considerations are another commonly cited reason for restrictions on gathering, justified by instances of gatherers collecting every mushroom they could find, later identifying those they wanted to keep and discarding the remainder in the car park.

Source: Alison Dyke.

Figure 5.3 *Native bluebell* (Hyacinthoides non-scripta), *a controversial wild harvest*

Box 5.2 Bluebells

Native bluebells (*Hyacinthoides non-scripta*) have iconic status in the UK, with their swaying blue flowers that carpet native woodlands in the spring. Britain is home to somewhere between 25 and 49 per cent of the world's population of the species. Bluebells are also a popular garden plant: there is a sizable trade in their bulbs as well as transplantation from the wild to domestic gardens. Both of these activities are illegal under Schedule 8 of the Wildlife and Countryside Act, which prohibits the 'sale or advertising for sale' of listed species. One solution to the demand for bluebells in the light of the ban on harvesting native bulbs has been the import of both native and non-native bluebell species, mainly from Spain.

Another solution would be the cultivation of native bluebells from seed. In theory, it should be possible to obtain a licence to gather bluebell seed, and such licences have been issued in England in the past. However, no licences for seed collection are available currently. It is difficult to distinguish native bluebells from non-native Spanish bluebells (*Hyacinthoides hispanica*) and hybrids of the two by any readily visible physical characteristics. As a result, the in-county agencies responsible for issuing seed collection licences have suspended doing so until a reliable genetic test is available. Meanwhile, planting projects that require native provenance seed are unable to obtain bluebells, and illegal gathering from the wild, of both native bluebell and Spanish bluebell, continues.

Poor labelling of marketed bulbs is an additional problem. Considerable efforts are being made to educate the public about the difference between native, Spanish and hybrid bluebells. But this understanding is hampered by plant breeders' poor labelling. Bulbs are sold with obsolete Latin names, illustrated with pictures of the wrong species, or simply under the wrong name. This poor labelling has resulted in some unfortunate errors. The village of Clent in Worcestershire spent £1000 on bluebell bulbs only to discover that they had bought (and planted) 7000 Spanish bluebells owing to misleading labelling. Obtaining sustainably sourced supplies of native bluebells is only half the problem, then. Ensuring that those supplies are, indeed, the native species is an even greater task.

Under common law, access and harvesting rights appear relatively straightforward: landowner permission is required to undertake harvesting for any purpose. Formal legal codes make these terms less clear. As previously noted, the Land Reform Act specifically excludes commercial harvesting from rights of access while making no mention of gathering for non-commercial purposes. The Wildlife and Countryside Act is silent on the issue of landowner permission for harvesting that does not involve the destruction of the organism. There is, for example, no explicit requirement to obtain permission prior to harvesting reproductive parts or foliage, which are the most commonly gathered materials. Understandably, a majority of harvesters have not made the considerable effort required to establish the exact legal position of their activity. Landowners are no better informed, and considerable confusion exists.

The arguably artificial distinction between much commercial and non-commercial gathering activity in the Land Reform Act may reflect a desire to preserve customary rights to gather for personal use. But the failure to provide explicitly even for non-commercial harvesting aggravates the murky, tenuous status of customary rights, which go largely unrecorded and receive little consideration in formal venues. Government publications describing the rights and responsibilities of access for the general public illustrate this tendency. For example, a Scottish Natural Heritage document on the implications of the Wildlife and Countryside Act states: 'Technically, wild flowers belong to the owner of the land and taking them may be theft' (Reid, 1998). Here the use of the word 'may' indicates the uncertainty of the position and the extent to which it may or may not be overridden by customary rights.

There are those who argue, on a variety of grounds, that commercial gathering is also a customary right. Some gatherers reject the notion of private landownership with exclusive rights to any and all resources on that land. These individuals may go so far as to view their commercial gathering as a form of protest against what they regard as an unjust distribution of benefits from the land. Others contend that ownership is defined by use and that landowners who do not use or manage a product themselves forfeit their right to it. Indeed, landowners themselves, when questioned, profess little concern about the harvest of fungi by unauthorized individuals provided the proprietor has no wish to exploit the resources and there is no long-term damage associated with the activity (Dyke, 1998a).

GATHERING PRACTICE AND LEGALITY

Contemporary NTFP-gathering and business practices fit uneasily into current law in Scotland. Official publications notwithstanding, customary rights often override formal legal measures, not least because the spatial and technological patterns of most harvesting make it hard to police and the laws themselves are difficult to implement. Fundamentally, the legal acceptability of gathering is a function of landowner permission. In practice, the need for explicit permission tends to increase with the degree of commercialization, the scale and mechanization of harvests and the distance gatherers travel to a harvesting site.

Degrees of commercialization

Much of the NTFP literature defines commercial activity in opposition to non-commercial harvesting (Schreckenberg and Marshall, 2006), implying a series of related binaries: commercial/non-commercial, cash exchange/personal consumption, economic/cultural. The assumption of these hard-boundaried distinctions is effectively codified in Scottish law. However, as illustrated above, in Scotland, as in most of the world, including the United States, a majority of gatherers' activities fall somewhere between strictly commercial and non-commercial activities (Carroll et al, 2003). One individual may gather for income, household consumption, gift-giving and trade or barter, even in a single outing. Collected materials may have considerable cultural importance while simultaneously providing much-needed cash.[3]

Scottish law governing NTFP enterprises is predicated on formal business models and exhibits a mismatch between the intent of legislation and the characteristics of many, if not most, of the enterprises to which it applies. NTFP businesses are often marginal in conventional economic terms, though they may be quite important socially, culturally and in terms of their contributions to household livelihoods. Research experience suggests that the great majority of NTFP-based businesses are cottage industries and microenterprises, employing no more than two or three individuals, if any, aside from the owner (Paul and Chapman, 2007). The volumes of plant material and fungi tend to be small and economic activity is frequently intermittent. As with gatherers, much of the exchange may take place in informal economic venues such as local fairs and personal social networks. However, a move to promote NTFP commercialization as an economic development initiative might well result in the formation of more formalized enterprises of the sort envisioned in legislation.

Scale and mechanization of gathering

The scale, or volume, of harvesting is often assumed to be a function of the degree of commercialization, but this is not always the case. An individual or family gathering fungi to make up a substantial portion of their diet may harvest more than someone seeking pocket money for a weekend outing, for example. Nevertheless, there is a widespread assumption among landowners and managers that commercial harvesters gather large quantities. For this reason, some commercial mushroom gatherers report using 'decoys'; they may carry binoculars so as to appear to be birdwatching tourists while in the harvest area, transferring their mushrooms to the crates that make them identifiable as commercial gatherers only when at a safe distance (Dyke, 1998a).

The degree of mechanization also is assumed to be a function of commercialization, but actual practices are considerably more complex. In general, NTFP harvesting is a manual process using only the most basic tools, if any, regardless of the livelihood use to which gathered materials are put. However, there are instances of highly mechanized gathering in Scotland. In particular, some commercial operations seek moss and bulbs in such large quantities that bulldozers are used for harvesting and tracked vehicles for removing plant materials from the site. The potential for damage to the resource and to the habitat as a whole from such mechanized harvesting is obviously high, and there have been examples of this type of harvesting activity being carried out without the permission of the landowner. But the use of mechanization also increases the visibility of harvesting, and, as a consequence, moss harvesting is one of the few areas where the use of harvesting agreements is relatively common.

Insider/outsider status and access

Formal law does not distinguish between gatherers on the basis of their place of origin or residence, but common practice often does. Many landowners support the tradition of local people gathering products for their own use but object to what they see as the appropriation of resources by non-local gatherers, especially when commercial sale is involved. These objections are based on twin apprehensions: first, that local people will suffer from the competition, and second, that outsiders' harvesting practices may be less sustainable. Harvesting by non-local people tends to be more visible by virtue of the identity of the individuals involved. A local car parked in a lay-by is generally unremarkable and could as easily be a dog walker's as a mushroom picker's. In contrast, an unfamiliar vehicle, especially if driven by unknown or 'foreign-looking' people, is more likely to be noticed and elicit some sort of response.

Numbers of non-local harvesters are likely to increase as markets for wild foods grow and increasing numbers of immigrants from Eastern Europe settle in urban Scotland, bringing strong traditions of fungi harvesting with them. The issue of gatherer identity and access will probably become more active in response to these changes. At present, the regulation of access based on place of origin depends on individual landowners, and no clear customary practice has emerged. There is, however, an obvious and as yet unanswered ethical question: is it acceptable to exclude people because they are not long-time residents of a local area, or can the concerns that might lead landowners to attempt to exclude non-local gatherers be adequately addressed through the regulation of harvesting practices and measures to address potential competition?

PROCESSES AND MECHANISMS OF ACCESS

While legal and customary rights constitute the formal basis on which individuals may access resources, actual practices may bear little resemblance to the law, reflecting a wider range of factors. The main processes and mechanisms that influence access to resources in the Scottish context are explored below in accordance with Ribot and Peluso's theory of access (2003).

Knowledge and lack of knowledge

Knowledge and lack of knowledge are perhaps the main factors determining access to NTFPs. Gatherers, business people and landowners tend to possess distinct bodies of knowledge that reflect their personal involvement with NTFPs. Gatherers' knowledge derives from ongoing interaction with fungi in the landscape. The specialist knowledge that distinguishes them and is essential to successful gathering gives them privileged access to fungi. Not uncommonly, it also conveys special social status through gifts to family and friends that embody the knowledge and effort that has gone into gathering and processing.

In addition to what might be thought of as traditional knowledge, new patterns of use and management are emerging around products that were not widely gathered in the past. Fungi are particularly noteworthy in this regard. With the increase in commercial and domestic culinary use, growing numbers of people are learning to identify and harvest edible fungi. Over the past 20–30 years an international network of craft workers who use fungi for dyeing has developed, including a strong contingent in Scotland. Through the sharing of knowledge within countries and across continents, it is now possible to produce almost the full spectrum of colors from fungi dyes. This type of expertise is termed 'new expert knowledge'.

Traditional and new expert knowledge often overrides legal and customary rights in conveying the ability to access resources. At present, individual gatherers are aware of the advantage their knowledge provides them in obtaining and using NTFPs. That authority enjoys no formal status, however, and there is little awareness on the part of gatherers that by combining efforts they might influence the legal terms of access.

Market knowledge also provides an advantage in gaining access to and making use of fungi. The owners and managers of more commercial businesses tend to be one step removed from the practice of gathering and, as a result, may lack the knowledge base that gatherers possess. But buyers and wholesalers often enjoy strong market knowledge. This expertise affords them a privileged position as middlemen when dealing with landowners and facilitating negotiations for access on favourable terms. Indeed, many well-established fungi buyers in Scotland have used their market knowledge to access international markets, importing fungi from other parts of the world when they are unavailable in Scotland, in order to provide a year-round supply to their customers.

In contrast, landowners generally lack knowledge of the NTFPs on their property and the markets that may exist for them. Landowners are accustomed to using professional advice with regard to other resources on their land, but at present there is little comparable data on NTFPs and there are few professionals to consult. Further, efforts to systematize scientific information on NTFPs or gatherers' traditional and new expert knowledge have been limited. Without such knowledge landowners are at a comparative disadvantage with regard to making use of fungi and other NTFPs. We note, however, the risk that the professionalization of this knowledge might entail for gatherers and buyers. The power conveyed by their gathering and market knowledge would probably be eroded if it were widely available to landowners, although it is not clear that landowners would wish to undertake gathering for their own or commercial purposes.

Representation

Representation in deliberative processes is closely related to questions of knowledge and power. Gatherers, buyers and processors are accustomed to operating as independent individuals and generally do not see themselves as constituting a group (or groups) joined by common interests. There is little sense of the potential for cumulative effects from their actions or the possibility that policy changes could impact on them individually. Not surprisingly, then, there is no organizational structure representing gatherers, buyers and/or processors in Scotland and there was no particular effort to engage these groups in consultations on the Land Reform Act. The absence of representative bodies for gatherers in particular also means that there are no clear groups for landowners to interact with and involve them in resource management.

Other stakeholders are also affected by the lack of any mechanism for communicating among interest groups. Landowners have no means of representing themselves and their perspectives to gatherers, and this contributes to their inability to benefit from the resources in their possession. Non-governmental organizations, which played a significant role in consultations on the Land Reform Act, are also hampered in their attempts to disseminate information by the lack of an easily defined and accessible audience. On an institutional level, the lack of representation aggravates the dissonance separating legislation and regulation from actual practices.

Scales of policy and implementation

NTFPs are a largely unmanaged resource in Scotland, in part because no one is quite sure whose responsibility they are. NTFPs are addressed in international policy documents such as the Convention on Biological Diversity (UNEP, CBD 1992), the Helsinki General Declaration (MCPFE, 1993) and the Global Strategy for Plant Conservation (UNEP, 2002). At the level of the European Union, their importance is well noted in the Pan European Criteria for Sustainable Forest Management (MCPFE, 2002). At the UK level, however, NTFPs scarcely receive policy mention. There is a time lag, as decisions taken at the international and European levels are subject to a series of delays in translation to national policy and legislation, and then into procedures for institutional implementation prior to being put into practice at the local level.

In addition, there is a disparity between the spatial scale of authoritative policy and that of allocative management (Geores, 2003). The legal structures discussed earlier set out to govern authoritative rights over resources nationally, but policy documents fail to make provision for implementation and allocation at the local level. These spatial and temporal mismatches are manifested in a missing policy and management layer that makes it difficult to govern resources at an operational level. At a very practical level, the size of units used for managing forests is too large to be of much use in the management of sites where NTFP species are found. Additionally, gatherers work on a wider scale than do woodland managers or landowners, often ranging across woodlands and ownerships in search of a variety of products from diverse habitats.

RECOMMENDATIONS

In practice, there is strength in the flexibility that the current system of ad hoc unregulated access affords gatherers. But it is also extremely vulnerable to unanticipated changes in circumstances. The increasing popularity of gathering, the involvement of urban and new immigrant populations, and the development of new products and appetites suggest that it may be time to consider changes in the legal instruments and management practices applied to NTFPs and their gathering in Scotland. The rights of old and new gatherers, the sustainability of NTFP species and their habitats may not be well served by the current system over the long term. Access rights and perceptions of access rights will probably be forced to evolve – not for the first time – to take account of changing circumstances.

One recent adaptation to changing NTFP dynamics is the development of gathering guidelines, initially for fungi and more recently for mosses and bulbs. Box 5.3 offers further discussion on guidelines for good practice and the process by which these may be developed, with reference to the Scottish Wild Mushroom Code. Our research suggests a number of additional actions that would provide for continued benefits from Scottish NTFPs while contributing to their sustainable management.

Box 5.3 The Scottish Wild Mushroom Code

In 1999, increased mushroom picking for domestic and commercial purposes was generating controversy in Scotland. Commercial gatherers tend to stick to species they know they will be able to sell in good condition. In contrast, those gathering for their own use may collect a wide range of species, in an attempt to improve their identification skills, before deciding what to keep. This practice generated the concern on the part of conservation organizations that an absence of fungi harvesting expertise could endanger particular species and habitats. Another concern was the potential for increased conflict between gatherers and landowners as more people went into woodlands in search of fungi. With consultations for the Land Reform Act getting under way, a code was commissioned with the aim of clarifying the confused picture of legal and customary rights to gather wild mushrooms.

The Scottish Wild Mushroom Code was developed by a group known as the Scottish Wild Mushroom Forum, which included gatherers (both recreational and commercial), landowners, mycologists, entomologists, conservation organizations and regulatory bodies (Scottish Natural Heritage, 2003). The code was developed prior to the Land Reform Act, but is based on similar principles, which include placing the responsibility for good conduct and personal safety on the gatherer and emphasizing courtesy and communication with other woodland users, particularly the landowner.

Guidelines are only a starting point; they also need to be implemented. Broad acceptance of the guidelines in the gatherer community will aid in this process, but the guidelines have no real teeth. In order to effectively influence gatherer and landowner practices, guidelines need to be accompanied by measures that allow gatherers to police one another's conduct and ensure that landowners consider the impacts of broader land and forest management practices on NTFP species.

Guidelines for gathering practice

As would be expected, there are examples of both good and bad gathering practice for each NTFP. Bad practice often takes the form of accidental damage by poorly prepared new gatherers and, in some cases, is due to the absence of long-term interest in productivity. Good practices tend to result from the knowledge and personal investment of gatherers and can form the basis for the development of guidelines on gathering practice for both new gatherers and land managers.

Guidelines are only as good as the process that creates them, however. The cooperation of many groups – gatherers, buyers, wholesalers, landowners and non-governmental organizations – is required to produce workable approaches that will be regarded as legitimate. The paucity of biometric data on NTFPs increases the importance of drawing on gatherers' knowledge of what has proven to be sustainable and unsustainable over time. Invoking their experience also heightens the likelihood that guidelines will be regarded as accurate and reasonable by the individuals who are asked to follow them. This, in turn, improves the chances that such guidelines on gathering will be accepted and voluntarily observed, factors that are crucial to their success.

However, guidelines alone are not enough to influence practice. The factors identified in the section above on 'Processes and mechanisms of access', including the impact of broader resource management decisions on NTFPs, must also be addressed in order to achieve broad acceptance and compliance. The process of creating guidelines initiates coordination and communication across the range of stakeholders, but more robust systems are needed to realize the potential for self-governance, including a way of monitoring the use of guidelines. Incorporating guidelines into broader resource management systems would provide mechanisms for enforcement beyond peer pressure and ensure that land managers take the actions and interests of gatherers into account.

Representation and communication

Many of the challenges to the sustainable management of NTFPs in Scotland stem from a lack of communication between parties. The fact that NTFPs rarely feature on the agenda of forestry policy processes exacerbates this problem, and a concerted effort is needed to ensure that NTFPs are addressed as part of broader forest policy. It is also important for the various stakeholder groups – gatherers, buyers, processors, landowners, non-governmental organizations and statutory bodies – that currently lack a forum in which they can communicate to develop methods and processes for doing so.

Systems of governance

As interest in managing NTFPs increases, the question of *how* these resources should be managed arises. In posing this question, it is imperative that we ask not only what changes might be made to current practices, but also according to what judgements and values any changes would be made and through which forms of governance. In particular, how can the governance of NTFPs be integrated into existing processes and mechanisms for natural resource management?

Accreditation and certification

In order to be effective, guidelines for good practice and sustainable gathering require enforcement. An accreditation scheme for gatherers would establish a public standard of good gathering practice that could be enforced by consumer demand and subject to peer monitoring by other gatherers. Accreditating gatherers rather than products has the advantage of being relatively easy to set up and administer. Similar schemes have been developed in Finland and are applied there as voluntary standards.

Ensuring that woodland management benefits NTFPs should go hand in hand with insisting on good gathering practices. At present, the certification of NTFPs under available schemes (such as the Soil Association's Woodmark scheme) is not practicable, since all such schemes require separate certification for each species gathered at each site. The ecological information needed to assess the impacts of gathering generally is not available, and the cost of administering such certification schemes is borne by the landowner and would be difficult to recover. New methods of certification, using groups of products and groups of landowners, might improve the situation. Standards requiring consideration of the impacts of forestry operation on NTFP species under the UK Woodland Assurance Standard would be more accessible and widely applied.

Subsidy

Without financial incentives, it is unlikely that forest managers will engage with the management of NTFP species. At present, however, there are no government programmes to encourage the management of these resources, although they would fit within the objectives of forestry grant aid. The Scottish Forestry Grant Scheme currently includes an element of support for recreational use, but NTFP gathering for personal consumption – a major recreational use of woodlands for many – receives no recognition in the way of grant aid. As a consequence, public and private landowners receive no financial incentive to incorporate NTFP gathering into woodland management. Grant aid programmes could create incentives for the management of NTFPs by requiring that all woodlands receiving subsidies be in compliance with the criteria for forestry practices set out in the UK Forestry Standard and by including consideration of the impacts of forestry operations on NTFP species in the standard.

CONCLUSION

NTFP gathering is widespread in Scotland today, and the practices associated with it reflect the vibrancy of contemporary culture in the nation. Gathering also offers benefits that address several national policy concerns. The micronutrients obtained and physical activity involved in harvesting wild plants and fungi contribute to health and well-being. By providing livelihood resources and cultural satisfaction, NTFPs contribute to the sustainability of rural lives. Gathering also involves people in environmental protection through direct engagement with the natural world.

In spite of these contributions, NTFPs are poorly incorporated into policy, in part due to conceptual boundaries between timber and non-timber, woodland and non-woodland, commercial and non-commercial. These distinctions do not reflect actual practice and serve to increase the difficulties that NTFP practitioners have in engaging with the development of national policy.

The most effective policy approaches to NTFP sustainability will be those that work with gatherers and incorporate their passion for the activity, allowing them to use their knowledge and building on the majority's desire to ensure that resources are available in perpetuity. This requires an element of trust that the state may find difficult to muster. Legislation will be of value in dealing with the extreme cases of bad practice that occur from time to time. Nevertheless, given the dispersed, low-technology nature of the vast majority of gathering, long-term prospects will be best served if gatherers retain the responsibility for managing NTFP resources and police themselves. Like the history of gathering in Scotland, present practices are dynamic and policy solutions will need to be flexible, accommodating the fluid and evolving nature of these activities.

NOTES

1   The ten vascular plant species mentioned most frequently by interviewees in the ethnographic study are all used as edible and/or beverage ingredients. Seven of the 'top ten' are berries: brambles (*Rubus fructosis*), raspberry (*Rubus idaeus*), blaeberry (*Vaccinium myrtillus*), elder (*Sambucus nigra*), sloe (*Prunus spinosa*), rowan (*Sorbus aucuparia*) and rose hip (*Rosa* spp.). The ten most frequently mentioned fungi are the easily identifiable chanterelle (*Cantharellus cibarius*), bolete (*Boletus edulis*), field and horse mushroom (*Agaricus campestris*, *A. arvensis* and allies), hedgehog fungus (*Hydnum repandum*), puffball (*Lycoperdon* spp., *Calvatia gigantea* and close allies), parasol (*Macrolepiota procera*), ink cap (*Coprinus comatus*) and wood blewitt (*Lepista nuda*). Holly (*Ilex aquifolium*), ivy (*Hedera helix*) and conifers find use as Christmas decorations. Mosses and bulbs (winter aconite (*Eranthis hyemalis*), snowdrop (*Galanthus nivalis*) and bluebell (*Hyacinthoides non-scriptus* and *Endymion hispanicus*)) are employed for horticultural purposes, while the young growth of various tree species is used in basket-making and a wide variety of vascular species and fungi find their way into dye pots. Chanterelles, boletes, hedgehog fungi and winter chanterelles (*Cantharellus lutescens*) are the most widely traded of the expanding variety of fungi for which commercial markets have developed.

2   Fungi are not directly referred to in the Wildlife and Countryside Act, but may, for purposes of the Act, be considered plants.

3   For examples from the United States, see Anderson et al (2000) about ferns, Carroll et al (2003) about huckleberries and Hinrichs (1998) about maple syrup. Wynberg and Laird (2007) provide a similar example on marula in South Africa.

REFERENCES

Anderson, J. A., Blahna, D. J. and Chavez, D. J. (2000) 'Fern gathering on the San Bernadino National Forest: Cultural versus commercial values among Korean and Japanese participants', *Society and Natural Resources*, vol 12, pp747–762

Carroll, M. S., Blatner, K. A. and Cohn, P. S. (2003) 'Somewhere between: Social embeddedness and the spectrum of wild edible huckleberry harvest and use', *Rural Sociology*, vol 28, pp319–342

Collis, P. (2005) *Land Registry Practice Guide: Prescription Act 1832*, Land Registry, London

Dyke, A. J. (1998a) Unpublished field notes

Dyke, A. J. (1998b) 'Wild edible mushrooms as a non timber forest product: The sustainability and potential of the harvest in Scotland', master's thesis, University of Edinburgh, Scotland

Dyke, A. J. (2006) 'The practice, politics and ecology of NTFPs in Scotland', PhD thesis, University of Glasgow, Scotland

Dyke, A. J. and Newton, A. C. (1999) 'Commercial harvesting of wild mushrooms in Scottish forests: Is it sustainable?' *Scottish Forestry*, vol 53, pp77–85

Emery, M. R., Martin, S. and Dyke, A. J. (2006) *Wild Harvests from Scottish Woodlands: Social, Cultural and Economic Values of Contemporary Non-Timber Forest Products*, Forestry Commission, Edinburgh

Food and Agriculture Organization (FAO) (2003) *State of the World's Forests*, Table 2: Forest area and area change, United Nations Food and Agriculture Organization, Rome

Forestry Commission (2001) *National Inventory of Woodland and Trees – Scotland*, Forestry Commission, Edinburgh

Geores, M. (2003) 'The relationship between resource definition and scale: Considering the forest', in N. Dolsak and E. Ostrom (eds) *The Commons in the New Millennium: Challenges and Adaptation*, MIT Press, Cambridge, MA

Gillam, S. (1999) *Public Opinion of Forestry 1999*, Forestry Commission, Edinburgh

Heggie, B. (2001) *Public Opinion of Forestry 2001*, Forestry Commission, Edinburgh

Hinrichs, C. C. (1998) 'Sideline or lifeline: The cultural ecology of maple syrup production', *Rural Sociology*, vol 63, pp507–532

Mabey, R. (1972) *Food for Free*, Collins, London

Ministerial Conference on the Protection of Forests in Europe (MCPFE) (1993) Second Ministerial Conference on the Protection of Forests in Europe, 16–17 June, Helsinki, Finland. General declaration. www.mcpfe.org/conferences/helsinki, accessed 7 February 2009

Ministerial Conference on the Protection of Forests in Europe (MCPFE) (2002) Pan-European Criteria for Sustainable Forest Management. Liaison Unit Vienna, Vienna, www.mcpfe.org/files/u1/publications/pdf/improved_indicators.pdf, accessed 7 February 2009

Nurse, A. (2003) *The Nature of Wildlife and Conservation Crime in the UK and Its Public Response*, University of Central England, Birmingham

Paul, J. and Chapman, E. (2007) *Scottish NTFP Sector Research and Development Project: Final Report*, Reforesting Scotland, Edinburgh

Reid, C. T. (1998) *Scotland's Wildlife: The Law and You*, Scottish Natural Heritage, Edinburgh

Reid, K. G. C. (1996) *The Law of Property in Scotland*, Butterworths, Edinburgh

Ribot, J. C. and Peluso, N. L. (2003) 'A theory of access', *Rural Sociology*, vol 68, pp153–181

Schreckenberg, K. and Marshall, E. (2006) 'Access rights and resources: The impacts of NTFP commercialisation', in E. Marshall, K. Schreckenberg and A. Newton (eds) *Commercialization of Non-Timber Forest Products: Factors Influencing Success, Lessons Learned from Mexico and Bolivia and Policy Implications for Decision Makers*, World Conservation Monitoring Centre, Cambridge

Scotland, Parliament (1995) Criminal Law (Consolidation) (Scotland) Act 1995, Chapter 39, www.opsi.gov.uk/ACTS/acts1995/ukpga_19950039_en_1, accessed 7 February 2009

Scotland, Parliament (2003) Land Reform (Scotland) Act 2003, asp 2. The Stationery Office Limited, Queen's Printer for Scotland, UK

Scotland, Parliament (2004) Nature Conservation (Scotland) Act 2004, asp 6. The Stationery Office Limited, Queen's Printer for Scotland, UK

Scottish Natural Heritage (2003) *Scottish Wild Mushroom Code*, Scottish Natural Heritage, Inverness, Scotland, www.snh.org.uk/publications/on-line/NaturallyScottish/fungi/wildmushroom.asp, accessed 7 February 2009

Snowley, H. and Daly, C. (2005) *Scottish Public Opinion of Forestry Survey 2005*, Forestry Commission, Edinburgh

TNS Global (2003) *Woodland Research: Results of an Omnibus Survey into Non Timber Forest Product Use in Scotland*, TNS Global, Edinburgh

UK Parliament (1968) Theft Act 1968. Chapter 60. HMSO, London, www.statutelaw.gov.uk/content.aspx?activeTextDocId=1204238, accessed 7 February 2009

UK Parliament (1981) Wildlife and Countryside Act 1981. Chapter 69. HMSO, London, www.statutelaw.gov.uk/content.aspx?activeTextDocId=809266, accessed 7 February 2009

United Nations Environment Programme (UNEP) (1992) *Convention on Biological Diversity*. Secretariat of the Convention on Biological Diversity, Montreal, Canada, www.cbd.int/, accessed 7 February 2009

United Nations Environment Programme (UNEP) (2002) *Convention on Biological Diversity. The global strategy for plant conservation*. Secretariat of the Convention on Biological Diversity, Montreal, www.cbd.int/doc/publications/pc-brochure-en.pdf, accessed 7 February 2009

West, V. and Smith, S. (2003) *Public Opinion of Forestry 2003*, Forestry Commission, Edinburgh

Wynberg, R. P. and Laird, S. A. (2007) 'Less is often more: Governance of a non-timber forest product, marula (*Sclerocarya birrea* subsp. *Caffra*) in southern Africa', *The International Forestry Review*, vol 9, no 1, pp475–490

From Barter Trade to Brad Pitt's Bed: NTFPs and Ancestral Domains in the Philippines

Yasmin D. Arquiza, Maria Cristina S. Guerrero, Augusto B. Gatmaytan and Arlynn C. Aquino[1]

INTRODUCTION

There is a scene from the popular American reality TV show *Queer Eye for a Straight Guy* that comes to mind whenever the topic of NTFPs is raised. In one episode, the main host looks around the room targeted for a makeover and says, 'You can't be in New York and not have wicker.'

Indeed, wicker baskets and furniture have become as common as bagels and cream cheese in New York. Yet not many people know that wicker comes from a vine called rattan, which grows in abundance in Philippine forests. As one of the leading rattan furniture and handicraft makers in the world, the Philippines is certainly a major source of wicker products that find their way to New York. In 2006, about US$60 million worth of rattan furniture from the Philippines went to international markets, principally the United States, which accounts for 70 per cent of all such exports (BETP, undated). In 2006, the versatile vine received a big boost in the global scene when *Time* magazine reported that actor Brad Pitt had bought a rattan bed from a well-known designer in Cebu, the furniture capital of the Philippines.

But few people probably know how rattan is harvested. In the Philippines, it is one of the major sources of livelihood for indigenous communities who toil silently in the forests, cutting the thorn-covered vine and carrying the heavy loads for many hours until they reach the lowlands. In the shadow of their thatched roof huts, they still have to scrape off the thorny covering and dry the rattan poles before these are transported to furniture makers in urban centres.

In addition to rattan, many indigenous Filipino communities have been economically dependent for many centuries on other commercially traded NTFPs, as well as a wide range of subsistence NTFPs, such as almaciga resin, abaca and wild honey. In

Source: Mangyan Mission.

Figure 6.1 *Alangan Mangyan rattan gatherers hauling bundles of split rattan*

1575, when the Spanish were just starting to colonize the country, up to 93 per cent of the Philippines' 30 million hectares of land was covered with forests (PAWB, 1998). These forests provided a rich source of natural resources that indigenous communities utilized for local as well as commercial purposes. For instance, when the Portuguese explorer Ferdinand Magellan and his companions arrived in Cebu in 1521, they found the island's inhabitants wearing clothes made from the fibre of the abaca plant, according to one account (FIDA, undated). The European contingent observed that abaca weaving was done everywhere on the island, indicating its importance in the daily lives of the people. Similarly, early explorers in the province of Palawan documented the trade in almaciga resin between indigenous peoples on the island and with seafarers from other places. Indigenous Tagbanua communities traded NTFPs for items such as metal and brass gongs from Chinese and Muslim merchants around the turn of the 20th century (Conelly, 1996).

Over the years, however, indigenous communities have gradually lost control of forest resources due to the increasing migration of lowlanders into upland areas, as well as government policies that tend to favour the moneyed elite in the granting of concessions for commercially important products such as rattan and almaciga resin. These developments have emerged as concerns in several research studies commissioned by the NTFP Task Force (NTFP-TF), a collaborative network of non-governmental organizations and people's organizations working to build the capacity of forest-dependent groups in sustainable livelihoods, conservation and land rights. The NTFP-TF is especially concerned with the resource rights of indigenous communities.

This paper provides a synthesis of NTFP-TF's research studies, with particular emphasis on the impact of new legislation regarding ancestral domains on the utilization of rattan

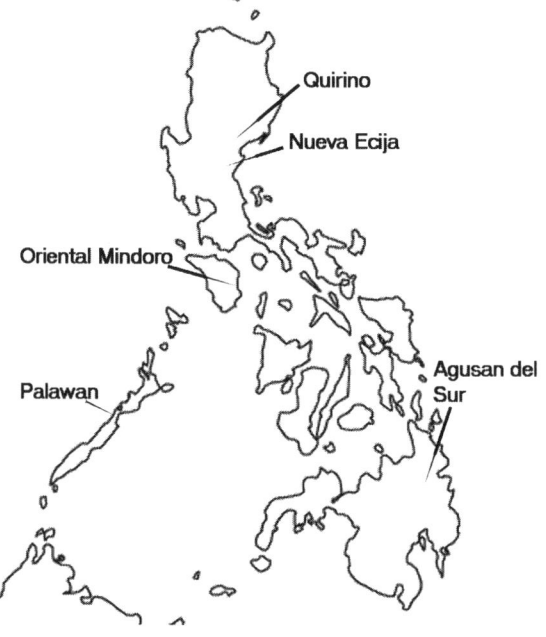

Source: Based on public domain map from www.lib.utexas.edu/maps/cia08/philippines_sm_2008.gif, modifications by Elizabeth Skinner.

Figure 6.2 *Map of the Philippines*

and almaciga resin among several indigenous communities. With the increasing importance of NTFPs in the manufacturing of products geared for the export market, there is hope that indigenous communities could regain control of natural resources in their home territories, and that society at large could better appreciate their role in forest conservation.

NTFPS AND THE ANCESTRAL DOMAIN LAW

Along with timber and other natural resources, NTFPs are considered part of the country's patrimony, and the government therefore exacts forest charges on their collection and trade. Regulations on NTFPs have been revised several times through the years, in response to changes in economic patterns and political trends.

From commodities in the barter trade in previous centuries, NTFPs have steadily grown in economic value, both as raw materials and as components of export products. Prior to the 1960s, most NTFPs were exported in their raw form, but the creation of the National Cottage Industries Development Authority in 1962 spurred the development of small- and medium-scale enterprises utilizing these resources in making furniture and other household products (Neri, 1994). Latest government data show that

Table 6.1 *Major NTFPs in the Philippines and their uses*

Type of NTFP	Commercial uses	Indigenous/local uses
Rattan: 12 out of about 62 species have commercial value; the most commonly used are palasan (*Calamus merrillii*), limuran (*C. ornatus var. philippinensis*), tumalim (*C. mindorensis*) and sika (*C. caesius*)	Furniture: living room sets, beds, tables, hammocks Handicrafts: baskets, hampers, bags, hats, walking sticks, carpet beaters, whips, twines, toothpicks, novelty items	Fruits: food and medicine Vine: fish traps, housing material, roosting area for chickens, rat traps, stakes for swidden farms, traditional calendar to mark the passing of seasons, conflict resolution mechanism for averting tribal wars, indigenous justice system (whoever cuts last strand is deemed guilty of certain offences)
Almaciga resin from *Agathis philippinensis* tree	Ingredient in the manufacture of high-grade glossy varnish, lacquer, paint, soap, plastic and waterproofing materials, ink, linoleum, floor wax, shoe polish	Incense in religious ceremonies, torches, starting fires, caulking boats, smudge for mosquitoes
Wild honey from *Apis dorsata* and *A. cerana* hives	Sweetener, health-food supplement	Vitamins for children, cure for colds, teething aid for babies, food supplement for fighting cocks, side dish for root crops and bananas, sauce for roast meat

Source: Florido and Arcillas, 1996; FDC, 2005; Arquiza, 2006.

almaciga resin exports for 2004 were valued at US$222,000 (DENR, undated) while rattan furniture exports for the same year reached US$98.2 million (BETP, undated).

The government started granting concessions for gathering rattan and almaciga, the two major NTFPs in the country, in the 1940s (PCSD, 1995). Records from the Department of Environment and Natural Resources (DENR), which is in charge of regulating NTFPs, show that most concessions were awarded to traders and politically well-connected individuals in urbanized settlements near forested areas. This has engendered resentment among upland dwellers, principally indigenous communities, and raised complaints that only those who could afford to pay application costs and were capable of going through the tedious process of following up the required documents could get concessions.

Major changes in government policy that had a tremendous impact on indigenous communities only came about after the overthrow of the former dictator, Ferdinand Marcos, in 1986, and the subsequent ratification of a new Philippine Constitution in 1987. For the first time, the country's basic law recognized the right of indigenous peoples to their ancestral lands, as well as the state's responsibility for protecting their culture and institutions within the context of national development (Republic of the Philippines, 1987).

The reforms could not come too soon, as Marcos had granted logging concessions to many of his personal friends and business associates during the years of martial law, resulting in the rapid loss of forests (Vitug, 1993). A comparison of DENR and National Commission on Indigenous Peoples (NCIP) statistics shows that many of the provinces with the thickest forest cover also have high concentrations of indigenous communities, so this loss of forest lands meant a shrinking domain for the indigenous peoples.

Beginning in the late 1980s, there has been a growing trend towards the granting of tenure over forested lands and management of NTFP resources to local communities in response to the new Constitution's thrust to promote social equity. In 1988, the DENR started awarding rattan concession contracts to people's organizations, cooperatives and indigenous communities that supplied processing facilities in areas where rattan was abundant. In addition to privately held concessions, the government also began to grant community-based forest management agreements that allowed local organizations to harvest NTFPs on a limited scale while conserving the resources in ecologically sensitive zones, such as primary forests, at the same time.

For indigenous communities, the DENR also came up with Administrative Order No. 2 in 1993, which granted Certificate of Ancestral Domain Claims (CADCs) in forest land areas classified as government land. The policy, popularly known as DAO 2, became the initial mechanism for some indigenous communities to harvest and trade NTFPs legally for the first time. The certificates and their accompanying Ancestral Domain Management Plans (ADMPs), which spelled out the resource use and conservation strategies of indigenous communities, functioned as permits for collecting and transporting NTFPs to lowland traders.

In 1997, the Philippine Congress passed the Indigenous Peoples Rights Act (IPRA) to fulfil its constitutional mandate. The central feature of the law is the formal recognition of the ancestral domains of indigenous peoples in various parts of the country, and their right to utilize the resources found in these territories according to their customs and traditions. Hailed as one of the most progressive laws for indigenous peoples anywhere, the IPRA strengthened the claim of indigenous communities to their long-held territories. The government created the NCIP to implement the new law. The agency's principal task is the granting of Certificate of Ancestral Domain Title (CADT), which, together with the Ancestral Domain Sustainable Development and Protection Plan (ADSDPP), became the new instrument for indigenous communities to collect and trade NTFPs. These two documents replaced the CADC and ADMP, which had served the same purpose prior to the IPRA.

However, legal challenges from business interests derailed the implementation of the IPRA for many years.

With the logging industry declining due to diminishing timber stocks, a new threat that has emerged in the past decade is the increase in mining claims that overlap with ancestral domain claims. According to the environmental network Alyansa Tigil Mina, the Philippine government is promoting 23 priority mining projects that would encroach on 53 per cent of ancestral domain claims (Alyansa Tigil Mina, 2006).

In 1998, the constitutionality of the IPRA was challenged in the Supreme Court on the grounds that the state, and not indigenous peoples, should have sole ownership and control of mineral wealth. The petitioners argued that whereas the Constitution absolutely prohibited the ownership of natural resources, this was allowed under the IPRA. The government froze all ancestral land claims until the matter could be resolved (AHRC, 2005).

The delay in fully implementing the IPRA resulted in many problems for indigenous communities. Difficulties lay in the conflicting mandate and scope of the two government agencies that shared the responsibility for NTFP utilization from ancestral domain areas. Although the law gives the mandate for ancestral domain recognition to

Table 6.2 *Timeline: NTFP utilization and ancestral domain legislation in the Philippines*

Year	Legal instrument	Coverage/subject	Major requirements
September 1970 (Neri, 1994)	Revised Forestry Licensing Regulations	Guidelines for the issuance of forestry licences, leases or permits for the extraction of NTFPs	Forest resource inventory to determine the amount to be extracted and ensure sustained yield capacity
1987	Philippine Constitution (after ousting of former President Ferdinand Marcos)	Recognized indigenous peoples' rights to ancestral domains for the first time	Congressional Act needed to define the scope of Indigenous community rights
29 March 1988	DENR Administrative Order No. 21: Revised Regulations Governing Rattan Resources	Competitive bidding for ten-year rattan licences per region: 55% for applicants with capitalization of P100,000 or less and 45% for applicants with capitalization of more than P100,000; includes exclusive cutting rights and transport	Supply contract, proof of availability of capital; maximum area is 5000ha for individuals and 30,000ha for groups; only 25 linear metres or longer can be cut, maximum allowable cut to be set by DENR; special rattan deposit for replanting obligations
1988 (Neri, 1994; IRG, 2006)	Negotiated rattan cutting contracts and indigenous rattan cutting concessions granted by the Bureau of Forest Development	10-year permit given to peoples' organizations, individual tribal people and cooperatives that have supply agreements with licensed processing plants, near forest areas where rattan is plentiful	Annual allowable cut, payment of forest charges and local government taxes; licensees have to plant at least ten rattan seedlings for every 100 linear metres harvested
10 January 1989	DENR Administrative Order 89-04: Revised Regulations Governing Rattan Resources	Competitive bidding for 10-year rattan licences per region (same as DENR Administrative Order 21 of 1988 except that capitalization was increased to P250,000)	Annual allowable cut, payment of forest charges and local government taxes; special rattan deposit to be used for planting seedlings to ensure sustainability of rattan
12 January 1989	DENR Administrative Order 04-01: Special provisions for the processing of rattan applications within area reserved/occupied by cultural communities	Negotiated rattan cutting contracts for indigenous communities not exceeding 10% of total production blocks per region; rattan production blocks to be modified if they encroach on indigenous communities' reserves; Indigenous communities' occupants given priority in bidding	Indigenous communities' offer should not be lower than floor price of rattan; Indigenous communities' consent is needed for non-indigenous communities to cut rattan or put up plantations in indigenous community areas; compensation package for Indigenous community if licence is awarded to outsider
1993	DENR Administrative Order 2	Granted ancestral domain certificates that indigenous communities can utilize as resource use permits for NTFPs	Ancestral Domain Management Plan; payment of forest charges to DENR and local government taxes; certificate of origin and permit to transport from DENR

Table 6.2 *Timeline: NTFP utilization and ancestral domain legislation in the Philippines* (Cont'd)

Year	Legal instrument	Coverage/subject	Major requirements
1996 (IRG, 2006)	Community-based forest management agreement DENR Administrative Order 1996-29: Rules and Regulations for the Implementation of Executive Order 263, otherwise known as the Community-Based Forest Management Strategy	50-year agreement between the DENR and a people's organization that gives usufruct rights to forest dwellers; includes rehabilitation and protection of forest land; allows harvesting of certain resources	Community resources management framework, annual work plan and resource use plan affirmed by DENR; payment of forest charges to DENR and taxes to local government; certificate of origin and permit to transport from DENR
1995	PCSD Memorandum Circular 01 Series of 1995	SEP clearance requirements and process for new licences/permits in the province of Palawan	Payment of fees and DENR endorsement, in addition to standard DENR requirements
1997	CADT and CALT through the IPRA	Property rights given to indigenous people who have lived in a given area for long periods and established their cultural affiliation to the land	ADSDPP, annual work plan, and resource use plan affirmed by DENR; payment of forest charges and local government taxes

Source: DENR Administrative Order 21 dated 29 March 1988: Revised Regulations Governing Rattan Resources; DENR Administrative Order 89–04 dated 10 January 1989: Revised Regulations Governing Rattan Resources; DENR Administrative Order 04–01 dated 12 January 1989: Special provisions for the processing of rattan applications within area reserved/occupied by cultural communities; DENR Administrative Order 1996–29: Rules and Regulations for the Implementation of Executive Order 263, Otherwise Known as the Community-Based Forest Management Strategy (CBFMS); Neri, 1994; PCSD Memorandum Circular 01 Series of 1995, Francisco and Segayo, 2000; IRG, 2006.

the NCIP, the task of regulating the collection and trade of NTFPs remains with the DENR. Since the NCIP could not grant titles to indigenous community applicants while the IPRA was being debated in court, some indigenous communities resorted to other legal instruments for NTFP utilization, such as rattan-cutting contracts and almaciga concession licences. Such instruments entailed a heavy financial burden for them, as only a few could afford the expensive forest inventory and other requirements of the DENR (see Table 6.2.) They also had little capacity and capital to manage such arrangements. As a result, some indigenous communities operated at a loss instead of benefiting from the resources found in their ancestral domains (Gatmaytan, 2004).

In 2000, the Supreme Court dismissed the petition and ruled that indigenous peoples maintained community ownership rights to their land. However, the court also gave the state control and regulatory powers over the resources found in these areas (AHRC, 2005). This meant that NTFP gatherers still had to overcome bureaucratic hurdles. Several case studies by the NTFP-TF have shown that different DENR offices have varying interpretations on whether indigenous communities may utilize the CADT and ADSDPP as a resource use permit under the IPRA law or still have to apply for a separate 'ordinary minor forest products' (usually shortened to OM) licence from the DENR (Gatmaytan, 2004; Arquiza, 2006).

In addition to inter-agency conflicts, indigenous communities also have to cope with territorial issues arising from overlapping legislation. Two laws in particular are often cited as major causes of conflicting interpretation in ancestral domain areas: the National Integrated Protected Areas System (NIPAS) Act and the Strategic Environmental Plan (SEP) – a law applicable only to Palawan Island. With most indigenous communities found in forest environments that they have protected for many years, it is no coincidence that many of the protected areas on a preliminary list of the NIPAS Act are also claimed as ancestral domains. Meanwhile, in the province of Palawan, the area-based SEP law was promulgated to conserve what is considered the Philippines' last major ecological frontier. However, Palawan also happens to have the highest forest cover in the areas where many indigenous communities that depend on the harvest of rattan, almaciga and wild honey are found (Arquiza, 2006).

As it stands, the passage of the IPRA may have strengthened the indigenous peoples' hold on ancestral domains, but many problems persist, especially in the area of NTFP utilization. These are shown, for example, in the state of rattan and almaciga resin enterprises. The following sections demonstrate how many indigenous communities receive the least financial benefit despite their back-breaking work and crucial role in harvesting the resources.

THE CASE OF RATTAN

Government statistics show that rattan permits for indigenous communities peaked at 121 contracts, covering nearly 1.4 million hectares of forests in 1998 (DENR, 1998). By 2004, however, the number of active permits had dropped to 29 and the area covered had shrunk to about one-third of the 1998 figure, according to the Indigenous Cultural Affairs Division of the DENR's Special Concerns Office (DENR, 2004).

In 2004, NTFP-TF commissioned two studies on rattan utilization in ancestral domains to find out if landmark achievements in recognizing community ownership of land had also led to economic and ecological benefits for indigenous peoples.

One study was conducted among six indigenous communities in different areas across the Philippines that had rattan operations in their ancestral domains. These were the Tagbanwa-Batak in Kayasan, the Tagabinet of Puerto Princesa City and the Pala'wan of Punta Baja in Rizal municipality, both in the province of Palawan; the Mangyan Alangan of Naujan in Oriental Mindoro; the Bugkalot in Giayan, Quirino; the Dumagat in Gabaldon, Nueva Ecija; and the Manobo in Latay, Agusan del Sur. Although all of them are CADC grantees, only three of the communities utilized their ADMPs as rattan licences. Two had no permits at all and operated on a smaller scale, as they found it less expensive and much easier to bribe government personnel rather than pay the required forest charges. One community, the Mangyan Alangan, obtained a ten-year negotiated rattan-cutting contract in 1991, when DAO 2 had not yet taken effect (Gatmaytan, 2004).

Typically, the rattan gatherers camp out in the forest and collect an average of 20 poles in two days, according to a separate and more detailed study on the Mangyan Alangan community (Aquino, 2004). Two production runs are usually made in a week, and during normal operations, a full-time rattan gatherer earns about 1500 pesos (US$30) a month (Aquino, 2004). In many communities, the local organization serves as the initial buying station at source, where the poles are scraped and cleaned. Local leaders may resell the rattan to financiers or transport the commodity themselves to buyers in the urban centres. Often it is easier for them to entrust the transport of rattan to financiers, mainly migrants with available working capital who have been in the trade for some time, as the requirements are cumbersome and forest charges too steep for the indigenous community to pay upfront. Financiers are generally traders who act as intermediaries between groups of rattan gatherers and manufacturers. They give advance cash payment and/or food supplies to gatherers during operations. Often they get advance payment from manufacturers and/or use their own money to finance rattan operations at the village level – the most critical expense being the forest charges that have to be paid in cash upfront when transporting rattan (Aquino, 2004). According to DENR Administrative Order 63, Series of 2000, forest charges for rattan vary depending on the species and diameter, with the fees ranging from P0.10 per linear metre for the smallest diameter to P5.50 per kg for split rattan (US$0.002–0.12). The rattan traders then sell the raw material to furniture manufacturers and exporters, mostly in the central island of Cebu (BCIC, 2004).

In a detailed analysis of value chain behaviour, the Mangyan Alangan case study showed that the gatherers benefited the least from the rattan trade. Based on a market price of P23.32 (US$0.50) per pole harvested from the forest, the study found out that rattan furniture manufacturers profited the most, cornering more than half its value at P14.17 (US$0.30). Rattan traders earned about P5.22 (US$0.11) per pole, while the government got P3.31 (US$0.07) in resource rent. On the bottom rung, even though they are in front line of the rattan industry, were the gatherers, who earned only P0.62 (US$0.014) for every pole after their labour costs were taken into account (Aquino, 2004).

Clearly there is a huge difference in benefits among the players in the rattan industry. A nationwide investigation of the rattan value chain (IRG, 2006) revealed

Table 6.3 *Institutional aspects of rattan operations*

Community	Beginning of involvement in rattan trade/year licensed operations began	Form of rattan licence held by community	Status of rattan operations (as of November 2003)	Entrepreneurial level of rattan operations	Interface between indigenous and local organization leaders
Giayan	1980s/part of expired rattan licence not controlled by community in 1990s	None	Ongoing, small-scale transactions	Individual/informal groups	n/a
Pinagkampohan	1930s–1940s, possibly earlier/1998	CADC/ADMP (1998)	Ongoing, active	Sitio (sub-village)-level organization (Mangayunan)	Clan or family heads or representatives represented in council, organization
SANAMA	1950s, possibly earlier/1990	Negotiated 10-year rattan-cutting contract (1990)	Suspended pending renewal of licence	Multiple-sitio organization (SANAMA)	Council of elders, *aplaki* integrated into management structure
Kayasan	1950s, possibly earlier/part of negotiated licence area in 1991, used CADC/ADMP when received in 1996	CADC/ADMP (1996)	Suspended pending submission of AWP, but ongoing small-scale transactions	Two-sitio organization (SATRIKA)	*Masikampo* and *orangkaya* (traditional leaders) marginalized
Punta Baja	1970s/part of negotiated licence area in 1989 or 1990, used CADC/ADMP when received in 1997	CADC/ADMP (1997)	Ongoing, inactive	Multiple-sitio organization (PINPAL)	*Mangungukom* integrated into management structure
Latay	1970s/part of expired rattan licence not controlled by the community in 1980s	None	Ongoing, small-scale transactions	Individual/informal groups	n/a

Source: Gatmaytan, 2004.

Table 6.4 *Rattan value chain: Custody and benefits*

Player	Value source	Value per pole (market price) (P)	Cumulative value chain (P)	% of total value of finished product
The government	Cost of rattan (resource rent)	3.31	3.31	4.7
IP gatherers	Direct labour (rattan gathering)	10.77	14.08	15.1
	Other costs	0.70	14.78	1.0
	SOP	0.24	15.02	0.3
	Profit	0.62	15.65	0.9
	Revenue (raw rattan poles)	15.64		
Trader	Cost of rattan	15.64		
	Other costs	6.73	22.38	9.5
	SOP	0.75	23.12	1.1
	Profit	5.22	28.34	7.4
	Revenue (raw rattan poles)	28.34		
Manufacturer/ exporter	Cost of rattan	28.34		
	Direct labour (furniture making)	14.17	42.51	20.0
	Other costs	14.17	56.68	20.0
	Profit	14.17	70.85	20.0
	Revenue (exported furniture)	70.85		100.0

Source: Gatmaytan, 2004.

that while the gatherers earned a 2 per cent return on their investment, traders got 87 per cent. Rattan gatherers provide the fundamental value of rattan as merchandise by covering the difficult stages involved in converting a thorny vine to a ready-to-use rattan pole. These rattan poles arrive in and leave the hands of the traders unchanged. Producing 10,000 poles requires the rattan gatherers to invest P100,000 (US$2200), while the traders would not need even half that amount. In this situation, the benefit to the gatherers and the traders respectively of participating in the value chain is not mainly defined by the transformation (value addition) applied to rattan as a product, or by the amount and profitability of their investments. Rather, it is often influenced by the ability to transport the rattan from the source to the buyers, traversing the range of forest charges and related fees charged by the DENR. This means that whoever has the cash to pay these fees up front gets more benefits. The DENR collects resource rent in the form of forest charges and grease money.[2] When the situation is such that only a few financier-traders and hardly any community groups are capable of paying this amount upfront, it becomes clear income for the DENR, arising from the fact that communities cannot compete with the financiers in making these payments.

The value chain study among the Alangan Mangyan showed that, despite their meagre income and their perception of their work as 'very difficult', rattan gatherers often disregard the value of their labour. Rattan is an important source of cash income that complements traditional swidden farms. While the latter provide a basic food supply, the former yields cash for basic household needs. Rattan also has many household uses, such as medicine, building materials, farm gadgets and garments. Measured in the context of the amount of effort used by indigenous people in various livelihood activities, rattan gathering is seen as a major activity. Hence many rattan gatherers regard the enterprise as their primary source of cash income (Aquino, 2004).

The depletion of rattan as an important NTFP is a growing threat in many ancestral domains due to overharvesting and habitat destruction. Both Giayan and Latay are areas that have been logged in the past, while Palawan's rattan stocks have suffered from unrestricted gathering through the years. Field data from the six indigenous communities taken together underscores the importance of land tenure. One finding is that local concepts and practices regarding land tenure and resource rights directly affect the performance of indigenous communities in sustaining their source of rattan. The variation is rooted in the fact that an entire CADC area is usually subdivided into sections owned by different clans, families or individuals. Those with the clearest notions of collective territory – either at the clan or community level – operate best in terms of area protection. However, in ancestral domains where land and resource ownership is allocated to families or individuals, it is difficult to act collectively to protect the area. This is because the family, clan or individual orientation of some tenure systems limits their ability to act at levels that transcend family, clan or individual interests (Gatmaytan, 2004).

Among the Alangan Mangyan, for instance, where landownership is vested in 59 communities, each with its own rattan *gŭtŭng* or territory, rattan cutters can harvest poles only within the home territory of their community of residence. They cannot harvest poles from other territories without prior permission from the *aplaki* (tribal elder) of the area where they intend to gather rattan. Indigenous values remain strong among these elders, whose protectiveness of their clan's property and interests tends to make them more effective in conserving their rattan resources. The tribal elders also function as a sort of 'environmental police' that enforces compliance with local rules within their territory (Gatmaytan, 2004).

Meanwhile, the communities with a weakly defined sense of collective ownership of territory fared poorly in terms of resource protection. In the Bugkalot and Manobo communities, there was a concept of community boundaries, but family and individual landholdings cut across the area. Their owners protected each landholding zealously, but were in no position to speak for other landowners in the territory on how to deal with outsiders. As a result, individual landowners sold, gave or bartered away their property to migrants, who gained control over much of the land within ancestral domains. They also had weaker mechanisms for enforcing rules, such that compliance with traditional methods for the appropriate harvest of rattan was left to each individual's discretion or conscience. Unlike other indigenous communities, in which customary leaders are still respected, these two study sites had no traditional institution that could ensure compliance with sustainable resource management policies (Gatmaytan, 2004).

Another observation was that there were problems in enforcement in the two communities that attempted to designate zones for NTFP management. In their respective management plans, these communities had decided to establish no-take zones and to replant zones, thus allowing resources to regenerate. The decision was based on their traditional practice; there is no government requirement for NTFP resource managers to establish no-take zones. However, the study (Gatmaytan, 2004) noted that in both cases rattan in the easily accessible areas was depleted, while that in more remote areas was abundant, suggesting that the no-take zoning system was ineffective. (In other sites, however, such as the Apurawan district (Virginio Tica, a Tagbanua elder, pers. comm.) in the Aborlan municipality, Palawan province, and the Hagpa district (Alvin Pantaon, pers. comm.) in the Impasugong municipality, Bukidnon province, no-take zones are common and tend to be effectively implemented, as the traditional control mechanisms in these communities are still strong.) From the indigenous community's perspective, it seems that replicating the practice in swidden farming of allowing a forested area to lie fallow for several years (i.e. rotational no-harvest zones) is the preferred method of sustaining renewable resources such as rattan, as opposed to the conservationists' proposal of establishing permanent reserves (Gatmaytan, 2004).

As for replanting, the lack of technical support and the novelty of the activity resulted in token rattan plantations covering only a minuscule portion of the ancestral domains. When interviewed, some communities pointed out that reducing extractive areas, which would mean adding more no-take zones, would allow the rattan populations to regenerate naturally anyway (Gatmaytan, 2004).

One of the significant findings of the rattan research is that there is no relationship between the form of licence used by a community and its performance as a resource manager. The Alangan Mangyan, the only group that had a rattan-cutting contract, performed their environmental protection duties just as well as the Dumagat and Pala'wan communities that made use of their CADC and ADMP as licences (Gatmaytan, 2004). Hence there is no reason for the government, through the DENR, to require a separate rattan licence from indigenous communities, as often experienced in many rattan-cutting areas.

In the same vein, the number of requirements imposed on a community organization, particularly the reports to be submitted and permits needed from various government agencies, does not affect its effectiveness in managing its resources. The Pala'wan and the Dumagat, with their light reporting requirements, did just as well in managing their resources as the Alangan Mangyans, who were burdened with heavy administrative demands. This means that the DENR can safely reduce the administrative requirements imposed on community organizations without endangering the forest or other resources under the latter's control.

For instance, the government requires NTFP harvesters to prepare a resource inventory as a tool for monitoring the sustainability of NTFP resources. However, it does not distinguish between resources that are harvested in a non-destructive manner, such as forest honey from indigenous bees and rattan from clustering palms, and those that are completely extinguished upon harvesting, such as single-stemmed rattan. With sustainable harvest methods, NTFPs that can be harvested non-destructively still have the ability to regenerate. Thus, the more important issue for such NTFPs is to determine the harvesting practices that cause the least damage to the resource, i.e. how to harvest, when to harvest, how much to harvest each time, etc., and then ensuring that these practices are used and that the NTFP resources remain healthy and productive. Another issue with the inventory requirement is the government's concept of an annual allowable cut (AAC), which is more useful to timber species that are felled and does not really ensure adequate regeneration of NTFPs.

The DENR also needs to enforce the uniform implementation of laws and regulations, particularly for securing transport permits for rattan pole deliveries. The Dumagat, for example, enjoy access to a 'one-stop shop' system for applicants that is largely localized in the district office of the DENR. Here the entire procedure takes a maximum of three days. In contrast, the Tagbanwa-Batak have to visit three separate DENR offices to request an inspection of the rattan stocks in their ancestral domain, arrange a schedule for the inspection, get the report, pay the assessed forest charges, get the transport form filled in and signed, and have the form notarized. The entire procedure takes from four days to a week, especially if key DENR personnel are unavailable (Gatmaytan, 2004). Such bureaucratic obstacles translate into heavy operational costs that reduce time, resources and the already marginal profits, which could rather have been used for community development.

ALMACIGA RESIN IN PALAWAN

Similar legal and institutional trends were discovered in a recent NTFP-TF study on the harvesting and trading of almaciga resin among indigenous communities in Palawan (Arquiza, 2006).

The almaciga tree (*Agathis philippinensis*) is found on other islands as well, but it is believed that much of the almaciga resin traded in the Philippines comes from Palawan province, where the drier climate and mountainous terrain are conducive to the growth of the valuable species (Callo, 1995). Known locally as *bagtik*, almaciga resin is the main source of cash income for several Batak, Tagbanua and Pala'wan indigenous communities in the central and southern parts of mainland Palawan (Arquiza, 2006).

Almaciga resin is known in the manufacturing industry as Manila copal. It is an ingredient in a wide range of commercial products such as high-grade glossy varnish, lacquer, paint, soap, plastic and waterproofing materials, ink, linoleum, floor wax, and shoe polish (Neri, 1994). Annual production figures from 1976 to 2004 ranged from a low of 191,000kg to a high of 1.4 million kg. Much of this was exported, with the highest receipts recorded at US$515,000 in 1979 (DENR, undated).

In 1995 the staff of the Palawan Council for Sustainable Development (PCSD), which implements the SEP law mentioned earlier, conducted a study on the almaciga trade and found that it commonly involved four major actors: the gatherer, the *kapatas* or field supervisor, the licence holder and the buyer. The main findings of the study, with additional data from a more recent NTFP study, are as follows:

• Most of the gatherers are indigenous people who perform the difficult task of harvesting the resin by tapping the almaciga tree in remote forested mountains. Typically, they load the resin on traditional rattan backpacks called *rarong* and haul them on foot until they reach a clearing where they can be loaded onto a buffalo-drawn cart or wooden raft. According to many gatherers, the harvesting of almaciga resin reaches its peak during the dry season. Tapping is usually done every two weeks, and each gatherer can carry up to 50kg of resin at a time.

- An average of 30 gatherers work under one supervisor, who often doubles as a gatherer as well. In ancestral domain areas, he is usually a local leader, but in concession areas, the supervisor may be an outsider. He shoulders the main responsibility of protecting the trees from destructive harvesting practices, and guards the almaciga concession against intruders who attempt to gather resin from trees that have been left to recover after a season of tapping. He is entrusted with cash to buy food and household goods that are handed to the gatherers as loans (which are later deducted from money owing to the gatherer). When the product is brought down, he weighs the resin, determines the price and pays the gatherer immediately if there is cash on hand. When the agreed volume has been harvested, he delivers the resin to the concession owner's warehouse or directly to the buyer.

- The supervisor reports directly to the licence holder, who may be a concession owner or a local people's organization with a CADC/ADMP or community-based forest management agreement. The licensee oversees the operations, shoulders all the costs for getting a permit from the DENR and often pays transport costs and forest charges. CADC holders say expenses normally include 'grease money' for government personnel manning checkpoints on the road, which affects the pricing of the commodity (Arquiza, 2006). The resin is brought to buying stations in the capital city of Puerto Princesa and other urban centres in Palawan, and the final price of the product is arrived at through negotiations between the buyers and licence holders.

A comparison of the 1995 PCSD study and the 2006 NTFP-TF study shows that the prices of almaciga resin have hardly changed. Gatherers received an average of P6 (US$0.13), while licence holders obtained about P12 (US$0.26) per kg of resin. In the PCSD study, concession owners were found to have gained the most from the industry, with an average net income of P412,820 (US$9000) per year. In contrast, gatherers earned a maximum of P6600 (US$145), while each supervisor received up to P36,000 (US$790) per year. Thus it appears that indigenous communities have not obtained substantial benefits from the enterprise despite their back-breaking work, and lowland traders have reaped much of the financial gains from the lucrative industry.

However, the situation has improved for at least one CADC holder that opted to sell directly to buyers. Almaciga resin tappers from the Pala'wan community in Rizal town who were interviewed in 2004 earned an estimated P390,000 (US$8500) a year as a result of their decision to transport and sell the product themselves. Bernardo Barahim, secretary of the Pinagtibukan et Palawan, which is a CADC holder, said the group had saved enough money for livelihood projects and monthly meetings. They also used some of the money to renovate their tribal centre-cum-storage area, which used to be an open space with a thatched roof. The group has since installed galvanized iron roofing to protect their NTFPs (such as rattan and almaciga) from the elements, concrete floors, and two rooms where members can stay while taking turns to guard the facility (Arquiza, 2006).

Table 6.5 *Almaciga resin prices*

Reference/IP community	Year	Amount paid to gatherer (on site)	Amound paid to kapatas (on site)	Amount paid by concession owner/ buyer	Amount paid by Manila buyer
PCSDS study	1995	P5.00–6.00	P6.00–8.00	P13.00–15.00	
Punta Baja, Rizal	2004	P7.00		P13.00	
Amas, Brooke's Point	2006	P8.00 – Brooke's Point buyers P12.00 – Muslim traders (direct selling by gatherers)			
Boong, Dumangueña, Narra	2006	P6.50–9.00	P7.00–9.50	n/a	
Daan, Apurawan, Aborlan	2006	P4.00–5.00		P8.00–10.00 (direct selling by gatherers)	
Marufinas, Puerto Princesa	2006	P5.00	n/a	n/a	
Ricky Magali, Las Insular Trading (varnish manu-facturer)	2006				Class A: P25/kg Class B: P23/kg
Anita Santos, Manaog Trading (also has concession in Samar)	2006			Palawan: Type A: P18.00–20.00 Type B: P15.00 Type C: P12.00 Samar: Type A: P13.00 Type B: P10.00	P23.00–24.00 P18.00–30.00 P15.00 P15.00 P15.00

Source: Arquiza, 2007.

Records from the Provincial Environment and Natural Resources Office (or PENRO, the local DENR branch) show that 56 licences were issued in 1994 for almaciga concessions in Palawan, covering a total of 368,695ha or more than one-fifth of the province's land area. Many of the licensees were heirs of the original concession owners, and they were granted an annual allowable quota of 1,588,634kg (PCSD, 1995).

However, the almaciga trade declined in Palawan after the PCSD issued a resolution granting preferential rights to indigenous communities in harvesting NTFPs. This was the response of the PCSD, the highest policy-making body on environment and development issues in the province, to the constitutional mandate to recognize indigenous peoples' rights. While indigenous communities welcomed the resolution, many lowland traders complained because they could not renew their concessions easily. The new directive required almaciga licence applicants who were not indigenous people to obtain free and prior informed consent from indigenous communities before they could get the permit, which is renewed annually.

As a result of the PCSD resolution, government records show, the number of licence holders had decreased by half in February 2006. Among them were six indigenous communities. One of them, the Nag-uyo-uyonon Tagbanua kat Mariwara in the district of Princess Urduja in the municipality of Narra, obtained a licence using the prescribed DENR procedures, and the other five used their ancestral domain and ancestral land certificates (PENRO-Palawan, 2006). The area covered by the concessions was reduced to 83,042ha, or about one-quarter of the 1994 figure. However, the decrease in the annual quota did not correspond to the decrease in area: it was set at 749,700kg, or just half the quota for 1994. These figures are significant, as it appears that almaciga licensees have been allowed to harvest a greater amount of resin from a smaller area, and this has long-term implications for the sustainable harvest of this commercially important commodity (Arquiza, 2006).

The NTFP-TF study on the almaciga trade in Palawan examined the situation of five indigenous communities that are highly dependent on the commodity for cash income. The study found that the communities with strong organizations and customary leadership practices made more headway in using their ancestral domain certificates to harvest and trade almaciga resin. However, only one group ventured into trading, while another opted for an informal arrangement that allowed an outside buyer to use the group's CADC/ADMP to transport the product outside the ancestral domain. At the time of the study, the three other groups were still going through the process of getting almaciga licences using their ancestral domain certificates, even as their members continued to harvest the resin and sell it surreptitiously, in small quantities, to buyers in the city. Their main stumbling block is lack of local capacity to meet government requirements such as local government endorsements and inventories, a problem that indigenous peoples' advocates and sympathetic DENR personnel have attempted to address through technical support (Arquiza, 2006).

In terms of conservation measures, customary law and indigenous knowledge have played an important role in sustaining almaciga resources. For instance, the Tagbanua communities in the municipalities of Aborlan and Narra either designated certain areas per tapper or shared the tasks of tapping and harvesting equitably. Bageral Ninge, a Tagbanua leader in Aborlan, where the sharing system is practised, explained that 90 per cent of the residents in his village were related to each other so they did not mind if someone else harvested from a tree that they had tapped. '*Marami naman kasing puno, at apo ko rin naman ang pakakainin sa kinuha ng iba kaya bakit ako magagalit?*' [There are many trees, and the tappers will be feeding my grandchildren anyway, so why will I get angry if someone else harvests the resin I have tapped?] he said (Arquiza, 2006).

Overall, communities with strong bonds and a high regard for traditional leaders were able to enforce boundaries and appropriate tapping practices better than those with fractured relations and more acculturated members. In contrast to the previously cited indigenous communities in Aborlan and Narra, where harmonious relations prevailed, the Pala'wan and Tagbanua local organizations near the more urbanized areas of Brooke's Point municipality and Puerto Princesa had serious problems in getting the support of their members for their almaciga licence applications. They also had more individualistic styles of harvesting almaciga resin and made hardly any collective effort to ensure that their sources of resin were not depleted (Arquiza, 2006).

For most communities, resource conservation does not seem a major concern, as the almaciga tree grows only in high-altitude forests where few lowlanders dare to venture. Tapping and hauling down the product are difficult tasks that only indigenous peoples are patient and persevering enough to undertake, as it has been their traditional livelihood for many decades. With their smaller populations, most indigenous communities are confident that, in terms of harvesting volume alone, their resources will not be depleted as long as they have security of tenure and their ancestral domain certificates (Arquiza, 2006).

However, government foresters have issued warnings about improper tapping practices that have led to the death of many trees in some concessions in Palawan. Almaciga resin exudes from the tree bark, the outermost part of the tree, where the cuts are made to extract the material. Between the bark and the wood is a thin intermediate layer called the cambium, which is responsible for the restoration of tissues, increase in wood diameter and formation of new bark. Government regulations forbid cuts that reach the cambium, as this is detrimental to the growth of the tree. Incisions that reach the wood are even more damaging, as they expose the tree to pests and diseases, ultimately resulting in its death (Callo, 1995).

The NTFP-TF study revealed that some indigenous communities do not conform to prescribed tapping regulations. For instance, one group's management plan allows incisions to reach the wood, which is not allowed by the DENR. However, indigenous tappers justify the practice as part of their traditional beliefs, and assert that none of their trees have died as a result of the practice. Various groups also had varying policies on the minimum tree size and size and type of cuts that can be done. (Arquiza, 2006).

However, in a workshop on best NTFP harvesting practices held in Palawan in 2006, indigenous gatherers from four provinces (Palawan, Mindoro, Negros Occidental and Bukidnon) agreed that the cuts should not be too deep. They also said it was important to determine the age and size of the tree before the first harvest to ensure the maturity of the tree, to limit the number of cuts per tree and the height of the first cut from the ground, the size of the cuts, as well as the frequency and positioning of the cuts. The participants felt that the whole package of best practices is crucial to the survival of the almaciga tree. Although their inputs are similar to the government regulations, most of them said they were unaware of the DENR's policies (NTFP-EP and NATRIPAL, 2006).

In areas where indigenous gatherers work for concession owners who are not members of indigenous communities, multiple cuts are made at depths harmful to the tree in order to reach the annual allowable cut specified in their permit. If the indigenous gatherers refuse or fail to meet the quota, non-indigenous gatherers are brought in, often resulting in harm for almaciga trees: non-indigenous gatherers have been blamed for many destructive tapping practices (Callo, 1995; Arquiza, 2006).

In the 1995 PCSD study, it was estimated that about 2000 tons of almaciga resin were collected every week from a 10,000ha area during the peak season. At this rate, damage to the tree may be inevitable. Through the years, DENR personnel have admitted that the lack of personnel has prevented the agency from effectively monitoring compliance with tapping guidelines (Arquiza, 2006).

Considering the unique characteristics of the source of almaciga resin, conservation measures would need to go hand in hand with the pursuit of the indigenous peoples' right to the utilization of resources inside their ancestral domains. Community leaders interviewed have said it is difficult to plant almaciga trees as their seedlings are hard to find. Also, trees only grow at higher elevations, and it takes many years for them to mature and become suitable for tapping (Arquiza, 2006). Therefore greater care is needed to ensure that appropriate strategies are incorporated in the management plans of Palawan's indigenous communities to ensure sufficient regeneration of almaciga populations.

THE SEEDS OF EMPOWERMENT

At the signing of the IPRA law on 29 October 1997, then-president of the Philippines Fidel V. Ramos summed up its significance with these words: 'Only a law of such breadth, depth and scope as R.A. 8371 can provide our indigenous peoples with the seeds of their empowerment and social equity' (CIPRAD, 2004).

It may be too early, a mere ten years on, to expect the seeds to have borne fruit, but nevertheless much more could have been done if the government's policy environment had been overhauled to keep pace with the progressive features of the IPRA. Clearly there is a pressing need for reforms in the Philippine government's policies towards natural resources in ancestral domains so that indigenous peoples can fully realize the benefits of the IPRA.

As of January 2007, the government had awarded 56 CADTs throughout the country covering some 1.1 million hectares (NCIP, 2005). Out of this figure however, no community has made use of its ADSDPP as an NTFP licence. Even with ADMPs, only the Tagbanua and Pala'wan in Palawan and the Dumagat are known to have used it as a licence, according to NTFP-TF partner communities.

Studies done by the NTFP-TF show that indigenous communities face many challenges in utilizing the resources, as summarized below.

Bureaucratic complexity and inconsistency in implementing laws

Government personnel are more likely to enforce laws in relation to commercially important products such as rattan and almaciga resin. Several indigenous communities have reported that DENR regulators rarely check for other NTFPs such as wild honey, bamboo, fibres and vines used for handicrafts. However, this seems to be changing, as the DENR is reported to be getting strict with large shipments of honey to Manila, for instance. With the IPRA in place, the communities' major complaint is the refusal of many DENR local offices to recognize the ADSDPP as a permit, which would spare them the bureaucratic maze of applying for a licence. For one thing, the process of applying for or renewing a licence for ordinary minor forest products (OM) is so lengthy that it can take up to a year, as a Tagbanua community in Palawan found out on renewing its almaciga licence. The Alangan Mangyan of Mindoro had the same experience: it took them more than four years to renew their rattan concession. With its area-specific SEP law, all applicants in Palawan have to comply with an added layer of bureaucracy in the form of a licence from the PCSD before the DENR approves its permit. Obviously there is a need to streamline the institutional set-up for NTFP development towards a more efficient delivery of services.

Lack of clarity on policies for indigenous communities

At the national level, the DENR has several unclear policies. Some policies relax licensing procedures for NTFP utilization and recognize one consolidated resource use plan as a permit, whereas others have strict and individual procedures for each NTFP. Indigenous communities are therefore often confused regarding which procedures to follow. Several requirements for an OM licence are either inappropriate or not required of indigenous communities, such as the necessary business capital, performance bond, income tax returns and financial statements for two years prior to the date of application.

Reduction in bribery

At all levels of the NTFP enterprise, unofficial payments to government personnel are one of the most commonly mentioned problems, and they are more acutely felt among cash-strapped indigenous communities. Known euphemistically as 'SOPs' (for 'standard operating procedures'), these additional expenses siphon benefits away from indigenous peoples, further reducing their meagre cash income. Some of these bribes are paid during the processing of government requirements, but most are collected at checkpoints along the route from the upland source of NTFPs in rural areas to the buying centres in urban areas. Table 6.6 illustrates how much traders and indigenous communities in three provinces usually pay to checkpoints.

Expensive requirements

Indigenous communities have to grapple with steep transaction costs as a result of the lengthy process, forest inventory and annual permit requirements, and forest charges. Ironically, many NTFP gatherers resort to higher-volume extraction in order to cover the administrative costs, despite the fact that government regulations are intended to support, rather than discourage, sustainable harvesting. One requirement in particular – resource inventory – constitutes a major expense that reduces the financial resources available for resource management activities such as reforestation and regeneration. One of the partner organizations of NTFP-TF, the Broad Initiatives for Negros Development in Negros Occidental province, spent over P160,000 (US$3500) on a 5 per cent inventory of a 240ha community-based forest management area (BIND, 2004). The cost would be much higher for larger ancestral domains. In several workshops

Table 6.6 *Unofficial payments in the NTFP Trade (for checkpoints only)*

NTFP site	Amount paid
San Mariano, Isabela province (rattan, trader)	P18,000
Nagtipunan, Quirino province (rattan, trader)	P60,000 (from Quirino to Manila Nueva Ecija)
Nagtipunan, Quirino province (rattan, trader)	P20,000 (from Quirino to Angeles City)
Rizal, Palawan (almaciga resin, IP community)	P5,000 (up to Puerto Princesa City)
Narra, Palawan (almaciga resin, IP community)	P1,000 (up to Puerto Princesa City)

Source: Key informant interviews in Arquiza, 2007.

organized by NTFP-TF, indigenous communities have repeatedly pointed out that the DENR's inventory system is impractical, too costly, too difficult and too rigorous.

Mining

The national government's vigorous campaign to attract mining investors has resulted in conflicting claims to forest areas, especially in ancestral domains (Alyansa Tigil Mina, 2006). In a recent video documentary, indigenous leaders in Palawan expressed concern about the negative impact of the resurgence in mining activities on their efforts towards sustainable forest resource use and management (NTFP-TF, 2006).

Unsustainable harvesting practices

Across the country, several groups have mentioned the problem of overharvesting in rattan and almaciga areas due to the encroachment of outsiders, mostly migrants from lowland areas. For rattan in particular, one of the common trends is that the harvesters have to go much deeper into the forest to collect the commodity than they had to in the past (Arquiza, 2006, 2007).

In some ancestral domains, the breakdown of traditional leadership has resulted in the unsystematic gathering of NTFPs by members of indigenous communities. Although the problem may be seen as difficulty in self-regulation, constraints in the trade regime that are beyond the control of indigenous communities may also be a factor in the overharvesting that occurs in these areas, especially if they do not have a licence.

Social equity

Lastly, as the research studies have shown, there is a need to attain social equity for indigenous peoples so that they can truly benefit from the IPRA. In general, the middle level and end of the value chain are receiving much more than indigenous communities at the starting point of the NTFP industry. Apart from some organized groups such as the Pala'wan in Rizal, Palawan, who ventured into the trading business, most indigenous communities remain dependent on migrant entrepreneurs for the sale of NTFPs. Mechanisms have yet to be put in place so that indigenous communities can share information among themselves on how to get better prices, or how to forge agreements between the harvesters and traders to fix a more equitable profit arrangement with NTFPs.

RECOMMENDATIONS

Support groups in the Philippines have been working for many years to address these challenges and fully realize the constitutional mandate to promote indigenous peoples' rights. Efforts are under way to come up with a unified and long-term policy that would harmonize the functions of various government agencies such as the NCIP,

Box 6.1 Draft administrative order for NTFP utilization in ancestral domains

In 2005, the NTFP-TF, together with consultant Augusto Gatmaytan, prepared a draft joint administrative order between the NCIP and the DENR in response to the need to harmonize utilization requirements in these government agencies. The PCSD was provided with copies and asked to comment on the draft, but no comments have been received so far.

The draft joint administrative order is entitled 'Rules and regulations for the extraction and marketing of non-timber forest products in ancestral domain areas'. Its objectives are:

- to simplify the administrative requirements for the extraction and marketing of NTFPs in ancestral domain areas;
- to provide adequate safeguards against the abuse or non-sustainable utilization of NTFPs in ancestral domain areas; and
- to clarify the respective roles and responsibilities of the DENR and the NCIP in relation to the extraction and marketing of NTFPs in ancestral domain areas.

The principal features of the proposal are the following:

- The ADMP, ADSDPP or five-year work plan shall be recognized as a resource use permit in order to 'enhance the indigenous peoples' or communities' access to, extraction and beneficial use of non-timber forest products, without impairing the state's interest in monitoring such resource utilization for purposes of regulation and ensuring long-term sustainability'.
- An ADMP or ADSDPP may qualify as a permit if it contains the following elements:
  a a duly prepared resource inventory or an equivalent participatory resource assessment for each NTFP that the holder of the CADC or CADT intends to utilize and/or market;
  b a duly determined AAC or an equivalent utilization limit based on a participatory resource assessment for each NTFP;
  c a statement onf who can utilize and/or market each of the NTFPs covered by the application;
  d indigenous or other rules on the appropriate extraction and marketing of the NTFPs in question, and a commitment to implement these rules and penalize violations thereof; and
  e a commitment to ensuring that the extraction of NTFPs does not adversely affect existing timber or non-timber resources, and to maintaining or ensuring adequate stocks of the NTFPs.
- The DENR shall explore the possibility of recognizing the traditional sustainable harvesting and management practices of certain commercial NTFPs and of establishing participatory monitoring mechanisms to ensure the regeneration of the NTFPs in the ancestral domain. Any such alternative mechanisms found mutually acceptable by the DENR and the indigenous community concerned may be substituted for the requirement under (a) and (b) above.

• The DENR shall explore the possibility of recognizing the traditional sustainable harvesting and management practices of certain commercial NTFPs and of establishing participatory monitoring mechanisms to ensure the regeneration of the NTFPs in the ancestral domain. Any such alternative mechanisms found mutually acceptable by the DENR and the indigenous community concerned may be substituted for the requirement under (a) and (b) above.

On 16 November 2006, the DENR invited the NTFP-TF to a meeting of the technical working group of the DENR and the NCIP regarding the harmonization of the implementation of indigenous peoples' rights and environmental policies. The NTFP-TF was asked to present this draft policy proposal. Both government offices recognized its importance, and they immediately established a working committee to come up with policy recommendations on the matter. The NTFP-TF was invited to join the policy discussions, and initial meetings have been positive. The bureaus concerned have submitted their comments on the draft and the NCIP has expressed eagerness to sign the joint administrative order. The DENR is working to integrate the policy proposal of the NTFP-TF with its own initiatives on this issue.

DENR and PCSD. Box 6.1 shows the salient points of a draft administrative order that contains proposed rules and regulations for NTFP utilization in ancestral domains.

The proposal and other recommendations from NTFP-TF studies seek to address the following action points.

Streamline government processes and procedures for applicants from indigenous communities

The main thrust of the draft administrative order is to formalize the recognition of ADMPs and ADSDPPs as multi-year permits in order to greatly ease the reporting and renewal requirements for affected communities. Aside from reducing red tape, this would also leave fewer opportunities for unofficial fees to be collected from vulnerable indigenous groups, especially at government checkpoints. On this point, one recommended action is to have a range of sectors represented at checkpoints in order to lessen the temptation for bribery. Community leaders and civic groups could be represented at the checkpoints, along with government personnel.

Document and respect the strengths of customary law on forest land tenure and NTFP resource management

One of the lessons of the rattan research is that tenure affects the implementation of utilization zones and forest reserves. For instance, these are not very viable when the territory has been subdivided among different families, clans or communities. If they were more thoroughly documented, the information could be shared among concerned agencies and thus enable policy-makers to make better and more informed decisions on NTFP development.

Local land and resource tenure concepts that work best for indigenous communities are already integrated into the formulation of ADMPs and ADSDPPs, so there is no need to override these with government requirements and statutory law. Instead, government agencies could incorporate indigenous resource management in crafting regulations for NTFP utilization within ancestral domains.

Institute more cost-effective and appropriate monitoring mechanisms

While the specific concern regarding conservation of NTFP resources is valid, the DENR's requirement of an inventory is not the only means of ensuring sustainability. Box 6.2 shows one example in the province of Bukidnon, where the indigenous Higaonon community uses traditional practices to ensure a bountiful supply of wild honey.

Box 6.2   Sustainable harvest of wild honey in Bukidnon: The role of customary law and resource management

During an NTFP Exchange Programme workshop in May 2006, the Higaonon participants from the village of Hagpa in the municipality of Impasugong, Bukidnon province, shared their management practices and regulations governing the sustainable harvest of wild honey, particularly *Apis dorsata*. They explained the environmental conditions that were important in keeping wild bees in an area and producing wild honey there.

First, explained the Higaonon, wild bees are attracted to certain trees with angular branches that lend themselves to hive-making. These trees included balete (*Ficus balete Merr.*), lauan (*Shorea contorta*), almaciga (*Agathis philippinensis*), hangilo (*Michelia platyphylla Merr.*), tongog (*Rapanea apoensis Elm.*) and malugo or maluko (*Pisonia alba Span.*). The community takes particular care that these trees are not harmed or chopped down.

In addition, they ensure the abundance of flowering trees where the bees get their nectar. In Hagpa, these are the salin-ubod or salingobad (*Saurauia macgregorii*), bita-ug (*Calophyllum inophyllum L.* or *Calophyllum megistanthum*), anie or anii (*Erythrina fusca Lour.*) and other trees known only by their local names: gitaan, kaitum-itum, luban, kabiti-biti and kadugi. These trees have a fragrance that attracts bees.

It is also important that no swidden farming is done in the areas where 'honey trees' and flowering trees are found. Preserving the vines and leaves in the forest canopy is also necessary, as theyse provide the proper shade and light intensity for the bees and the hive. They shield the bees and the hive from the rays of the sun so that the hive does not melt away. The vines also protect the hive from strong winds. Finally, the Higaonon are also aware that the bees need ample sources of water, so it is important to keep streams clean and flowing freely for the bees, as well as other forms of life in the forest.

The Higaonon believe there is a 'goddess of the bees' named Palayag. She is the protector of the bees and she must be respected. Rituals are performed before and after harvesting wild honey to ask Palayag, or thank her, for an abundant harvest. In the bee areas of the forest, loud noise and laughter have to be avoided, because Palayag should not be disturbed.

Anyone failing to follow the rules of the community must perform a ritual asking for forgiveness, because if the goddess gets angry, there is risk that all the bees and honey will be lost and the whole community will not have any harvest at all. Thus there is an incentive for the entire community to put pressure on the erring individual to seek pardon or atonement. Punishment is necessary, as otherwise, so the Higaonon believe, the bees may not return to the forest.

Through the local monitoring of environmental conditions and the strong community regulatory system for the protection of this area and these species, the Higaonon are able to conserve the honey bee and its habitat. This ecological monitoring is vastly different from DENR's requirement that the abundance of hives and volume of honey alone be assessed to measure sustainability, which does not take into account the cycle of the flowering trees and other environmental conditions. These processes, as related by the Higaonon, could help indicate where the ecological balance should be restored to preserve the habitat of the wild bees and improve the ecological health of the area.

Sources: Salvosa, 1963; documents on the 'Workshop on Guidelines for the sustainable management of three important NTFPs in the Philippines: rattan, almaciga resin and honey'; personal communication with Ino Pantaon, Higaonon community member of Sitio Pulahon, Bgy Hagpa, Impasugong, Bukidnon, May 2006.

Forge meaningful partnerships with indigenous communities

In the area of social equity, support groups have a large role to play in helping indigenous communities across the country enhance their capability to take active roles in marketing their products. As various studies have noted, NTFP gatherers, who are mostly members of indigenous communities, often get short-changed, and their conditions need a lot of improvement. One positive trend cited in the rattan research is that manufacturers are actually willing to pay up to 12 per cent more than the current price of the raw material, which means the gatherers can demand better prices (Aquino, 2004).

Reduce forest charges or provide resource rent to indigenous communities

In keeping with their economic status, indigenous communities could be charged less for harvesting NTFPs so that they can give rattan gatherers better wages and earn more profits as operators in the rattan-cutting industry. A more radical step would be to pay resource rent to indigenous communities as administrators of their ancestral domains. This is discussed in the next recommendation.

Transfer the administration of forest charges to the community

As stewards and owners of ancestral domains, indigenous communities could be entrusted with responsibility for managing the fees imposed for the extraction of resources in their territories. One model for this is found in Brazil, where officials in the north-western state of Acre provide incentives to NTFP gatherers in order to stimulate the industry. This paradigm shift would require a huge leap in perspective for Philippine government officials, but is not entirely impossible, as experience in Latin America shows.

For now, the Philippine government appears willing to take smaller steps in providing better economic benefits to indigenous peoples for the resources they have long depended on for survival. This is demonstrated in the openness with which key officials involved in natural resource management are pressing on with the proposed reforms in NTFP development.

With the shifting economic trends, it is indeed high time for the Philippine government to examine how it can work better with indigenous communities in the area of NTFP utilization. The total ban on log exports, which started in 1987 and was followed by the restriction of lumber exports in 1989, has led to a decline in the importance of the forestry sector to the country's economy. In 1973, when wood products were a major export commodity, the forestry sector's share of the gross national product was 3.93 per cent. By 1990, this had dropped to 1.1 per cent (Neri, 1994) In contrast, the value of NTFPs in the export market remains high and the demand for raw materials has outstripped supply.

According to Marlene Gatpatan-Bedia, head of the membership and cluster development unit of the Cebu Furniture Industries Foundation, many of the foundation's members have started importing rattan from Indonesia due to restrictions on local supply. It is worth noting that two of the top rattan users in Cebu, Pacific Traders & Manufacturing Corporation and Mehitabel Furniture Inc., reported annual exports of US$20 million and US$14 million in 2005, respectively (CFIF, 2005). These figures indicate the huge potential for NTFP utilization, and represent a giant leap from the days when rattan was bartered for gongs.

As of 2002, the Philippines' forest cover had diminished to 18 per cent due to logging and the conversion of forests to farms and commercial estates (PAWB, 1998). If the country's indigenous communities gain more control over their ancestral domains and obtain more favourable policies on NTFP development, there is hope that more forests will be protected in the years to come.

NOTES

1   Maria Cristina S. Guerrero is the deputy director of the NTFP Task Force. The three co-authors are independent researchers commissioned by the NTFP Task Force to undertake studies on various NTFPs in the Philippines.

2   Grease money: Bribes seen from the angle of the briber and alluding to the drop of oil given to a squeaky wheel of excessive bureaucracy to make the things move smoothly again (U4, undated).

REFERENCES

Alyansa Tigil Mina (2006) *Organizational History*, www.alyansatigilmina.net/about-us, accessed 2 October 2008

Aquino, A. (2004) 'Report on Labor Cost Analysis of Rattan Gathering in a CADC area: The Case of SANAMA of Paitan, Naujan, Oriental Mindoro', unpublished paper, Non-Timber Forest Products Task Force, Manila

Arquiza, Y. (2006) 'Bagtik kat Rarong: The Almaciga Trade and Palawan's Indigenous Peoples', unpublished paper, Non-Timber Forest Products Task Force, Manila

Arquiza, Y. (2007) 'Food, Shelter, Clothing and Cash: The Economic Importance of NTFPs in the Philippines', unpublished paper, Non-Timber Forest Products Task Force, Manila

Ateneo Human Rights Center (AHRC) (2005) *A Case Study on Indigenous Peoples' Claim to Parts of Reservations*, Manila

BCIC (2004) *State of the Sector Report – Philippine Furniture*, Pearl 2 Project Technical Paper No. 2, British Columbia Innovation Council, Vancouver

BETP (undated) *Tradeline Philippines*, Bureau of Export Trade Promotion, www.tradelinephil. dti.gov.ph/betp/trade_stat.main, accessed 21 August 2008

BIND (2004) *Financial Report to IUCN-Netherlands for Project Entitled: Community Property Rights and Forest Conservation Through NTFP Development 6AS00110A*, Bacolod

Callo, R. (1995) *Damage to Almaciga Resources in Puerto Princesa and Roxas, Palawan Concessions*, Ecosystems Research and Development Bureau, Department of Environment and Natural Resources, Manila

Cebu Furniture Industries Foundation (2005) Company Profile of Cebu Manufacturers (CFIF members), Schackant, S. (ed)

CIPRAD (2004) *Guide to RA 8371*, Coalition for Indigenous Peoples' Rights and Ancestral Domains, International Labour Organization and BILANCE Asia Department, Manila

Conelly, W. T. (1996) 'Strategies of indigenous resource use among the Tagbanua', in J. Eder and J. Fernandez (eds) *Palawan at the Crossroads: Development and the Environment on a Philippine Frontier*, Ateneo de Manila University Press, Manila

DENR (1998) *List of Negotiated Rattan Contracts Issued from 1989–1998*, Department of Environment and Natural Resources, Manila

DENR (2004) *List of Negotiated Rattan Cutting Contracts to Indigenous Communities*, Indigenous Cultural Affairs Division, Department of Environment and Natural Resources, Manila

DENR (undated) 'Environmental and Natural Resources Accounting Project', CD-ROM

FIDA (undated) *Market Report*, Fiber Industry Development Authority, www.fida.da.gov.ph/ About%20Fiber%20Industry1.html, accessed 21 August 2008

Florido, H. B. and Arcillas, R. (1996) *Resin-Producing Tree Species: Research Information Series on Ecosystems 8*, Ecosystems Research and Development Bureau, College, Laguna. Quoted in Diaz, C., Florido, L., Lapis, A., Sy, M. and Faraon, A. (2004) *Determination and Assessment of Current Practices in the Production and Harvesting of Rattan and Almaciga Resin within the Samar Island Biodiversity Project*, Final Report, February, Ecosystems Research and Development Bureau, Department of Environment and Natural Resources, UPLB, College, Laguna

Forestry Development Center (FDC) (2005) *Analysis of Policies and Regulations Governing Non-Timber Forest Products (NTFPs) in the Philippines*, Forestry Development Center, College of Forestry and Natural Resources, University of the Philippines in Los Baños, College, Laguna

Francisco, G. J. and Segayo, M. L. P. (2000) *Forest Resources Accounts Update: 1989–1997 Rattan Resources Accounting Update*, Department of Environment and Natural Resources, Manila

Gatmaytan, A. (2004) 'Case Studies in Rattan Utilization in Ancestral Domain Areas', unpublished paper, Non-Timber Forest Products Task Force, Manila

IRG (2006) *FRAME: Philippines Rattan Value Chain Study*, International Resources Group, US Agency for International Development, Manila

National Commission on Indigenous Peoples (2005) Status Report of the 56 Priority CADT Applications, www.ncip.gov.ph/downloads/StatusReport-PDAP.pdf, accessed 2 October 2008

Neri, B. S. (1994) 'Philippines', in P. B. Durst, W. Ulrich and M. Kashio (eds) *Non-Wood Forest Products in Asia*, Regional Office for Asia and the Pacific, Food and Agriculture Organization of the United Nations, Bangkok, www.fao.org/docrep/X5334e/x5334e09.htm, accessed 2 April 2009

NTFP-EP and NATRIPAL (2006) *Workshop for developing guidelines for sustainable management of three important NTFPs in the Philippines: rattan, almaciga resin and honey*, workshop proceedings, Non-Timber Forest Products Exchange Programme, Manila.

NTFP-TF (2006) 'Message from the Forest: Developing Guidelines for Sustainable Harvesting of Three NTFPs: Rattan, Honey and Almaciga Resin', video documentary, Non-Timber Forest Products Task Force, Manila

PAWB (1998) *Implementing the Convention on Biological Diversity in the Philippines: The First Philippine National Report to the Convention of Biological Diversity*, Protected Areas and Wildlife Bureau. Quoted in Bugna, S. C. and Blastique, T. (2001) 'Description and analysis of the protected area system in the Philippines', *ASEAN Biodiversity*, January–June, pp28–32

PCSD (1995) *Report on Almaciga Resin Gathering and Marketing*, Planning and Technical Services Department, Palawan Council for Sustainable Development staff, Puerto Princesa City

Provincial Environment and Natural Resources Office (PENRO) – Palawan (2006) *List of Almaciga Licensees in Palawan*

Republic of the Philippines (1987) *The 1987 Constitution of the Republic of the Philippines*, www.gov.ph/aboutphil/constitution.asp, accessed 15 August 2008

Salvosa, F. M. (1963) *Lexicon of Philippine Trees*, Forest Products Research Institute, Laguna, Philippines

U4 (undated) *Corruption Glossary*, U4 Anti-Corruption Resource Centre, www.u4.no/document/glossary.cfm, accessed 19 August 2008

Vitug, M. D. (1993) *The Politics of Logging: Power from the Forest*, Philippine Center for Investigative Journalism, Manila

From Indigenous Customary Practices to Policy Interventions: The Ecolocal and Sociocultural Underpinnings of the NTFP Trade on Palawan Island, the Philippines

Dario Novellino

INTRODUCTION

Contemporary features of Batak food-procurement strategies include the harvesting and trade of commercially valuable non-timber forest products (NTFPs). Often such strategies are perceived by conservationists, the government and non-governmental organizations (NGOs) alike as inherited traits of the Batak 'mode of subsistence', but it is difficult to label current Batak NTFP management strategies as primarily 'customary' and distinctively 'indigenous', as such practices have developed and continue to develop as micro-responses to government programmes (Bryant et al, 1993) and to other unpredictable factors such as ecological and climatic changes. Because of deforestation, land-use changes, demographic pressure, increasing market demand, competition with non-indigenous collectors, environmental policies restricting the use of traditional resources and NGO approaches to conservation, the Batak of Palawan receive few economic benefits from the sale of their NTFPs, especially if one considers the time and physical exertion required to pursue these activities.

This paper assesses the interlocked events, circumstances and policies influencing Batak involvement in the trade of rattan and almaciga resin in the social, cultural and historical context of the NTFP trade. Specifically, we will examine the central and underlying factors determining the effectiveness of policies and laws on NTFP use, management, and trade. The main lesson drawn from the case studies is that small-scale indigenous communities such as the Batak have great difficulties in dealing with and responding to market forces and to the complex bureaucratic procedures underlying the implementation of NTFP policies. As a result, such communities – because of insufficient managerial experience and a lack of credit facilities and support services – are unable to profitably engage in the trade of NTFPs and to free themselves from long-standing patron-client relationships.

THE PEOPLE

The Batak are currently scattered across the north-central portion of Palawan Island in the Philippines. They have a heterogeneous mode of food procurement, mainly centred on shifting cultivation, but also including hunting and gathering, the commercial collection of NTFPs, and wage labour. They move from one activity to another according to ecological and economic circumstances, but often pursue them simultaneously. At the close of the 19th century, approximately 20–50 Batak families were associated with each of the nine river valleys that made up their territory (Eder, 1987). As of 2005, however, there were only 155 individuals with two Batak parents, a decline in the Batak 'core' population of almost 57 per cent within a period of 33 years (Novellino, 2007b).

Most of the information presented in this chapter concerns the Batak community living in the territorial jurisdiction of Barangay Tanabag in the north-central portion of the island. It consists of 30 families with a total population of 153. Aspects of the

Source: Based on public domain map from www.lib.utexas.edu/maps/cia08/philippines_sm_2008.gif, modifications by Elizabeth Skinner.

Figure 7.1 *Map of Palawan, the Philippines*

discussion also relate to Batak and Tagbanua communities settled further south in the villages of Kayasan.

Contrary to the standard description of Batak as 'pure' hunters and gatherers, they do engage in upland farming. Batak have a very complex and detailed mythology involving rice and elaborate swidden rituals. Numerous legends trace the origin of rice to their remote past. They name and recognize over 70 varieties of upland rice, of which 44 are said to be *dati* (old) and *tunay* (original) to the area. Batak fallows include a higher number of useful species than primary forest (Novellino, 2007b). The Batak envision a cyclical system in which the seasonal production of honey and rice depends upon the flow of bees and of the life-forces *(kiaruwá)* of rice from *gunay gunay*, a mythical place at the edge of the universe where important resources are concentrated. Access to bees and rice depends on the Batak ability to enhance their dispersal through shamanic practices.

Source: Dario Novellino.

Figure 7.2 *Tapping resin from an Agathis tree, Palawan*

Batak livelihoods also include harvesting NTFPs for sale and subsistence. The resin from *Agathis philippinensis* (*bagtik* or almaciga) is gathered for sale. Rattan canes (*Calamus, Daemonorops* and *Korthalsia* spp.) and wild honey are gathered for both domestic consumption and commercial sale. This paper focuses on commercially traded NTFPs, but subsistence use of NTFPs by the Batak is broad and complex (see Novellino, 1999).

CHANGES AFFECTING FOREST USE AND NTFP MANAGEMENT

After World War II: The beginning of migration and the intensification of trade

Between 1945 and 1960, the Tanabag Batak used lowland areas extensively, as well as the nearby coral reefs and mangrove forests that provided important fishing grounds and additional sources of protein for the people. According to elders, wild honey was collected and stored for periods of several months to support them during seasonal food shortages. *Agathis* resin was harvested from tree branches sporadically and bartered with local traders. Domestic root crops and upland rice sustained the people during their expeditions in search of *Agathis* resin, in contrast to today, when they work with middlemen and patrons, and purchase rice to feed themselves on expeditions. In addition, Batak elders in Tanabag claim that their swiddens were much more productive in the 1950s than today. This is because their ancestral territory was not yet occupied by migrants and thus sufficient land was still available for long fallow periods. Batak swiddens cut from secondary forest regain fertility after a period of 7–18 years on average (Cadeliña, 1985, p25).

It is only after World War II that the migration of Filipino settlers seeking new agricultural land increased significantly, and roads did not enter Tanabag until 1956. In the 1950s the national demand for NTFPs (especially *Agathis* resin) also intensified. Migrant concessionaires moved into the region, and the Tanabag Batak began to acquire new tapping techniques from them (e.g. the skill of using tapping knifes). As a result, the Batak became increasingly involved in the commercial trade and collection of resin (Novellino, 2007b).

Increased migration into Batak territory, competition for land and resources, increased indebtedness and nutritional decline

In the early 1960s, the Batak traditional coastal areas were more intensively occupied by settlers, and barrios and municipalities were established. Concessions to extract forest resources (including *Agathis* resin and rattan) were given to influential politicians, and numerous illegal and unauthorized concessionaires also operated in the area. During the next 20 years, the area between the lowland coastal zone and the present Batak settlement of Kalakuasan was heavily deforested by migrants and logging companies.

Eder (1978) reports that the Batak suffered hunger more frequently during these years, because of the loss of traditional food niches, and were chronically undernourished. At the same time, particularly in the late 1970s, the external demand for NTFPs grew exponentially. As a result, and to compensate for the decline in customary food sources, more people decided to prioritize the collection of NTFPs over other traditional activities. The transformation of the landscape at the hands of migrants not only produced 'spatial disorientation' (Kirsch, 2001, p249), but also dislocated memories of the past. Lowland areas include traditional graveyards and sacred sites that the Batak regard as physical evidence of mythological events and associate with important cosmological principles. The people see the destruction of these historical and natural landmarks as an obliteration of their history.

During the 1980s, Palawan underwent dramatic political change. Nationwide, this period was characterized by a democratic revival leading to a proliferation of NGOs and peoples' organizations. More importantly, there was a radical restructuring of the development paradigm: NGOs were no longer seen as a threat to the elite and bureaucracies, but rather as organizations providing services, especially to the poorest sectors of society (Contreras, 2000, p146). They became 'the missionaries of the new [neoliberal] era' (Tandon, 1996, p182). In these years, the Batak came to be seen as the epitome of a vanishing Filipino culture needing to be saved from imminent extinction, and thus an ideal target for so-called integrated conservation and development projects (ICDPs).

It was at this very time, however, that the dependence of Batak on lowland Filipino society increased. In these years the gathering of *Agathis* resin, rattan and honey (all male activities) acquired a central role in people's livelihood. These activities became the primary way to obtain cash for necessary purchases. However, the reliance on middlemen often results in increased indebtedness because Batak gatherers have to borrow money for food to sustain themselves and their families while collecting *almaciga* and rattan canes. The Batak usually 'borrow' rice and other commodities from Filipino traders and middlemen. But even while repaying their debts in rattan and *almaciga*, they have to continue to borrow food, thus trapping themselves in a vicious cycle of indebtedness. Additionally, as Wakker notes, 'credit is also a source of conflicts, such as when the gatherer does not cut enough rattan to pay back the advance or when the creditor does not want to give advances' (1993, p20). In addition, migrants are generally more skilful than Batak at trading forest products and often have a better understanding of forest laws, as well as closer connections with local politicians and patrons, which aids their control of the NTFP trade.

Increased NTFP gathering in the 1970s and 1980s, even as part of ICDPs, did not improve Batak nutrition, but rather contributed to its decline. Diets became less diverse and more dependent on retailed rice obtained through the sale of *Agathis* resin. The integration of traditional foraging and farming practices with the commercial gathering of NTFPs, wage labour and other options resulted in gross caloric decline. Although this multi-pronged strategy increased the amount of food produced, it appeared to be less efficient than traditional subsistence strategies in terms of calorie and protein intake (Cadeliña, 1985, p119). In fact, the work of collecting resin and transporting it to the hauling points left the Batak with little opportunity for other livelihood activities. Certain agricultural practices such as weeding were often neglected, and this resulted in poorer rice yields.

Logging also eroded the benefits of commercial NTFP gathering. In 1986, when I first visited the Tanabag Batak, the community in the settlement of Tina (six hours' walk from the nearest Filipino settlement) demonstrated a high degree of social cohesion. In 1987 a logging company reached their upstream settlement and advanced further into the interior. The ancestral territory of the Tanabag Batak was by then criss-crossed by logging roads. In the locations of Kapuyan, Kapisan and Maniksik the *Agathis* trees on which Batak depended for commercial resin were felled. As a result, the Batak lost most of their extractive reserves closer to the coast, and were forced to harvest resin in the far interior. Consequently, the energy and time needed for transporting resin to the coast increased by up to six times – an unprecedented level (Novellino, 1999). To cope with this new crisis, the Batak managed to enter into informal agreements with logging company truck drivers to transport resin to the coast.

In addition to felling valuable NTFP species, logging also opened up more remote forest areas to migrants who competed with Batak for NTFP resources. The Batak were forced to the fringes of their territory to look for new sources of NTFPs. The non-aggressive Batak were easily intimidated by migrants; rather than confront them, they attempted to withdraw physically, even to the point of abandoning their own resources.

The rise of environmental conservation and indigenous peoples' rights in 1990s policy and law

In the 1990s, the national government enacted measures to protect the environment and path-breaking legislation to safeguard indigenous rights to land and resources (Novellino, 1999, 2000a, 2000b). Politicians well known for their ties to destructive logging operations now turned 'green'. Environmentalists, policy-makers and even many businessmen claimed to embrace the 'sustainable development' paradigm (Bello, 2004). Examples of these policy efforts, which combined sustainability objectives with indigenous peoples' rights, were the community-based forest management agreements (CBFMAs) and the proliferation of ICDPs carried out by NGOs. As we shall see, these projects and programmes had a significant impact on people's ability to manage NTFPs.

The coming into being of 'negotiated' contracts

In the early 1990s NGOs in Palawan began to support and facilitate the shift of NTFP licences from private concessionaires to indigenous communities. So-called 'negotiated' contracts were now concluded with associations of indigenous peoples mainly from Tagbanua communities. In 1990, some Tagbanua formed their own legally registered association known as SAMAKA (Samahan sa Maoyon ng mga Katutubo, or Association of Indigenous People in Maoyon) and obtained from the Department of Environment and Natural Resources (DENR) a contract for the extraction of rattan in a concession area that included their traditional territory as well as that of other groups. For the first time in the history of Palawan, a concession was released directly to indigenous communities, and this appeared to be a turning point on the way to liberation from patronage and exploitation by middlemen.

Soon, however, the initiative began to exhibit some controversial features: the concession released to SAMAKA encompassed the area inhabited by Batak communities that were not members of the organization. At that time, the Tagbanua and Batak resolved the matter peacefully and SAMAKA agreed not to extract rattan from the ancestral territory of the Tanabag Batak. However, after a few months of operation, the absence of credit facilities and insufficient managerial experience forced SAMAKA to borrow money from the Chinese businessmen in control of the rattan trade. To pay back these debts, the SAMAKA gatherers had to increase their rattan production, and thus encroach again on Batak territory. This was the cause of severe social tension between the two groups. Eventually, SAMAKA had no option but to sell its invoices for rattan shipments to moneylenders. As a result, its associates went from being resource managers to labourers in their own concession.

The Strategic Environmental Plan

Republic Act 7611, also known as the SEP (Strategic Environmental Plan), was enacted in June 1992. It established the legal basis for the protection and management of the environment in Palawan. Protective measures proposed by the law include the demarcation of areas as either off-limits to the human population or reserved for local 'indigenous cultural communities' (ICC), or both (Novellino, 2000a, 2000b).

The SEP law provides a comprehensive framework for sustainable development and contains a package of strategies to prevent further environmental degradation. The centrepiece is the establishment of the Environmentally Critical Areas Network (ECAN), which places most of the province under controlled development. The law establishes that core zones (e.g. habitats of endemic and rare species on steep slopes, primary forest and areas above 1000m elevation)

> *shall be fully and strictly protected and maintained free of human disruption …. Exceptions, however, may be granted to traditional uses of tribal communities of these areas for minimal and soft impact gathering of forest species for ceremonial and medicinal purposes.*
> *(Congress of the Philippines, 1992, p101)*

The ECAN core zones, however, coincide with large portions of the Batak hunting and gathering ground. The resin of *Agathis* trees, for example, is usually extracted in commercial quantities from forest around 1000m above sea level, now classified as 'core zones' (Novellino, 2003b). Having been pushed to the fringes of their territories over the preceding decades, the Batak were now informed that these very areas – remote, steep and high altitude – were the priorities for conservation in the region and activities within them should be carefully controlled.

Community-based forest management agreements

The CBFMAs are part of a policy of the DENR that allows local communities to manage forests that have been converted to non-timber uses. One of its objectives is to develop self-sustaining production systems in the uplands by replacing indigenous swidden practices with permanent forms of agriculture (Novellino, 2003a). However, despite their seemingly lofty objectives, including more participatory approaches to forest management, CBFMAs appear to deny indigenous peoples' rights to their ancestral land, reducing them to stewards and guards of public land.

For example, in the agreement entered into between the Provincial Environment and Natural Resources Office and the Association of Batak of Tina, it is specified that the indigenous beneficiaries should 'immediately assume responsibility for the protection of the entire forest-land within the CBFM area against illegal logging and other unauthorized extraction of forest products, slash-and-burn agriculture (*kaingin*), forest and grassland fires, and other forms of forest destruction, and assist DENR in the prosecution of violators of forestry and environmental laws' (Novellino, 2007a). In effect, the contract requires the Batak to guard their area from their own practices, such as swidden cultivation. The CBFMA does not recognize the claims of indigenous communities over their ancestral domain and instead places indigenous forest management under government control, using the people, in effect, as subcontractors of the DENR.

With a CBFMA in place, things turned out to be even worse for the Batak than they expected. They were unable to fulfil most of the bureaucratic obligations associated with their CBFMA, and did not submit their Annual Work Plan and Community Resource Management Framework to the Community Environment and Natural Resources Office. These reports have to be written according to strict government standards, but the Batak are illiterate. Because they did not produce these reports, the DENR withdrew the permits the Batak needed to sell NTFPs (Novellino, 2007a). Communities in Palawan are often illiterate, and lack of managerial and administrative experience is common.

In addition, the Tanabag Batak were unable to control the entry of illegal gatherers from the neighbouring Tarabanan valley into their area, resulting in the overharvesting of important species. The rights included in CBFMAs proved worthless since communities were not sufficiently empowered to defend such rights within their territories.

In the early 2000s, the drastic reduction of agricultural production caused by the combined effect of El Niño and La Niña (see 'External factors with major impacts on NTFPs' below) and the sudden collapse of copra prices in the national and international market (Novellino, 2007b), followed by the economic uncertainties of the Asian financial crisis, forced lowland migrants and coastal residents to increase the collection of NTFPs on indigenous land. The destructive tapping techniques employed by migrant Filipino gatherers, involving cuts exceeding the thickness of the bark, resulted in the destruction of the cambium (the thin layer between the wood and the bark), exposing the tree to attack from termites and fungi (Callo, 1995; Novellino, 1999). As a result, many *Agathis* trees became unproductive and died, and the most important source of Batak income (almaciga) was severely depleted. All this was happening at a time when agricultural production had collapsed after years of city government prohibition on swidden cultivation (see 'Policies to replace shifting cultivation with "alternatives"' below).

Because they lack financial capital and have limited technical skill in dealing with buyers, the Batak also struggle to create beneficial deals. It is difficult for communities to participate in the NTFP trade equitably if credit is not available to underpin their bargaining power with traders and allow them to respond to market cycles. In some older and established cases, there have been advantages for the Batak in their patron–client relationship with middlemen, particularly during the cyclical and seasonal periods of food shortage. As Platteau (1995, p767) notes, 'patron–client ties are not limited to transaction of economic goods and services but also include symbolic exchange of personal favours and obligations'. Although this relationship is inequitable and in need of reform, the Batak tolerate a certain level of inequity in order to avoid the worst-case scenario: total exclusion from local and regional networks.

The rise of environmental law and indigenous peoples' rights in the 1990s also created a situation in which NTFP law became inherently confusing. The state has made little effort to harmonize overlapping and contradictory laws that are often implemented at the same time, in the same region and within the same community. Such laws, on the one hand, restrict people's access to protected areas and, on the other, pursue a community-based approach to the management of natural resources. This contradictory and ambiguous situation results in a confused understanding of policies on the part of local communities and fosters increased disenchantment towards state agencies.

Policies to replace shifting cultivation with 'alternatives'

In 1992, a new goal of the DENR was to reforest 600,000ha in five years. Little information existed and less effort was expended, however, on identifying how much of this area consisted, in fact, of indigenous swiddens under fallow. At this time, the replacement of shifting cultivation with alternative livelihood practices (e.g. the sustainable harvesting of NTFPs for the market) became one of the cornerstones of DENR community forestry programmes, as it was of environment departments around the world.

The result in Palawan was that indigenous peoples were no longer able to practice their traditional forms of agriculture. By the late 1990s, several members of the Tanabag Batak complained that their fields had become *maniwang* (thin), in the sense of being infertile and with poor yields For instance, according to Ubad, the eldest Batak in Tanabag,

> because of government restrictions to cut old fallow forest, the people clear their swidden plots after three to five years, when trees have not even reached the size of a leg. When you burn them, little ashes are produced – not enough to make your rice healthy. (Interview with author, 15 August 2005)

At the local level, Edward Hagedorn, city mayor of Puerto Princesa municipality, enforced a ban against shifting cultivation in 1994. In the same year, the rice yields of Batak and Tagbanua communities fell dramatically and the people faced severe hardship. The ban altered the entire indigenous agricultural system, and local varieties eventually became rare or even extinct. Such a prohibition flagrantly violated major tenets of the Indigenous Peoples Rights Act of 1997 (Republic Act 8371, (Congress of the Philippines, 1997)) that recognized, protected and promoted the rights of indigenous cultural communities. However, as a result of Survival International's campaign in 1996, the Mayor of Puerto Princesa City allowed indigenous communities to cultivate small swiddens using controlled burning methods, but this arrangement has not been formalized and recent evidence indicates that city government anti-shifting cultivation measures are still enforced with vigour. Ultimately, hundreds of indigenous people had little choice but to exponentially increase the collection and sale of rattan, almaciga resin and honey to compensate for the loss of agricultural production, with significant negative impacts on these species' populations (Novellino, 1999, 2007b).

During these years, the 'alternative to shifting cultivation' paradigm was also embraced by local environmental organizations. In 1994, the NGO Haribon-Palawan implemented an ICDP among the Tanabag Batak that was financed through the technical assistance of the International Union for Conservation of Nature (IUCN). A major objective of the project, much like the government's at the time, was to 'shift from *kaingin* [swidden] to sustainable upland agriculture' (Haribon-Palawan and IUCN, 1996). A technical evaluation of the project in 1997 found that the lack of legal recognition of Batak resources was a major cause of low motivation among the beneficiaries. The report further stated that 'as long as the local communities do not have control of the NTFP resources, other planned project activities such as Community Based Sustainable Resource Management (CBSRM), processing and marketing are interesting (theoretical) studies but remain meaningless' (Bech, 1997, p10).

In another example, a memorandum of agreement was signed in 2003 between the European Commission, the United Nations (UN) Development Programme, the Small Grants Program for Operations to Promote Tropical Forests and the TagBalay Foundation Inc. to finance the Community Development and Mobilization for Forest Development and Protection project in Bayatao, Barangay Tagabinet, central Palawan, in order to develop the ancestral domain title among the Batak and Tagbanua and to provide alternative livelihood opportunities. The project had offered little in representative participation or consensus, as community members complained that the prominent Tagbanua person chosen by Edward Hagedorn (who, besides being mayor of Puerto Princesa, is chairman of the TagBalay Foundation) had misrepresented their interests. The project, further supported by NPO2050 and Cosmo Oi in Japan, invested considerable energy in educating and training indigenous women on eri-silkworm rearing. Despite these investments, however, women – because of cultural prohibitions – refused to bring the rearing cages and silkworms into their households to engage in family-based silk production. Rather than invest in a new form of livelihood with uncertain economic outcomes, they preferred to continue their daily subsistence activities. As a result, production of quality cocoons and finished products remained low and most of the material for the project (spinning wheels, boiling equipment, etc.) remained unused.

In addition – in order to show Tagbalay's commitment towards conservation – a nursery of the useful species ipil (*Intsia bijuga*) was established, but failed. However, a few days before the arrival of the donors' delegation, the indigenous members of the Bayatao and Kayasan communities were asked to collect wildlings of ipil, narra (*Pterocarpus indicus*), and almaciga (*Agathis philippinensis*) for the nursery. In less than 15 days a project nursery was created with almost 10,000 plants, most of which also failed within days of the donor delegation's departure (Novellino and Dressler, in press).

These and similar projects in the region are designed to support environmental protection and the business of conservation, rather than indigenous communities' livelihoods, and suffer from similar institutional and bureaucratic problems to those impairing government programmes, since most of the funds are available only to legally recognized entities with at least two years' experience. As a result, most indigenous communities are excluded from applying, unless an NGO helps by acting as project proponent and administrator. While NGOs can assist their indigenous partners with the operational aspects of project implementation, too often the result is projects that further disempower local communities. Both NGO and government programmes in the region are largely conceived by external agents, and local people are asked to participate with little genuine consultation. This kind of 'participation' often interferes with traditional Batak patterns of food procurement and makes the community more vulnerable to outside forces over which they have no control. As Contreras has argued,

> *Fund-driven programming has bureaucratized participatory efforts and has somewhat eroded the potential for nonbureaucratic modes of organizing. The unwarranted appropriation of participatory approaches has led to the proliferation of programs and strategies which confuse, rather than induce, meaningful empowerment. (Contreras, 2000, p145)*

We have good reason to believe that in the 1970s, at the height of the Marcos dictatorship, the Batak were much better off, and were still able to carry out most of their swidden practices undisturbed. Yet it was in 1975, during Ferdinand Marcos' time, that the state prohibition on slash-and-burn cultivation was reinstated through Presidential Decree No. 705; and it was in 1976 that one-third of the total land area of Palawan was given to timber concessions (Conelly, 1996). Nevertheless, in Marcos' day the state had limited capacity to control remote communities, partly because, unlike today, it could not obtain the collaboration of non-government and people's organizations. The latter were perceived as enemies of the state and, in many instances, banned and suppressed. Furthermore, the Batak were too geographically marginal and politically insignificant to warrant attention. More importantly, northern Palawan was not a site of insurgency, and thus the state did not try to gain firm control over the province.

It was only in the late 1980s that the Batak fully emerged from their 'political isolation' and, particularly in the 1990s, began to interact 'freely' with government and non-government agencies. As Foucault (1982, p221) puts it, not only is freedom the precondition for power, but 'power is exercised only over free subjects and only insofar as they are free'. In the 1990s, through devolution, government programmes and NGO projects, the Batak were no longer displaced outside the boundaries of the state. Rather, they became recipients of external assistance and were invited to 'participate' in meetings and seminars and to settle down closer to the coast. Thus they become 'locatable' and 'being locatable, local peoples are those who can be observed, reached and manipulated as and when required' (Asad, 1993, p9).

EXTERNAL FACTORS WITH MAJOR IMPACTS ON NTFPS

NTFPs and climate change

In addition to legal and policy developments promoting the environment and indigenous rights, the late 1990s were characterized by climatic changes and unpredictable seasonal fluctuations that had a dramatic impact on people's livelihood. In some ways, the increase in temperature registered during El Niño encouraged the harvesting of certain NTFPs. For instance, according to the Batak, the dry weather improved the production of *Agathis* resin. In Palawan, resin production decreases during the wet season, when the rain dilutes and washes away the exudates from tree trunks, and increases in hot, dry weather conditions. The dry weather experienced during El Niño also simplified the collection of rattan canes. According to the Batak, tree trunks and vines were less slippery during El Niño and could be climbed more easily to within reach of the terminal part of the rattan palms. However, these benefits were largely offset by a multitude of negative effects affecting other spheres of people's livelihoods. For example, the soil became hard and dry, so cassava plants grew higher but produced small or no tubers at all, and upland rice production dropped dramatically. Wild fruit trees and banana bore little or no fruit, which also affected the population of game animals (e.g. boars and monkeys); pollen-producing vines and trees did not bear flowers, causing honey production to collapse. Starvation reduced resistance to disease among the Batak, and gastroenteritis decimated the infant population.

La Niña followed hard upon El Niño, and was felt until late April 1999 and again in 2000. The continuous rain prevented gatherers from drying lengths of rattan, which were damaged by fungus and thus unmarketable. Moreover, the rain stopped the Batak from burning more than small portions of their swiddens. The result was crop failure. To cope with the new food crisis, Batak became involved for the first time in alternative livelihood strategies such as the collection and sale of small trees to be used in charcoal-making (ten pieces were sold for P100 – less than US$2).

The rise of mining in 2008

Despite having ratified a range of international treaties – such as the Convention on Biological Diversity, the Basel Convention on Hazardous Wastes, the UN Convention on the Law of the Sea, the Convention on Marine Dumping and the Convention on Wetlands of International Importance, as well as the recent UN Declaration on the Rights of Indigenous Peoples – the Philippine government under President Gloria Macapagal-Arroyo is calling for a revitalization of the mining industry that relegates environmental protection, the sustainable use of forest for NTFPs and other purposes, and indigenous peoples' rights to a position of secondary importance. As much as 30 per cent of the country's land area has already been opened to mining, and 2000 mining permit applications are pending nationwide (more than 300 in Palawan alone), some in core zones, protected areas, watershed areas, fertile agricultural land, NTFP extractive reserves, the ancestral domains of hundreds of indigenous communities and CBFMA areas. At present, the ancestral land of the Tanabag Batak is not directly threatened by mining operations because the local government of Puerto Princesa has banned mineral extraction in its territory. However, this might change after the next election.

This is a telling example of the way non-timber values, and the rights of indigenous peoples, are quickly discarded when powerful corporate interests arrive in forest areas. It also demonstrates the institutional confusion found in many governments, and the contradictory nature of policies and rules emanating from a single government agency. In this case, the DENR, the agency in charge of signing CBFMAs and enforcing regulations for the protection of the environment, also approves mining applications. Resolving this conflict of interest requires clarification of the different roles played by the DENR, which one imagines should focus on its mandate to protect the Philippine environment and renewable natural resources, leaving other agencies such as the Department of Mines, Hydrocarbons and Geosciences to deal with the licensing of mining permits, ensuring compliance with the highest international technical standards (Doyle et al, 2007). This is particularly true since mining brings few benefits to local populations, or even the country as a whole, and results in hundreds of local communities being deprived of livelihoods based on farming, fishing and the collection and trade of NTFPs (Doyle et al, 2007).

CONCLUSION

There are a number of key challenges to the effective implementation of NTFP policies in Palawan:

1   the impact of socioeconomic and environmental changes on 'sustainable' patterns of NTFP extraction;
2   poorly formulated government and NGO interventions that are intended to promote equity and sustainability in the NTFP trade, but fail to address the true livelihood needs of indigenous groups;
3   the lack of technical capacity within indigenous communities to comply with the legal and bureaucratic procedures governing the harvesting, transportation and sale of NTFPs;
4   the knock-on effect of the government ban on shifting cultivation, climate change and macroeconomic factors affecting both the availability and management of NTFPs; and
5   the threats posed by the new state policy calling for a revitalization of commercial mining.

Undoubtedly the combined effect of these factors suggests that a holistic, multidisciplinary and multi-stakeholder approach is needed in order to harmonize NTFP policies with Batak livelihood needs and also incorporate managerial skill.

Historically, the combination of poorly conceived laws and policies with complex socio-political, economic and climatic factors has meant that the increasing involvement of Batak communities in the trade and harvesting of NTFPs has effectively further disempowered them. Ironically, while forest cover is decreasing at an alarming rate, Palawan Island continues to be publicized as the last green frontier of the Philippines. The Puerto Princesa City tourist brochure promises travellers 'a journey abounding with breathtaking scenarios, distinct sights, a rich cultural heritage, and the warmth of the people', and the motto now becoming popular among the 'greens' of Palawan is recited: 'Take nothing but pictures, leave nothing but footprints, kill nothing but time.' However, as a Batak leader told me, 'From the plane tourists can still look over the forest, but what they cannot see is that below the standing trees there are starving people'.

ACKNOWLEDGEMENTS

This article is based on fieldwork carried out while I was a visiting research associate of the Institute of Philippine Culture at the Ateneo de Manila University. A special debt is owed to my Batak friends for their warm hospitality. I acknowledge invaluable funding (grant no. 7136) from the Wenner-Gren Foundation, the Urgent Anthropology grant of the Royal Anthropological Institute in 2007/08 and the Christensen Fund (grant 2007–03068).

REFERENCES

Asad, T. (1993) *Genealogies of Religions: Discipline and Reasons of Power in Christianity and Islam*, Johns Hopkins University Press, Baltimore and London

Bech, J. (1997) Project evaluation: 'Sustainable utilization of non-timber forest products in Palawan', January–February, unpublished manuscript

Bello, W. (2004) *The Anti-Development State: The Political Economy of Permanent Crisis in the Philippines*, Department of Sociology, College of Social Sciences and Philosophy, University of the Philippines Diliman, Quezon City, Philippines

Bryant, R., Rigg, J. and Stott, P. (1993) 'Forest transformation and political ecology in Southeast Asia', *Global Ecology and Biogeography Letters*, vol 3, pp101–111

Cadeliña, R. V. (1985) *In Time of Want and Plenty: The Batak Experience*, Silliman University, Dumaguete City, Philippines

Callo, R. A. (1995) *Damage to Almaciga Resources in Puerto Princesa and Roxas, Palawan Concessions*, College, Laguna, Philippines, Report to Ecosystems Research and Development Bureau, Department of Environment and Natural Resources, Los Baños, Philippines

Conelly, W. T. (1996) 'Strategies of indigenous resource use among the Tagbanua', in J. F. Eder and J. O. Fernandez (eds) *Palawan at the Crossroads: Development and the Environment on a Philippine Frontier*, Ateneo de Manila University Press, Manila, pp71–96

Congress of the Philippines (1992) Republic Act 7611, Manila, Philippines

Congress of the Philippines (1997) Republic Act 8371, Manila, Philippines

Contreras, A. P. (2000) 'Rethinking participation and empowerment in the uplands', in P. Utting (ed) *Forest Policy and Politics in the Philippines: The dynamics of Participatory Conservation*, Ateneo de Manila University Press and United Nations Research Institute for Social Development, Manila, pp144–170

Doyle, C., Wicks, C. and Nally, F. (2007) *Mining in the Philippines: Concerns and Conflicts*, Society of St Columban, Solihull, UK

Eder, J. F. (1978) 'The caloric returns to food collecting: Disruption and change among the Batak of the Philippine tropical forest', *Human Ecology*, vol 6, pp55–69

Eder, J. F. (1987) *On the Road to Tribal Extinction: Depopulation, Deculturation, and Maladaption among the Batak of the Philippines*, University of California Press, Berkley, CA

Foucault, M. (1982) 'The subject and power', in H. L. Dreyfus and P. Rabinow (eds) *Beyond Structuralism and Hermeneutics*, University of Chicago Press, Chicago, IL

Haribon-Palawan and IUCN (1996) *Sustainable Utilization of Non-Timber Forest Products*, Phase I: Final Report, Palawan, Philippines

Kirsch, S. (2001) 'Lost worlds: Environmental disaster, "culture loss", and the law', *Current Anthropology*, vol 42, no 2, pp167–198

Novellino, D. (1999) 'The ominous switch: From indigenous forest management to conservation – the case of the Batak on Palawan Island, Philippines', in M. Colchester and C. Erni (eds) *Indigenous Peoples and Protected Areas in South and Southeast Asia*, IWGIA Document No. 97, International Work Group for Indigenous Affairs, Copenhagen, pp250–295

Novellino, D. (2000a) 'Recognition of ancestral domain claims on Palawan island, the Philippines: Is there a future?' in *Land Reform: Land Settlement and Cooperatives 2000/1*, Food and Agriculture Organization, Rome

Novellino, D. (2000b) 'Forest conservation in Palawan', *Philippine Studies*, vol 48, pp347–372

Novellino, D. (2003a) 'Miscommunication, seduction and confession: Managing local knowledge in participatory development', in J. Pottier, A. Bicker and P. Sillitoe (eds) *Negotiating Local Knowledge*, Pluto Press, London, pp273–297

Novellino, D. (2003b) 'Contrasting landscapes, conflicting ontologies: Assessing environmental conservation on Palawan Island (the Philippines)', in D. Anderson and E. Berglund (eds) *Ethnographies of Conservation: Environmentalism and the Distribution of Privilege*, Berghahn, London, pp171–188

Novellino, D. (2007a) '"Talking about kultura and signing contracts": The bureaucratization of the environment on Palawan Island (the Philippines)', in C. A. Maida (ed) *Sustainability and Communities of Place*, Berghahn, London and New York

Novellino, D. (2007b) 'Cycles of politics and cycles of nature: Permanent crisis in the uplands of Palawan (the Philippines)', in R. Ellen (ed) *Modern Crises and Traditional Strategies: Local Ecological Knowledge in Island Southeast Asia*, Berghahn, London and New York, pp185–219

Novellino, D. and Dressler, W. (2010) 'The role of "hybrid" NGOs in the conservation and development of Palawan Island, the Philippines', *Society and Natural Resources*, vol 23, no 2, pp165–180

Platteau, J. P. (1995) 'A framework for the analysis of evolving patron-client ties in agrarian economies', *World Development*, vol 23, no 5, pp767–786

Tandon, Y. (1996) 'An African perspective', in D. Sogge (ed) *Compassion and calculation: The business of private foreign aid*, Pluto Press with Transnational Institute, London/Chicago

Wakker, E. (1993) *Towards Sustainable Production and Marketing of Non-Timber Forest Products in Palawan, The Philippines*, Tropical Social Forestry Consultancies, Haarlem, Netherlands

Case Study D
Overregulation and Complex Bureaucratic Procedure: A Disincentive for Compliance? The Case of a Valuable Carving Wood in Bushbuckridge, South Africa

Sheona Shackleton

INTRODUCTION

Pterocarpus angolensis DC., commonly know as kiaat, African teak or wild teak, is one of the most valuable hardwood species in the dry forests of east and southern Africa (Vermeulen, 1990). Indeed, its excellent quality wood and high market value have led to its overexploitation in many regions (Vermeulen, 1990; Mushove, 1996). In South Africa this species has a relatively restricted geographic distribution, occurring only in eastern KwaZulu-Natal and narrowly delimited parts of Limpopo and Mpumalanga provinces in the north-east of the country. In Limpopo it provides the basis for the local woodcraft industry. Carvers and furniture makers in the rural municipality of Bushbuckridge have been harvesting kiaat for decades from the extensive communal lands surrounding their villages, turning it into a range of utilitarian goods that they sell in external markets (Shackleton and Steenkamp, 2004; Shackleton 2005a). For these few hundred entrepreneurs this species forms the mainstay of their livelihoods and is critical to their ability to earn an income. The industry is not a lucrative one and generally incomes are modest[1] with many producers only just getting by (Shackleton, 2005a). The high costs associated with harvesting and marketing are the main factors limiting profits.

Within South Africa, concern regarding the vulnerability and potential overuse of this important species has prompted a profusion of legislation to control its exploitation. Regulations on kiaat harvesting and transportation can be found at all levels of government and in a variety of government institutions. For example, in regions where this species occurs, most provincial departments and agencies responsible for conservation (e.g. the Mpumalanga Parks Board and Ezemvelo KwaZulu-Natal Wildlife) have their own legislation on the use of this species. Moreover, the traditional authorities (chiefs) in some regions may also assert control over important species such as carving woods, as well as the cutting of live wood in general.

In Bushbuckridge, kiaat has been protected for decades by laws and ordinances established during the apartheid era in Transvaal, Lebowa and Gazankulu, and more recently by Limpopo provincial legislation (Limpopo Environmental Management Act 7 of 2003, Schedule 12). This means that a permit from the nature conservation

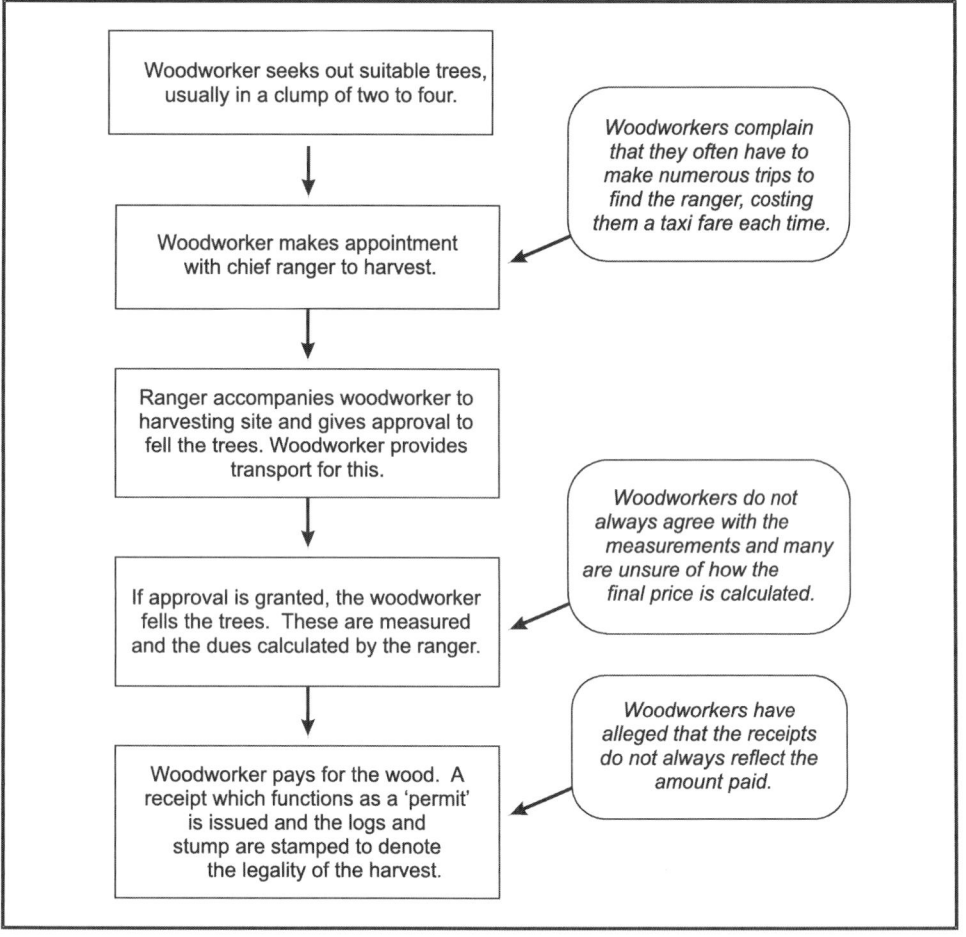

Source: Shackleton (2005b).

Figure D.1 *Harvesting P.* angolensis *in Bushbuckridge – procedure and problems*

authorities is required to fell kiaat. Furthermore, within the former 'homeland' of Gazankulu, woodworkers were required to pay for *P. angolensis* at a cost of R6 (just under US$1) per running metre. Apparently the charge was instituted to encourage producers to use this valuable wood responsibly (Tsweni, pers. comm.).

The steps employed by woodworkers for procuring[2] *P. angolensis* were as follows (Shackleton 2005b, Figure D.1). Two to four suitable trees were selected from the communal lands and approval to fell these was granted by the chief ranger following a site visit. After felling, the logs were measured by the ranger who accompanied the woodworker on the harvesting trip and calculated the amount owing. This was usually between R200 and R450 (US$29–65), and was paid in cash with a receipt issued. The stumps and logs were stamped to denote the legality of the harvest and the revenue was banked at a provincial level (i.e. it was not reinvested in the management of kiaat) (Tsweni, pers. comm.).

In interviews conducted by the author (Shackleton, 2005b), woodworkers identified a variety of problems with this system, which relied heavily on law enforcement by rangers and tribal authorities and was extremely bureaucratic. Factors causing discontentment included:

- the exclusion of producers from resource management decisions;
- the tedium and cost of the process, mainly due to rangers often being unavailable or not having transport;
- poor communication about the method for calculating payment;
- harassment and corruption (e.g. rangers taking wood or issuing incorrect receipts); and
- a lack of credit facilities to purchase wood.

Consequently woodworkers sometimes bypassed the system, especially if they became frustrated when trying to find and pin down the ranger. At the same time, however, they were concerned about the decline in the resource base and the increased appropriation of wood by outside groups. Anecdotal evidence suggests that many previous controls had broken down (Macleod, 1999; Shabangu, pers. comm.), mainly as a result of institutional confusion and a lack of clarity regarding which authorities at provincial and local level (local government and/or chiefs) were responsible for this function, as well as budgetary and capacity constraints. There have, however, been recent efforts by Limpopo to address this by supporting traditional leaders in reasserting their customary control over the natural resource base. For example, a Deforestation Liaison Committee was formed for Bushbuckridge. This appears to have had some effect, as all woodworkers stated that the chiefs were becoming increasingly vigilant regarding the use of carving timbers, and were issuing high fines to illegal harvesters. One producer mentioned how in his tribal authority a woodworker could be banned from the trade if charged for more than three offences.

In addition to being covered by provincial legislation, kiaat has been listed as a protected species at national level under the Department of Water Affairs and Forestry's (DWAF) National Forests Act 84 of 1998 (amended and gazetted in 2006). In terms of this Act, protected tree species may not be cut, disturbed, damaged, destroyed or their products possessed, collected, removed, transported, exported, donated, purchased or sold except under licence granted by DWAF (or a delegated authority). However, at the time of writing, it remained unclear how these regulations would practically relate to current provincial legislation, or how they would be implemented and enforced.

It appears that any users of protected species will need to apply to both national and provincial authorities for a permit or licence, unless there is a formal arrangement (such as a memorandum of agreement) to coordinate applications (Van der Merwe, pers. comm.). No such agreements have been established yet. It is also still unclear how this new legislation will apply in the case of communal land users. A licence application process exists for private landowners wishing to cut protected species on their land as well as for transporters and wholesalers of products derived from these species, but the application form as it stands is unsuitable for local, often poorly educated, woodworkers. To harvest legally it is likely that woodworkers will need to deal with another layer of bureaucracy, but at this stage few are probably aware of this. Ideally, a simpler system is required that is compatible with the long-standing procedure for kiaat harvesting.

In response to an enquiry, a DWAF official suggested that, for political and pragmatic reasons, the protected species regulations were currently unlikely to be enforced in the case of small-scale, low-income, informal-sector users and that, certainly for the foreseeable future, the traditional authorities would remain the primary regulators of resource use in communal areas. He said that action would, however, be taken if any large-scale harvesting from communal areas for the benefit of formal commercial concerns came to the attention of the department (citing the example of a barbecue wood business) (Van der Merwe, pers. comm.).

To date all management efforts have been targeted at controlling and curtailing harvesting through law enforcement, with little concern for the ecological management of the wild resource. No attempt has been made to establish proactive, ecologically sound management guidelines for the species to encourage regeneration and growth, and prevent local extinction. Potential for this appears to exist, though. Van Daalen (1990), Vermeulen (1990), Holmes (1995) and Krynauw (2004) describe various approaches, encompassing selective harvesting, the nurturing of seedlings and the creation of suitable conditions for regeneration, that could contribute to improved management of natural populations of *P. angolensis*.

Since it is unlikely that woodworkers will substantially curb their use of mature trees, it is critical that such an approach be incorporated into any broader management strategy

---

Box D.1 Carvers' and furniture makers' comments on the harvesting system

It (the kiaat) would disappear quickly if there were no control. People would cut the wood and sell it outside of Gazankulu.

The rangers are not straight – they steal the wood from the bush when the owners go away. They then approach other woodcarvers and offer to sell it.

The government does not help the woodcarvers. We promote the area with our skills but they are unhelpful. The rangers are unhappy with what we are doing. We don't have the money to buy logs – but we must pay cash. It would be better if we can get the wood and then pay later.

As a carpenter I am constantly being followed by nature conservation while other people are cutting and carrying logs away. Rangers are always after bribes.

Source: Shackleton, 2005b.

for this species. In the long run this may prove more effective than strict regulations that are difficult to implement and enforce, that producers circumvent because of the complex and costly nature of the process, and that may ultimately alienate local users.

NOTES

1   These amounts are in South African rands. US dollar equivalents are as follows: $29–142 (mean = $65 ± 95) for carvers and $116–348 (mean = $179 ± 384) for furniture makers in 2000.
2   Although this procedure was operational in 2003, it is not known whether this is still the case.

REFERENCES

Holmes, J. (1995) *Forest Ecology and Management*, vol 1, Natural Forest Handbook for Tanzania series, Faculty of Forestry, Sokoine University of Agriculture, Morogoro, Tanzania, pp243–247

Krynauw, S. (2004) 'The feasibility of harvesting *Pterocarpus angolensis* from Mawewe Nature Reserve', in M. J. Lawes, H. A. C. Eeley, C. M. Shackleton and B. G. S. Geach (eds) *Indigenous Forests and Woodlands in South Africa: Policy, People and Practice*, University of KwaZulu-Natal Press, Pietermarizburg, pp416–417

Macleod, F. (1999) 'Syndicates strip Bushbuckridge reserve', *Mail and Guardian*, 17 December

Mushove, P. T. (1996) 'Population dynamics of *Pterocarpus angolensis* stands growing on Kalahari sands', in P. T. Mushove, E. M. Shumba and F. Matose (eds) *Sustainable Management of Indigenous Forests in the Dry Tropics*, proceedings of an international conference, Kadoma, Zimbabwe, 28 May–1 June, ZFC and SAREC-SIDA, pp155–167

Shackleton, S. E. (2005a) 'Spoons, bowls and other useful items: The kiaat woodcrafters of Bushbuckridge, South Africa', in A. Cunningham, B. Campbell and B. Belcher (eds) *Carving Out a Future: Forests, Livelihoods and the International Woodcarving Trade*, Earthscan, London, pp81–102

Shackleton, S. E. (2005b) 'The significance of the local trade in natural resource products for livelihoods and poverty alleviation in South Africa', PhD thesis, Department of Environmental Science, Rhodes University, Grahamstown, South Africa

Shackleton, S. E. and Steenkamp, C. (2004) 'The woodcraft industry in the Mpumlanga-Limpopo Province lowveld', in M. J. Lawes, H. A. C. Eeley, C. M. Shackleton and B. G. S. Geach (eds) *Indigenous Forests and Woodlands in South Africa: Policy, People and Practice*, University of KwaZulu-Natal Press, Pietermaritzburg, pp399–430

Van Daalen, J. C. (1990) 'Survey of utilisable indigenous timbers in Kangwane', unpublished report, South African Forestry Research Institute, Pretoria

Vermeulen, W. J. (1990) 'A monograph on *Pterocarpus angolensis*', SARCCUS Technical Publication, Department of Environmental Affairs and Tourism, Pretoria

PERSONAL COMMUNICATIONS

Shabangu, Alfred. Woodcarver and furniture maker, Mkhuhlu, Bushbuckridge; first chairperson of the Mhala Woodworkers Association

Tsweni, T. Chief law enforcement officer for the Bushbuckridge region, former Northern Province Department of Agriculture and Environment

Van der Merwe, Izak. Official responsible for regulations, national Department of Water Affairs and Forestry

Chapter 8

Overcoming Barriers in Collectively Managed NTFPs in Mexico

Catarina Illsley Granich, Silvia E. Purata, Fabrice Edouard, Fernanda Sánchez Pardo and Citlali Tovar

OVERVIEW OF NTFPS IN MEXICO

Inhabitants of rural Mexico use non-timber forest products (NTFPs) for a wide variety of purposes. The species they use, the products they make and their local management systems are diverse, as are the social contexts. However, among all the variations in biological and social dimensions, there are similarities with regard to tenure and the legal context for most NTFPs: most products are extracted by poor people from commonly owned land and, by Mexican law, illegally.

Because NTFP extraction can be an incentive for forest conservation and important for rural livelihoods, a growing number of institutions and conservation organizations promote the sustainable management of these products as a means to conserve forests and biological diversity, safeguard environmental services and promote the socioeconomic development of rural populations. There are, however, a series of barriers of different kinds preventing this strategy from becoming truly effective.

In this chapter we will briefly describe the situations in which NTFPs are used and traded in Mexico, presenting the case of a wild agave as an example and focusing on the constraints imposed by an inappropriate legal framework and the ways some communities have found to overcome these barriers.

Overview of NTFPs in Mexico: Biology, management and economics

Due to the great biodiversity of Mexican forests and the country's rich cultural background, many species of NTFPs are used in many different ways. No official recent statistics exist, but according to some estimates (FAO, 1995) approximately 2000 species of plants are used by Mexican rural people either to cover basic subsistence needs such as food, fuel, medicine and construction materials or to make utensils and handicrafts which are sold on a small scale for income generation.

Figure 8.1 Agave cupreata *in its natural habitat*

Table 8.1 lists some of the main commercial and subsistence products and their uses (PROCYMAF, 2000).

Some NTFPs, such as *Chamaedorea* spp. leaves used in the floral industry and *Bursera* spp. wood for carvings, have a short history of commercial use. In other cases, commercial use dates as far back as pre-Columbian times, including palm leaves (*Sabal* spp., *Brahea dulcis*) for thatching and basket weaving, copal resin (*Bursera* spp.) for incense, bark fibres (*Ficus* spp., *Trema micrantha*) to make paper, and entire plants for food and drink (*Agave* spp.).

Some NTFPs are harvested for subsistence use from forests using traditional management systems, which in some cases have guaranteed sustainable use, favouring species conservation through norms and community institutions. Other NTFPs are extracted in response to external demands imposed by markets and poverty, resulting in overharvesting, overexploitation, scarcity and resource depletion, disregarding the natural capacity of the species to recover. This can be a consequence of a lack of knowledge about the species or due to the weakening of social organization at community level; in many cases overexploitation comes as a result of structural poverty and social marginalization.

In addition to the importance of NTFPs for basic family needs, the commercialization of NTFPs also plays an important role in people's livelihoods at many different

Table 8.1 *Examples of important NTFPs for market and subsistence economies in Mexico*

Common name	Scientific name	Uses
Oak	*Quercus* spp.	Fuel
Fungi	Species of the genus *Amanita, Tricholoma, Morquela, Cantharellus, Lactarius* and *Boletus*	Edible
Pine resin	*Pinus* spp.	Several industrial uses
Pinyon	*Pinus cembroides, among others*	Edible
Moss	*Hypnum* and *Polytrichum*	Christmas ornaments
Heno	*Tillandsia, Bryum, Morinia* and *Braunia*, among others	Christmas ornaments
Doradilla	*Selaginella lepidophylla*	Medicinal
Camedor palm	*Chamaedorea* spp.	Ornamental leaves
Guano palm	*Sabal* spp.	Rustic thatching
Hat palm	*Brahea dulcis*	Weaving baskets, bags, mats, hats, handicrafts, thatching
Gum latex	*Manilkara zapota*	Chewing gum
Allspice	*Pimienta dioica*	Condiment
Pita	*Aechmea magdalenae*	Fibre
Cuachalalate	*Amphipterygium adstringens*	Medicinal
Copal resin	*Bursera bipinnata, Bursera* spp.	Incense
Yucca species	*Yucca schidigera, Yucca* spp.	Edible and for industrial use
Cacti	*Mammillaria, Pereskiopsis, Hylocereus* and *Lophophora*, among others.	Ornamental
Candelilla	*Euphorbia antisyphillitica*	Wax for cosmetics, confectionary, electronics
Gobernadora	*Larrea tridentata*	Medicinal
Lechuguilla	*Agave lechuguilla*	Fibre
Zacaton roots	*Muhlenbergia macroura*	Fibre

levels in Mexico. Some NTFPs have been traded historically, while new products or new uses for old products are making their way to the local or the global market. Some NTFPs without long histories of traditional use have important commercial roles today, often driven by international demand, as in the case of certain wild mushrooms; this creates a policy gap, since there is no customary regulation as there is for species with long traditional use.

Even though revenues from NTFPs are not typically very high, their sale usually provides extra money that complements family income, critical in times of illness and during religious or family festivities. The income generated by NTFP trade is often the only source of income for some sectors of society, such as old people, unmarried women and women with small children and an emigrant spouse. This is the case for many species of mushrooms which are collected during the rainy season mainly by women and their families.

Another characteristic shared by most NTFPs in Mexico is lack of information regarding the rate of extraction and trade. Due to the type of use and the small commercial scale, most escape registration by forest authorities, so there is no official record of the amounts extracted or the impact that the harvest has on ecosystems.

Even though there is a government agency collecting data concerning NTFP extraction, there is no unified system to register this information. Only a handful of the multiple resources extracted, used and traded – those of economic importance internationally – are registered by forest authorities. Consequently, at present it is not possible to give realistic estimates of numbers of species used or to assess the ecological, social and economic impacts of NTFP extraction and commercialization. Hence, official data grossly underestimate the importance of the contribution of NTFPs to the local economy.

NTFP LEGAL FRAMEWORK

It is estimated that Mexico has 56.5 million ha of forest land (covering 29% of the country); 22.1 million ha more are considered to be deteriorated forest lands (SARH, 1994, cited in Bray and Merino, 2004). Mexico is unusual in that most of its forests are socially owned. Land reform started at the beginning of the 20th century and established two forms of collective property: *ejidos* and indigenous communities,[1] which today make up more than 50 per cent of the national territory. These agrarian nuclei, as they are called, have a certain degree of autonomy in the management of their natural resources, and so there is a rich variety of indigenous forms of community property and management. However, the state maintains a great deal of control through the legal system, especially through federal laws (Bray and Merino, 2004).

Mexican environmental legislation is recent: the first General Law for Ecological Balance and Environmental Protection (Ley General del Equilibrio Ecológico y Protección al Ambiente, LGEEPA) was enacted in 1988 and, for the first time in national history, an environmental office with a specific mandate was established. Since then, several laws have undergone important changes, among them the General Sustainable Forestry Law (Ley General de Desarrollo Forestal Sustentable, 2003), and new regulations have appeared, such as the General Wildlife Law (2000 Ley General de Vida Silvestre). This has led to frequent changes in how forest issues are treated in public policy, and in what is required to comply with various laws.

A confusing legal framework thus characterizes NTFP regulation in Mexico. Various laws and regulations sometimes contradict one another and are generally confusing. Ironically, overregulation includes great voids where exercising discretion can result in interpretations far removed from the legislators' initial objective of regulating NTFP access and use.

On the other hand, at the international level, the Convention on Biological Diversity (CBD) directs that each country regulate its operation by national legislation. Even though several Mexican laws consider the general principles of the CBD and incorporate them in their rules, application of the principles has been very limited and at times confusing due to the inconsistencies in the national legal framework.

Barriers within the legal framework

Environmental laws

The two laws that apply directly to all NTFPs are the General Sustainable Forestry Law and the General Wildlife Law, both enforced by the Ministry of the Environment and Natural Resources (SEMARNAT). Each law establishes its own legal and administrative procedures, causing a clear legal overlap and overregulation for NTFPs. Authorities tend to exercise discretion in the application of regulations, depending on the importance of the species to be harvested and extracted. Application of the law varies from one state office to another.

A general rule has been informally established: the Wildlife Law applies to NTFPs that are listed as species at risk, in danger of extinction or under a special protection regime; the Forestry Law covers all other NTFPs. For species that grow in humid tropical ecosystems or in natural protected areas or that do not regenerate with ease, there are additional legal requirements (see Table 8.2).

To confuse the picture even more, some NTFPs also are regulated by intellectual property laws for geographic indications (GI), under the Ministry of Finance and its Mexican Institute for Industrial Property (IMPI). In some cases, when NTFPs are part of what is protected, this regulation may overlap or duplicate others already established by the environmental sector. The most important form of recognition of GIs in Mexico is the appellation of origin (*denominación de origen*, DO).[2] For example, with the legal establishment of the appellation of origin for mezcal (governed by the norm NOM 070-SCFI 1994-Mezcal) in 1994, official regulation became even more complicated for wild species of *Agave*, which provide the raw material for mezcales, spirits distilled from these plants. This norm duplicates some of the procedures required by SEMARNAT and adds more paperwork and expense to the process of legalizing the extraction of this group of NTFPs.

The number of regulations and the studies and administrative procedures required make the process of legal extraction of NTFPs difficult and expensive, a great burden to communities and a disincentive to compliance, as we will see in the case study.

CASE STUDY: MEZCAL FROM *AGAVE CUPREATA*

This case study is based on the collaborative work of an NGO, Grupo de Estudios Ambientales, AC (GEA); a peasant organization, Sanzekan Tinemi; and 30 communities surrounding Chilapa, in the state of Guerrero, one of the poorest regions of Mexico. One objective of this collaboration is the sustainable production of mezcal, a traditional Mexican alcoholic beverage prepared by the distillation of agave plants that have been baked and fermented. The collaborators have been carrying out basic biological, ecological and ethnoecological research on *Agave cupreata* and mezcal, as a starting point for developing community-based sustainable agave management programmes, grounded on the understanding and further development of traditional practices within community institutions (Illsley et al, 2007).

Table 8.2 *Legal procedures for extraction of NTFPs in Mexico*

Group of species	Laws and regulations	Legal instruments required for extraction
NTFPs in general	Forestry Law and rules; Several Mexican official norms (NOMs)	Notice of extraction
Forest topsoil for gardening, species of genus *Yucca*, and families: Agavaceae, Palmae, Cactaceae, Cyathaceae, Dicksoniaceae, Nolinaceae, Orchidaceae and Zamiaceae	Forestry law and rules; Several NOMs	Simplified management plan
From tropical ecosystems or with difficult regeneration	Forestry Law and rules; Several NOMs; LGEEPA	Notice of extraction or simplified management plan and environmental impact assessment
At risk, endangered or under special protection regime	General Wildlife Law; NOM 059	Management Units for Wildlife Conservation (UMA) and management plan for the species
Within a natural protected area	General Wildlife Law	Opinion from the National Commission of Natural Protected Areas (CONANP); and Management Units for Wildlife Conservation (UMA); and Management plan for the species
Included in an appellation of origin (geographic indication)	Industrial Property Law (Ley de la Propiedad Industrial); NOMs	Registration of areas for NTFP extraction before the regulating council

---

Box 8.1 Wild mushrooms in Mexico

Wild mushrooms provide an excellent example of the need to manage timber with NTFP values in mind; mushrooms are valuable, but they are negatively affected by logging. Steps have been taken in certain regions, such as with Pueblos Mancomunados in Oaxaca, to exclude areas with high mushroom density from logging.

Like many NTFPs, mushrooms do not provide large incomes to harvesters, but they do provide income during times when agricultural production is low (for example, in the rainy season), helping households cover extra expenses.

Efforts to regulate mushrooms are overly expensive and confusing, favouring industrial agricultural production over small-scale harvesters: it is easier to obtain a timber authorization than to obtain authorization for the extraction of mushrooms in the same area (see Table 8.3).

Table 8.3 *Use of wild mushrooms in Mexico*

Species	Amanita caesarea (Tecomate, hongo amarillo, hongo de huevo)	Cantharellus cibarius (duraznito)	Boletus edulis (hongo de pan, boleto)	Morchella spp. (pancita, elotito)	Tricholoma magnivelare (Matsutake, hongo blanco, hongo de pino, hongo de venado)
Distribution in Mexico	Widely distributed in most temperate forests	Widely distributed in most temperate forests	Widely distributed in temperate forests, at altitudes over 2500m	Limited distribution in central areas; probably related to the Neovolcanic Belt (Mexico State, Morelos)	Limited distribution, related to ecological conditions such as type of soil, type of pine, slope
Legal status		Listed in NOM 059[a]	Listed in NOM 059	Listed in NOM 059	
Form of consumption[b]	Mostly fresh and more appreciated when young (buds)	Fresh (Mexico), dehydrated and pickled (Europe)	Dehydrated, fresh and pickled	Dehydrated and fresh	Fresh and processed for medicine
Traditional use	Traditional consumption in the whole country as food, probably the most popular wild species	Traditional consumption in the whole country as food	Traditional consumption, but more limited than the previous species	Traditional consumption limited to the areas of highest production	Limited traditional consumption. Not used traditionally by communities that currently collect and commercialize it
Commercial trade/primary	Strong local demand in rural areas and markets in bigger cities; strong international demand, particularly in Italy, but there is no trade with Mexico	Traditional consumption in the whole country as food	Low demand in local rural areas, good demand in restaurants with European cuisine, and dried for export (but Mexico does not export)	Market concentrated in Mexico City for European restaurants, and dried for export to USA and Europe	Most important market is fresh for export to Japan
Volume and size of trade[c,d]	No separate data for this species at national level	No separate data for this species at national level	No separate data for this species at national level	No separate data for this species at national level	30–40 tons per year

Table 8.3 *Use of wild mushrooms in Mexico (Cont'd)*

Species	*Amanita caesarea (Tecomate, hongo amarillo, hongo de huevo)*	*Cantharellus cibarius (duraznito)*	*Boletus edulis (hongo de pan, boleto)*	*Morchella spp. (pancita, elotito)*	*Tricholoma magnivelare (Matsutake, hongo blanco, hongo de pino, hongo de venado)*
Price per kilo (US$/kg)	2–3 to the producer, 4–6 in regional markets depending on quality	3–5 to the producer depending on quality, 5–7 in the regional markets	2–4 to the producer depending on quality, 4–6 in the regional markets, 120 dried in shops	6–8 to the producer, 10–14 in the regional market, 350–400 dried in shops	7–35 depending on quality, more than 150 fresh to the consumer in Japan
Legal requirements	Notice of extraction	Management Unit for Wildlife Conservation (UMA) and management plan	Management Unit for Wildlife Conservation (UMA) and management plan	Management Unit for Wildlife Conservation (UMA) and management plan	Notice of extraction

Notes: [a]Norma Oficial Mexicana (Mexican Official Norm) 059, meaning that the species is at risk, in danger of extinction or under special protection regime.
[b]This species is sold and eaten all over the world.
[c]National market of around US$450,000–850,000 per year for 3000 peasant families (including all main commercial species).
[d]For the wild species except matsutake: 100–120 tons (collected with a permit given by SEMARNAT). An important percentage of the production escapes this registration because it is commercialized in a regional market.

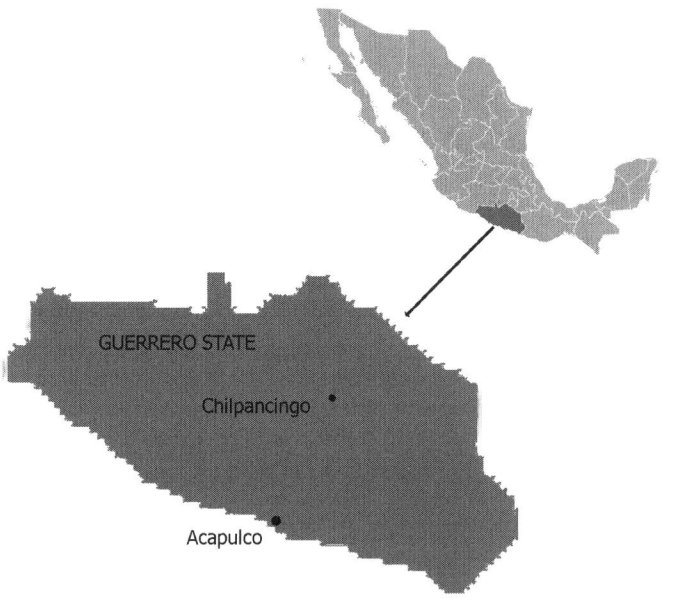

Figure 8.2 *Map of Guerrero State in Mexico*

Background

Over 40 different species of *Agave* are used for mezcal production in 26 of Mexico's 31 states. The genus *Agave* contributes to Mexico's great biodiversity, while the processes for making the various kinds of mezcal reflect the country's cultural richness. Ethnic groups relate to particular species of *Agave*, which they use not only for mezcal production, but also as food, building materials, medicine, fibres and materials for rituals (Colunga-GarcíaMarín et al, 2007).

Each mezcal is different, not only because of the particular plant species it comes from, but also because of the process used in its preparation. No other alcoholic drink in the world, as far as we know, comes from such great natural and cultural diversity. Making mezcal is a skill that has been passed down from one generation to the next for hundreds of years. It is generally done in small, rustic installations in remote, poor and often indigenous communities. Only a few mezcales are made industrially, the most famous being *tequila*.

Agave cupreata Trel. et Berger, Agavaceae, locally known as *maguey papalote*, is one of the main species used in the state of Guerrero. It is endemic to the Balsas River watershed, which extends throughout the centre of the states of Guerrero and Michoacán (García Meneses, 2004; Ocaña-Nava et al, 2007). The plant grows in the understorey of tropical dry oak and other deciduous forests and in greater densities in grasslands and palm–agave associations. It takes between 6 and 12 years for the plant to reach maturity. Mature plants reproduce only once in a lifetime, producing a large flowering stalk. After the seeds are dispersed, the plant dies.

Mezcal production has a direct impact on the sustainability of wild agave populations because the entire plant is extracted immediately before flowering, precluding the production of seeds. For agave species such as *A. cupreata* that only reproduce sexually, this may have negative ecological consequences, requiring careful management to avoid depletion of populations. Mezcal production directly affects the genetic variation of agaves. Large-scale sustained production of mezcal depends on the genetic vigour of the populations and on their pollinators (Illsley et al, 2007).

Role of local institutions in access to agaves

Agave extraction has been regulated for hundreds of years through local institutions, within the *ejido* and indigenous community structures. Like most NTFPs in Mexico, especially those that are local and of little interest to national markets, *Agaves* and their management have been ignored by the government and academic institutions until recently.

Local institutions have been responsible for regulating access, management practices and the distribution of benefits. The regulations each community has adopted derive from its particular conditions and history. A number of management practices have developed over time, based on traditional knowledge of the species. These are articulated by norms and agreements[3] established by each general assembly (the top authority in an agrarian nucleus) and are continually modified or replaced in a dynamic process of response to new situations and to the tensions of environmental, socioeconomic, cultural or technological origin. Generally it is the duty of the *ejido*'s vigilance committee to enforce the approved norms (Aguilar et al, 2002).

The specific combination of agreements and norms varies from place to place (see Table 8.4); a practice may be established as an agreement in one community and as a norm in the next, and not be present at all in another. The degree to which regulations are respected also varies. The result at the regional level is a patchwork of very different situations, where one community may have developed a de facto sustainable management plan for its agave and other resources (water, soil, forests), while its neighbour has completely depleted them and no longer has any agaves. Between the two extremes there are many intermediate conditions.

Ejido, indigenous community and private property coexist in this region, and access to agave is initially linked to land tenure. In *ejidos* and indigenous communities, agave is a common-pool resource (CPR), whereas on private property it belongs to the owner of the land. In *ejidos* and communities plots of communally owned land can be assigned to individuals, who act as owners as long as they tend the plots, which cannot be sold. If a plot is abandoned, it goes back to the community. The resources spontaneously growing on the plot, including agaves, may be treated either as private or as CPR, by decision of the general assembly.

The right to harvest agave is strictly regulated and, when it is treated as a CPR, arrangements vary: the general assembly may collectively sell the mature agaves and determine the use of the income (which can go to pay for community expenses or infrastructure), or the *mayordomos*[4] may be given the right to sell agave in order to help finance the annual religious *fiesta*. In other cases, agave is distributed among the members of the agrarian nucleus (*ejidatarios* or *comuneros*), each receiving an equal quota of plants.

Role of local institutions in the management of agaves

Management is also regulated by local institutions. *Agave cupreata* is considered a 'wild' species, but for hundreds of years local inhabitants have been managing it, increasing its density and selecting the best plants for seed, although without actually cultivating the plant. Again, different types of management systems have been developed in each community and can vary tremendously, even among neighbours.

Management practices include sparing plants for seed, guaranteeing seed maturity and dispersal, extracting only mature plants, preventing forest fires and excluding cattle from areas with agave. Ecological studies (Illsley et al, 2007) corroborate the soundness of these management practices, while pointing out the need for improvements, as some traditional practices are no longer effective; for example, the way of sparing plants for seed production.

One effective management system is practised in communities where, in addition to enforcing good management practice regulations, they have also developed unwritten de facto management plans. These plans include the systematic rotation of harvest areas, strict monitoring for the extraction of only mature plants and decreeing harvest bans for one to four years if the agave population drops visibly. Here the territory is divided by imaginary lines and harvest is carried out systematically, creating a three-year rotation cycle for every harvested plot.

Under this system, each *ejidatario* receives 140 agave plants, which can be harvested or sold to another *ejidatario*, but not to outsiders. The vigilance committee holds a list to control the order in which each *ejidatario* receives his or her quota of plants. Only one *mezcalero*,[5] who is also an *ejidatario*, is authorized to receive all the agave. *Mezcaleros* cannot buy agave from other communities and will be fined for every immature plant found on their premises before processing. If authorities, by visual inspection, determine agave populations are diminishing, they will impose a ban on harvesting for two to four years. Bans are also used as a way to force *mezcaleros* to increase the price they pay for agave plants. The firewood they use is also carefully monitored.

In some cases, especially on private property, another interesting traditional agroforestry system is used: inducing the growth of very dense patches of agave, selecting the plants spared for seed, and annually harvesting only ripe plants. These patches, though small, can reach densities of 3500–4000 plants per ha. Yet another system, recently introduced by the regional peasant organization Sanzekan Tinemi, is to collect seeds to produce plants in nurseries, nurture them for one or two years and then plant them into forests and agroforestry areas (Illsley et al, 2007).

Agave plants face serious challenges to their existence, including forest fires, which are especially damaging to immature plants, and grazing. The long dry season drives cattle to eat not only young, tender plants, but even the spiny, fibrous adult agave. Well-organized communities can almost always control fires before they cause too much harm, but local management has not been successful in finding the best solution for the coexistence of cattle and forests. This creates internal tensions that have not been resolved (Illsley et al, 2007).

The impacts of agave use on the environment are not limited to the *Agave* species. Mezcal production requires large quantities of firewood to attain the particular artisan quality desired in the drink. Intensive use of firewood in the distilleries threatens forest remnants. The negative consequences of loss of forest cover can be significant, including severe erosion and the reduction of water sources. Some communities are well aware of this and strictly regulate firewood extraction, while others no longer have any firewood left and are forced to buy from neighbouring communities (Aguilar et al, 2002; Illsley et al, 2007).

In short, the capacity of local institutions to sustainably manage agave as a NTFP is related to the traditional knowledge of the species, the type of de facto management plan implemented, and the capacity of the community to enforce its norms and agreements, to regulate the actions of different sectors, especially *mezcaleros*, and to manage conflicts (Aguilar et al, 2002). GEA and Sanzekan have organized workshops for community-to-community interchange about norms and agreements for the management of natural resources. The result has been cross-pollination and a series of changes in communities that decided to follow the lead of more efficient neighbours.

From agaves to mezcal: Transformation of the NTFP

Mezcal is produced in small rustic distilleries set out in the fields, close to water springs. Ripe agaves are harvested by cutting the leaves to expose the sugar-rich centre of the plant, transported by donkeys to the distillery and baked in pit-ovens over oak firewood and hot stones. The cooked agave is then crushed, fermented in wooden vats and finally distilled twice in copper distillation pots.

Generally, the *mezcalero* will buy the agave in the field and hire the harvesters. Sometimes other social arrangements occur; for example, an agave producer and a *mezcalero* will agree to go half-and-half (*a medias*), or an individual (agave producer or not) will rent a distillery from a *mezcalero*. In these cases they will share costs and divide the end product between them.

The role of *mezcaleros* in conservation is critical under CPR conditions, for unless the vigilance committee keeps a very close watch, *mezcaleros* will unilaterally decide what plants to harvest or spare. If they are conscientious and attentive to quality, they will harvest only mature agaves, but if they are looking to work less and to make cheap alcohol, they will take immature plants and thus deplete populations.

On the other hand, *ejidatarios* and *comuneros* who have land directly allotted to them, as well as private agave producers, will set their own conditions. They go out to their field with the *mezcalero* and point out the plants that can be harvested as well as those to be spared for seed, usually the strongest, best-coloured ones, ensuring conservation and genetic selection.

Mezcaleros are one of the few groups that need not migrate in order to maintain their families: the commercialization of mezcal allows them sufficient income to be among the wealthy of their communities. But they are not the only ones who profit, as

Table 8.4 *Role of local institutions in regulating the agave-mezcal production chain in Guerrero*

Production chain link	Ownership of resource	Regulating institution	Local regulation: Agreements and norms
Agave production	Common; private	General assembly in *ejidos* and indigenous communities; private	For sustainable management (regulation of ripeness for harvest, harvesting of flowers, flower stalks, leaving seeds in the field, rotating harvesting areas, harvest bans); for distribution of benefits (access to harvest, market destination, election of *mezcaleros* and exclusive agreements, price of agave)
Transformation into mezcal	Private; private under common surveillance	General assembly; private	For quality in mezcal (ripeness); for distribution of benefits (right to work in factories)
Commercialization of mezcal	Private; private under common surveillance	General assembly; private; local market	Distribution of benefits from selling mezcal; right to sell; price of mezcal

the benefits are distributed among many inhabitants in the region. Mezcal production provides work at all points in the commodity chain; it is one of the few sources of local employment.

Making mezcal is a way of adding value to agaves; aging mezcal adds even more value to the product. If the *mezcalero* has the opportunity to save, he will make more money selling aged mezcal than putting his money in the bank. Mezcal is often kept for years, buried in the ground in glass flasks, as 'savings' against the costs of special celebrations or emergencies such as illness, as well as schooling needs (Illsley et al, 2007).

Commercialization of mezcal

The typical mezcal commodity chain is very short. Often the *mezcalero* sells directly to the consumer or to a corner store or restaurant, which will in turn sell to customers. Over the years, *mezcaleros* acquire stable clients who buy part of their production. Many of the clients are influential people in the region: politicians, teachers or lawyers. There are some intermediaries who buy from *mezcaleros* and sell by the litre, in plastic Coca-Cola bottles, in regional markets and streets. In some cases, the right to be an intermediary of this sort is subject to certain control: for example, a general assembly can give an *ejidatario* in dire need the right to sell mezcal in the region.

Local markets exercise a certain control over *mezcaleros*: those that have won regional recognition and status set the standards for the quality and price of mezcal and agave in the region. Prestigious *mezcaleros* with high-quality mezcal sell at the best prices and generally have the most stable clients; lower-quality mezcal is sold in the streets and markets at lower prices. Adulterated mezcal has the lowest prices and a poor reputation, yet is bought by those who cannot afford anything better.

In the past several years mezcal has started to gain popularity among national and international spirit 'connoisseurs' and has been slowly reaching global markets. As a result, the government and investors have declared mezcal to be the upcoming agro-industry for Mexico, following the model of tequila, the most famous of all mezcales. Local producers are aware of this opportunity and some are scaling up their process and looking to enter wider markets, as is the case of a group of *mezcaleros* who have organized under Sanzekan Tinemi in Guerrero. The level of official regulation and control over mezcal has risen along with its popularity, as will be described.

Meeting official regulations

Environmental sector laws: regulations for agave extraction

In the past few years there has been growing pressure from government authorities to comply with national plant extraction regulations. Poor traditional peasants are being fined by the Environmental Protection Agency (PROFEPA) when they do not present a legal permit for the extraction of agave or firewood.

Independently of local institutions, the General Sustainable Forestry Law establishes an apparently simple general rule for the commercial extraction of NTFPs: compliance with a written 'notice of use' submitted to SEMARNAT's state delegations before extraction. However, the written notice must be accompanied by a complex technical study. Additionally, the Forestry Regulation and the technical Mexican Official Norms (NOMs) establish exceptions to the general rule and specify that certain NTFPs require both an authorization and the elaboration of a 'simplified management programme', as is the case for wild agaves.

To carry out the required studies, the community must hire a professional, generally a forest engineer or a biologist, to draw up the management programme and to prepare periodic reports of quantities extracted. Traditional de facto management plans, norms and agreements are invisible to these professionals and to policy-makers, who believe that poor people are ignorant and need to be taught. On the other hand, as happens with most NTFP species, there is very little published literature on which to base a management programme. Time and economic resources are scarce and so the professionals are not willing or able to do the basic research required for a sound management plan.

Technically, to comply with the law, there is no requirement to consider any of the practices, agreements and norms the community is already enforcing or wishes to institute. The professional needs only to roughly determine the number of plants in the field and, according to his or her best judgement, establish how many can be harvested each year. The professional must also state in writing that 20 per cent of mature individual plants will be spared for seed – as required by law – and hope nobody will come to check whether this is being followed in the field. The present proportion of mature plants spared is 1–5 per cent, and 20 per cent would never be accepted by the peasants, nor is it necessary, according to ecological studies. The management programme system suggested can be whatever plan occurs to the professional, most of whose energy will then be devoted to paperwork and expensive trips to and from the state capital to obtain legalization and to hand in reports.

At best, this legal plan will be paperwork that does not change the way things are done, but it may also impose changes in precisely those practices, such as rotation of extraction areas and harvest bans, that have ensured sustainable management over the years; and thus actually result in the depletion of plant populations.

To make things more complicated, the Ministry of Agriculture and some other government and academic agencies assume mezcal agaves should not be regulated by the environmental sector at all. They are pressing to change the traditional organic management of agave as an NTFP into an agricultural monocrop system with fertilization, chemical control of pests and, if possible, *in vitro* reproduction, based on the myth that only high-tech agriculture is productive (Barrios et al, 2006). This kind of monocrop production system for agaves has been shown to have heavy environmental impacts, including deforestation, erosion, soil and water contamination, and loss of diversity at ecosystem, species and genetic levels. The best example of this is found in the tequila region (Valenzuela-Zapata and Nabhan, 2003; Valenzuela-Zapata and Simoes, 2007; Martínez Rivera et al, 2007).

The economic sector laws: Regulations derived from the appellation of origin

At national level, the regulation of the appellation of origin for mezcal contains elements which directly and indirectly impact the conservation and management of agaves. The regulation does not consider the diversity of *Agave* species used for mezcal production, nor does it recognize the historical mezcal regions or consider the environmental impact of intensive agave production. No reference is made to the need to conserve biodiversity and to manage and use agave sustainably.

In addition, the regulation protects municipalities in only 7 Mexican states, although mezcal is produced in 26. This limited coverage highlights the lack of coordination among government sectors when the regulation was drafted – no members of the environmental sector were included in this process – as well as the lack of participation of mezcal producers.

In addition to having to comply with SEMARNAT's requirements, under the new regulation agave plantations must be registered with the Regulating Council for Mezcal, which has offices in the city of Oaxaca. It establishes only one such council for mezcal at the national level.

This council has developed one set of rules, based on the two commercial agave species (tequila and espadín) which are grown in monocrop plantations, to be applied to many species, regions and mezcales in the seven protected states. It does not consider the diverse schemes that exist for wild agave management, incorrectly assuming a situation similar to tequila. The regulation ignores the fact that agaves exist as NTFPs, even though over 35 species fall under this category or under low-intensity forms of management.

Following this pattern, the regulation does not consider specific traditional local methods used in mezcal production, such as fermentation in clay pots or animal hides and aging in glass or clay. Instead the norm only acknowledges fermentation in wooden vats and aging in wooden barrels, again as done in modern times with tequila. This imposes homogeneity and pushes aside traditional species, equipment and practices that otherwise might be factors in creating a niche market and higher prices.

Rather than promote differentiation, the regulation promotes uniform production, in the name of 'modernization', with the result that local know-how is being lost in the process. If uniqueness is not recognized, the taste derived from local species and methods will not be valued and the tendency will be to promote monocrop plantations and uniform mezcal that offers quantity over quality.

The only way to legally commercialize mezcal is to have it certified by the official DO regulating council (COMERCAM, Mexican Regulatory Council for the Quality of Mezcal). Certification under COMERCAM requires a great deal of paperwork and expenditure from each *mezcalero*. The cost of complying with this regulation can be four times the annual income of an average small-scale *mezcalero* (Illsley et al, 2007). The bureaucracy and costs of certification plus a 50 per cent tax on alcoholic beverages could push legitimate poor *mezcaleros* out of business and into migration. Some *mezcaleros* will probably remain, as they have done before, in clandestine resistance, but exposed to abuse and corruption.

Final thoughts around geographic indications

The agave case shows that even though GIs are a very interesting concept, merely declaring them is not enough to promote local development and local rights. The strategies for determining who has control and governance within a GI must be carefully constructed, guaranteeing equal rights of participation of all legitimate sectors in the commodity chain.

Recognizing the complexity of GIs, their planning cannot be left to one administrative sector, but must include the environmental, health, social development and cultural as well as economic sectors. This avoids the duplication of regulations and simplifies procedures. Lack of this integration can lead to a loss of biodiversity and of valuable traditional knowledge and sustainable practices.

With mezcales, Mexico has a unique opportunity to build and implement a GI system which could be a model for the world, if existing policies are improved and local participation guaranteed, thus supporting the recognition of local rights. Headway has been made, but changes are needed before interests become further entrenched. In the longer run, the benefits of a solid GI system for mezcales could highlight Mexico's natural and cultural diversity to the world, serving as a centrepiece for regional development with cultural identity and NTFP conservation.

GENERAL CONCLUSIONS AND RECOMMENDATIONS

Among the characteristics of NTFPs that make their sustainable management difficult is the great heterogeneity of products and the diversity of species, parts of plants used and management systems. This complexity makes the design of general mechanisms for their regulation quite difficult. It is necessary to treat each case separately, which at present prevents the application of general rules.

There is very little or no scientific biological and ecological information for most of the species, even the most conspicuous. This lack of knowledge makes it difficult to develop methods to reliably estimate sustainable harvest levels.

Respect for traditional knowledge and local regulations is established as a very general principle in the Forestry and General Wildlife Laws, yet it is impossible to give it effect within Mexico's extremely formal and bureaucratic legal system. In reality, Mexican official regulations generally ignore traditional knowledge and local practices, norms and regulations.

Compliance with the law becomes a slow and costly affair, generating paperwork that does not reflect the way communities relate to their NTFPs and does not necessarily enhance sustainable use and conservation. There is also no recognition or incentive for communities that make the effort to preserve biodiversity and produce NTFPs using sustainable practices. On the contrary, the burden is greater for them.

Moreover, due to the inappropriateness of the regulatory system that governs the extraction and commercialization of NTFPs, most extractors and traders stay out of the legal process, even where they are making efforts to improve their practices for sustainable use and management.

In any case, the government does not have the capacity to enforce its environmental laws. There are not enough PROFEPA technicians to monitor all agrarian nuclei and often the cases they do monitor are a response to conflicts between groups or individuals and do not necessarily promote the sustainable management of resources.

There is an urgent need for better understanding and recognition of the key role of NTFPs in local economies and as a tool for biodiversity conservation. As the agave case demonstrates, the obstacles and legal impediments that communities face when they want to use their resources legally have been resolved to a large extent, thanks to the support of institutions outside the communities. This represents a step forward in the development of alliances between groups of professionals and communities involved in natural resource management. However, these levels of external support are difficult to maintain on a large scale and in the long term. This is why it is necessary to develop regulatory schemes that remove impediments so that communities can autonomously use their natural resources in a sustainable way.

According to its promoters, the GI system should bring many benefits, not only to producers but also to consumers and local communities, as GIs are intended to add value and improve market access while providing for the protection of local know-how, diversity and natural resources; and as such, they claim, GIs can be a key development tool (OriGIn, 2003). Specialists have also pointed out that GIs can be useful for protecting biodiversity as well as certain forms of traditional knowledge and can help communities protect themselves from illegitimate patents (Waglé, 2004). However, as we have described, the present situation of mezcal in Mexico shows a number of ways in which GIs are not producing all the positive results identified by promoters of the international registration of GIs.

An interesting alternative to GIs for mezcales is being proposed by the Mexican National Commission for the Knowledge and Use of Biodiversity, CONABIO. It begins with the recognition and inclusion of diverse *Agave* species, mezcal regions, traditional processes and types of mezcales (Larson, 2005; Ocaña-Nava et al, 2007).

Recognizing that the generic term 'mezcal' cannot be applied to any one region, the system would have to identify 10–20 regions and allow each to characterize itself and establish its own GI, self-regulating council and set of rules, building from the bottom up. This would not preclude the existence of some basic general rules, but specific regions could develop unique environmental, cultural and technological spec-ifications. Each council could also have the right to monitor the sustainable manage-ment of the agave species in its region, decreasing the paperwork and costs (Larson, personal communication, 2004).

The legal framework for NTFPs needs to be reformed to do the following:

- Offer legal certainty to the owners of forest lands by establishing in a clear and coherent way the application of the legal instruments and administrative proce-dures of the Forestry Law and the Wildlife Law, while avoiding granting discretion to the authorities.
- Give true legal recognition to the local rules for sustainable management by inte-grating community management programmes through local institutions.
- Strengthen the legal framework for community level land-use planning.
- Offer legal recognition and validation to traditional knowledge.

As for public policies, much has to be done. Nevertheless, we would like to assert the importance of three strategic actions:

- Strengthen decentralization to the states of the power to authorize extraction, since on a smaller scale it is possible to make decisions in a more holistic way.
- Strengthen organizational processes within the communities, to achieve a higher and better level of regional participation by the communities, to increase possibili-ties for funding and support, and to develop production and commercialization projects based on their natural resources.
- Develop policies that allow access to legality with lower costs, considering legality as the door to commercialization.

NOTES

1   *Ejido* is a legal form of land tenure, recognized in the Mexican Constitution of 1917 as a result of the Mexican Revolution. Land owned by the state was given over to groups of peasants together with the rights over its use; this land could not be sold or rented. The *ejido* consisted of agricultural land as well as common forests and pastures. Due to the neoliberal policy measures introduced by the Mexican state in the 1990s, these peasant farmer lands can now be privately owned by the peasant families and can be sold to anyone who wants to buy them. Indian community land is also a legal form of land tenure. After the revolution, the Mexican state recognized the right of indigenous peoples to their traditional lands, the property that the Spanish Colony recognized as original Indian settlements and territories. The land is collectively owned by members of Indian communities and cannot be sold to outsiders. A community can give settlement rights to outsiders, however, without giving them formal land titles. 'Indigenous communities' must not be confused with the concept of Indian reserves or Indian territories as in the USA, Canada or other Latin American countries.

2   Mexican law defines a DO as 'the name of a geographical region of the country used to designate an original product whose quality and characteristics are due exclusively to the

natural and human environment' (IMPI). This form of GI was first employed in Mexico in 1977 to limit the use of the term *tequila* to the spirit produced with the blue variety of *Agave tequilana* Weber in a region where this species originated and is now cultivated.

3 We use the term 'agreements' for voluntary practices that are generally accepted; 'norms' for practices that are compulsory, carrying punishment (such as a fine, imprisonment, prohibition from taking part in festivities) for failure to to comply with them. Locally, the term used for both is *acuerdos* (agreements).

4 *Mayordomos* are a group of people responsible for organizing patron-saint festivities in each village. It is a position of status, changed every year or two.

5 A *mezcalero* is a person who owns the pots and pans and has the knowledge to transform the ripe agave plants into the distilled spirit called mezcal. His installation is called a 'factory'.

REFERENCES

Aguilar, J., Illsley, C. and Gómez, T. (2002) *Normas comunitarias campesinas e indígenas para el aprovechamiento de los recursos naturales de acceso común*, Grupo de Estudios Ambientales, AC, Mexico

Barrios, A., Ariza, R., Molina, M., Espinoza, H. and Bravo, E. (2006) *Manejo de la fertilización en magueyes mezcaleros* (Agave *spp.*) *de Guerrero*, Folleto Técnico 13, INIFAP, Mexico

Bray, D. and Merino, L. (2004) *La experiencia de las comunidades forestales en México: Veinticinco años de silvicultura y construcción de empresas forestales comunitarias*, SEMARNAT, INE, CCMSS, Mexico

Colunga-GarcíaMarín, P., Zizumbo-Villarreal, D. and Martínez Torres, J. (2007) 'Tradiciones en el aprovechamiento de los agaves mexicanos: Una aportación a la protección legal y conservación de su diversidad biológica y cultural', in P. Colunga-GarcíaMarín, A. Larqué Saavedra, L. E. Eguiarte and D. Zizumbo-Villarreal (eds) *En lo ancestral hay futuro: del tequila, los mezcales y otros agaves*, CICY, CONACYT, CONABIO, INE, Mexico

CONABIO (2002) *Mapas de distribución de agaves mezcaleros, escala 1:250,000*, Comisión Nacional para el Conocimiento y Uso de la Biodiversidad (CONABIO), Mexico

FAO (1995) *Memoria – Consulta de expertos sobre productos forestales no madereros para América Latina y el Caribe*, Serie Productos forestales no madereros – 1, Food and Agriculture Organization, Rome

García Meneses, P. M. (2004) *Reproducción y germinación de* Agave cupreata *Trel y Berger (Agavaceae) en la localidad de Ayahualco, Guerrero*, bachelor's thesis, Facultad de Ciencias, Universidad Nacional Autónoma de México

Illsley, C., Purata, S. E., Edouard, F., Hersch, P. and Robinson, D. (2005) *La normatividad oficial y los productos forestales no maderables en México*, paper presented at the 5th Iberoamerican Forest and Environmental Law Congress, Aguascalientes, Mexico, June

Illsley, C., Vega, E., Pisanty, I., Tlacotempa, A., García, P., Morales, P., Rivera, G., García, J., Jiménez, V., Castro, F. and Calzada, M. (2007) 'Maguey papalote: hacia el manejo campesino sustentable de un recurso colectivo en el trópico seco de Guerrero, México', in P. Colunga-GarcíaMarín, A. Larqué Saavedra, L. E. Eguiarte and D. Zizumbo-Villarreal (eds) *En lo ancestral hay futuro: del tequila, los mezcales y otros agaves*, CICY, CONACYT, CONABIO, INE, Mexico

Larson, J. (2005) *Collective Biological Resources: The Cases of Mezcales and Corn*, Latin American and Caribbean Regional Workshop on Sustainable Use, Buenos Aires, Argentina

Ley de Desarrollo Forestal Sustentable (2003) *Diario Oficial de la Federación*, Mexico

Marshall, E., Schreckenberg, K. and Newton, A. C. (eds) (2006) *Commercialization of Forest Products: Factors Influencing Success. Lessons Learned from Mexico and Bolivia and Policy Implications for Decision-makers*, UNEP World Conservation Monitoring Centre, Cambridge, UK

Martínez Rivera, L. M., Rosales, J., Gerritsen, P., Moreno, A., Iñiguez, L., Palomera, C., Contreras, S., Cárdenas, O., Rivera, L. E., Solís, A., Cuevas, R., García, E., Ramírez, M., Aguirre, A., Olguín, J. L., Santana, F., Carrillo, R., Pacheco, C. (2007) 'Implicaciones socioambientales de la expansión del cultivo de agave azul (1995–2002) en el municipio de Tonaya, Jalisco, México', in P. Colunga-García Marín, A. Larqué Saavedra, L. E. Eguiarte and D. Zizumbo-Villarreal (eds) *En lo ancestral hay futuro: del tequila, los mezcales y otros agaves*, CICY, CONACYT, CONABIO, INE, Mexico

Ocaña-Nava, D., García-Mendoza, A. and Larson, J. (2007) 'Modelación supervisada de la distribución de magueyes mezcaleros en México y sus posibles aplicaciones', in P. Colunga-García Marín, A. Larqué Saavedra, L. E. Eguiarte and D. Zizumbo-Villarreal (eds) *En lo ancestral hay futuro: del tequila, los mezcales y otros agaves*, CICY, CONACYT, CONABIO, INE, Mexico

OriGIn (2003) *Geneva Declaration*, Organisation for an International Geographical Indications Network, Geneva, www.origin.technomind.be/fileadmin/origin/PDFs/English/ORIGIN.Declaration.EN.pdf, accessed 21 May 2008

PROCYMAF (2000) *Base de datos de especies con usos no maderables en bosques de encino, pino y pino-encino en los estados de Chihuahua, Durango, Jalisco, Michoacán, Guerrero y Oaxaca*. Proyecto de Conservación y Manejo Sustentable de Recursos Forestales en México.

Reglamento de la Ley de Desarrollo Forestal Sustentable (2005) *Diario Oficial de la Federación*, Mexico

SE (1994) 'Denominación de origen mezcal', Secretaría de Economía, *Diario Oficial de la Federación*, Mexico

SECOFI (1994a) 'Norma oficial de bebidas alcohólicas, mezcal, especificaciones NOM-070-SCFI-1994-BEBIDAS ALCOHOLICAS-MEZCAL-ESPECIFICACIONES', Secretaria de Comercio y Fomento Industrial, *Diario Oficial de la Federación*, Mexico

SECOFI (1994b) 'Norma oficial de bebidas alcohólicas, tequila, especificaciones NOM-006-SCFI-1977-BEBIDAS ALCOHOLICAS-TEQUILA-ESPECIFICACIONES', Secretaría de Comercio y Fomento Industrial, *Diario Oficial de la Federación*, Mexico

SEMARNAP (1994a) 'Norma oficial mexicana de emergencia que establece los procedimientos, criterios y especificaciones para realizar el aprovechamiento de leña para uso doméstico', Secretaría de Medio Ambiente, Recursos Naturales y Pesca, *Diario Oficial de la Federación*, Mexico, 13 April, p35

SEMARNAP (1994b) 'Norma oficial mexicana de emergencia que establece los procedimientos, criterios y especificaciones para realizar el transporte y almacenamiento de corteza, tallos y plantas completas de vegetación forestal NOM-005-RECNAT', Secretaría de Medio Ambiente, Recursos Naturales y Pesca México, *Diario Oficial de la Federación*, Mexico, 20 May, pp9–11

Valenzuela-Zapata, A. G. and Nabhan, G. P. (2004) *Tequila! A Natural and Cultural History*, University of Arizona Press, Tucson

Valenzuela-Zapata, A. G. and Simoes, O. (2007) *Las denominaciones de origen y su importancia en la conservación de variedades de agaves y vides en México y Portugal*, paper presented at the 5th Congreso Europeo CEISAL de Latinoamericanistas

Waglé, S. (2004) *Geographical Indications Under TRIPS: Protection Regimes and Development in Asia*, SAWTEE Policy Brief No 8, Kathmandu, Nepal

Table 8.5 Appendix *Barriers within the national legal framework*

Key laws and implementing agencies	Activities regulated	Legal instruments and administrative procedures regarding NTFPs	Remarks
General Law for Ecological Balance and Environmental Protection (Ley General del Equilibrio Ecológico y la Protección al Ambiente) SEMARNAT	Environmental impact assessment, natural protected areas, hazardous waste, pollution and general activities related to forestry, water, wildlife and soil	Requires elaboration of an environmental impact assessment for NTFP extraction before extraction. The same assessment is needed for forest land-use change to other (non-forest) use, even in arid regions.	Within SEMARNAT's Department of Environmental Protection, there are three subordinate operating authorities that regulate NTFPs.
General Sustainable Forestry Law (Ley General de Desarrollo Forestal Sustentable) and its regulations SEMARNAT CONAFOR	Forestry for timber in natural forests and plantations Forestry for NTFPs for commercial use, domestic use and scientific research	Extraction for commercial use requires compliance with a written 'notice of use' submitted to SEMARNAT's state delegation before extraction. Additionally, the Forestry Regulation and the technical Mexican official norms (NOMs) establish exceptions to the general rule and require, for certain NTFPs, not only authorization but also a 'simplified management programme'. If the areas of extraction for domestic use are considered habitat for endemic, threatened or endangered wildlife, use of those products must not alter the necessary conditions for the subsistence, growth and evolution of the species in danger. Extraction for scientific research activities requires authorization. The extraction of timber and NTFPs for domestic use does not require authorization, but may be otherwise regulated under a NOM or legal regulation.	The written 'notice of use' has to be accompanied by a complex technical study that requires human and economic resources. For administrative procedures, all application forms and documents are received in one place. From there, each application is sent to the competent authority for resolution. Due to poor administrative coordination among the diverse authorities, this system does not simplify proceedings as it should. Domestic use, in legal terms, does not include low-profile commercial activities to provide the income necessary for the subsistence of poor families. SEMARNAT's authorities tend to use discretion in the application of regulations, depending on the importance of the species to be harvested and extracted. Application of the law varies from one SEMARNAT state office to another.

Table 8.5 Appendix *Barriers within the national legal framework* (Cont'd)

Key laws and implementing agencies	Activities regulated	Legal instruments and administrative procedures regarding NTFPs	Remarks
General Wildlife Law (Ley General de Vida Silvestre) SEMARNAT	Extraction of flora (which include NTFPs) and fauna, especially species listed in Mexican Official Norm NOM 059 that are at risk, in danger of extinction and under a special protection regime	Units for Environmental Management (Unidades para la Conservación, Manejo y Aprovechamiento Sustentable de la Vida Silvestre, UMAs) Implementing conservation activities in wildlife areas does not require authorization, but only notification of SEMARNAT. Those areas will be incorporated into the National System of Environmental Management Units (Sistema de Unidades de Manejo para la Conservación de la Vida Silvestre).	The UMAs are part of a national system of units that focuses on the development of productive activities for rural communities; the application of traditional biological knowledge; combating illegal traffic and appropriation of wildlife specimens; and the establishment of biological corridors between natural protected areas.
Federal Law for Establishing Norms and Standards (Ley Federal Sobre Metrología y Normalización) SEMARNAT SE	Mexican official norms (NOMs)	For NTFPs, NOMs establish the exceptions to the written notice required by law and determine specific procedures and criteria for commercial use, transportation and storage. NOMs also regulate domestic use and activities related to the scientific sampling of biological material of wildlife flora and fauna. SE creates NOMs for the regulation of GIs.	There are contradictions between some NOMs and the Forestry Law. NOMs created for GIs do not consider Forestry Law procedures; they duplicate and sometimes contradict them.
Industrial Property Law (Ley de la Propiedad Industrial) IMPI	Patents, marks, industrial and intellectual property disputes	GIs (denominaciónes de origen)	Has influence on certain NTFPs including wild agaves. Overlaps with environmental laws.

Notes: SEMARNAT: Federal Ministry of the Environment (Secretaría de Medio Ambiente y Recursos Naturales)
SE: Ministry of Economy (Secretaría de Economía)
CONAFOR: National Forestry Commission (Comisión Nacional Forestal)
IMPI: Mexican Institute for Industrial Property (Instituto Mexicano de la Propiedad Industrial).

Chapter 9

Fiji: Commerce, Carving and Customary Tenure

Francis Areki and Anthony B. Cunningham

INTRODUCTION

The relationship between forest resource commercialization, sustainable management and livelihoods is complex. On the one hand, expanded trade may improve forest management by increasing the value of forests, thereby providing incentives to local people to invest in long-term sustainable management strategies. This is particularly true when land tenure is strong, harvested tree species are used, and valued locally as well as commercially, and value-adding takes place. On the other hand, potentially negative effects of expanded trade in hardwoods for timber and non-timber use may include increased extraction rates from a declining resource base.

Widely distributed in South-East Asia, where it is heavily logged for commercial timber, with smaller populations in East Africa (Tanzania) and the western Pacific, *Intsia bijuga* is one of the most highly valued trees in the Pacific (Thaman et al, 2004). Due to a combination of habitat loss for farming, widespread use for traditional housing, commercial logging and commercial woodcarving, *Intsia bijuga*, a hardwood tropical tree, is now listed in Fiji under the Endangered and Protected Species Act. This chapter synthesizes lessons from the unsustainable logging of *Intsia bijuga* for use in woodcarving, not by a foreign company, but by local people on a remote Pacific island, Kabara, one of 100 small Fijian islands collectively known as the Lau Group. The Lau Group lies midway between Tonga and the main islands of Fiji (Viti Levu and Vanua Levu), with Kabara about 250km by sea from Fiji's capital, Suva. Kabara covers a total land area of 33km and is composed almost entirely of limestone, with the exception of a volcanic outcrop along the north-west side of the island. Due to the extremely rough terrain, this limestone area is still covered by natural forest, whereas the volcanic outcrop, with its better soils and more accessible terrain, has been cleared for cultivation.

This case study could be considered a 'natural experiment' focused on a timber species of outstandingly high cultural and economic value, found in a landscape where all land is under customary tenure, foreign logging companies are not present and loss

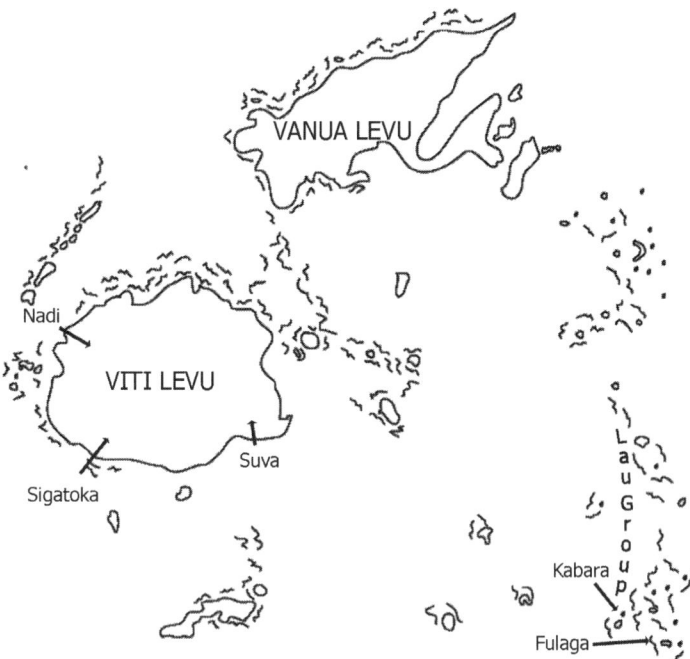

Source: Public domain base map from www.commons.wikimedia.org/wiki/File:Fiji-map-blank.png, modifications by Elizabeth Skinner.

Figure 9.1 *Map of Fiji*

of habitat due to agricultural expansion is minimal. The question this chapter addresses is: what happens in such an 'ideal' case from a conservation perspective, when people want to enter the cash economy, their income-earning options are limited and new logging and woodcarving technologies are introduced that make cashing in on existing resources possible on an entirely new scale? And following from this, what combination of customary and statutory policies could support sustainability and equity in the trade?

In this chapter, we first describe the cultural significance of *Intsia bijuga* in Fiji. We then highlight the extent to which households on Kabara rely on income from carving *Intsia*, and how this has changed since the 19th century through the adoption of new woodcutting and carving technologies. Finally, we examine *Intsia bijuga* use in terms of land tenure and forest legislation in Fiji and how such a high-value tree species is managed on remote islands like Kabara, far from state control, yet with strong customary law and cohesive communities.

CULTURAL VALUES OF *VESI*

The cultural significance of *Intsia bijuga* (known in Fijian as *vesi*) is diverse and is evident in many Fijian cultural expressions and beliefs. The tree itself was sacred among ancient Fijians as it was believed that the first Fijian man and woman had sprung forth and were reared beside the tree by the ancestor-god, Degei. Sacred *Intsia bijuga* groves were maintained by ancient Fijians near their temples for the purpose of performing rituals and sacrifices. *Intsia bijuga* was used for the main pole to hold up traditional temples and chiefs' *bures* (houses), to build the *drua* or *waqa tabu* (sacred canoe) reserved for those of noble birth and to make the *lali* (traditional gong) used to announce important events. *Intsia* is a totemic tree for several Fijian clans and its hard and seemingly indestructible wood is equated with hardiness as a desirable human quality. *Intsia bijuga* has long been an honoured tree, and many ruling noble families incorporated *Intsia bijuga* into their clan names, such as the family name Lalagavesi (meaning 'wall of *Intsia bijuga*') of the ruling dynasty of Cakaudrove in northern Fiji.

Many Fijian native expressions and traditional practices still in use today incorporate the word *vesi* to identify a person of noble birth or one of strong character. For instance '*Sa ciri na vesi* (The *Intsia bijuga* is afloat)' is used when bidding farewell to a paramount chief, '*kaukauwa vaka na vuni vesi* (strong as the *Intsia bijuga* tree)' describes a person of firm character, and '*sa bale na vesi levu*' (the great Intsia bijuga tree has fallen) refers to the death of a high chief. Important household items carved from *Intsia bijuga* have also made their way into native proverbs, such as the *kali* (traditional headrest) used as a taunt in quarrels: '*Na kali oqo, na kali oqori*,' (literally: 'The headrest here, the headrest there,' meaning, 'If you have power, so do I'). When *kava*, the ritually important drink made from *Piper methysticum* roots, is mixed in the *tanoa* (a kava bowl carved from *Intsia bijuga*) and presented during ceremonial occasions the mixer will call out, '*E saqa ena kuro vesi*' (It is cooked in the *Intsia bijuga* pot) (Davies and Bulicokocoko, 1960). Traditional medicines derived from *Intsia bijuga* are also widely used in Fiji. The bark and leaves are used in decoctions and infusions are used to treat various illnesses such as rheumatism, diarrhoea, arthritis, asthma, colds and headaches (Cambie and Ash, 1994).

However, the tree was and is primarily used for its wood in the manufacture of traditional handicrafts, canoe building and house construction. Woodcarving traditions in Fiji can be traced back to a single area of origin in the southern Lau Group. The art of fine woodcarving is considered to have been brought to the Lau Group first by the Samoans in the 18th century and then from Tonga in the 19th century (Thompson, 1940; Hooper, 1982). Woodcarving expertise existed prior to these arrivals, but the fine carving skills found today owe much to the influence of carvers from the Samoan Lemaki clan on Kabara and the Tongan Jafau clan on Fulaga. Carving is now no longer kept within these particular clans but shared among all clans from these two islands.

The carving and manufacture of cultural items, such as kava bowls, headrests and ocean-going canoes, is far more common in the Lau Group, in particular the islands of Kabara and Fulaga, than other parts of Fiji, primarily because these arid limestone islands harbour some of the richest stands of *Intsia bijuga*. So important was the timber from this tree that control of its stock and monopolizing the trade in crafts made from *vesi* impacted greatly upon Fiji's political landscape and history, including provincial demarcations, in the 19th century. As Banack and Cox (1987), page 161, note:

> *In a sense, the floristic resources of Kabara are an analogue of modern day strategic minerals. Just as the possession of titanium resources – important in aircraft*

> *construction – bestows strategic importance to a country possessing them, possession of large trees of Intsia bijuga useful in canoe construction conferred a political advantage on Kabara. Material constraints of ocean-going canoe construction led early Tongan shipwrights to capitalize on the floristic resources of Kabara.*

For example, the Tongan king Enele Ma'afu's ambitions and conquest of Lau were partly fuelled by his need to gain control of *Intsia* trees as a strategic material, for whoever controlled the *Intsia bijuga* forests and craftsmen in Lau monopolized South Pacific trade in war canoe construction and weaponry. So valuable were the double-hulled sailing canoes that historical anecdotes indicate Ma'afu traded a war canoe made from *Intsia bijuga* with the high chief from Cakaudrove for Vanua Balavu, an island about 53km^2 in size (Haddon and Hornell, 1936).

COMMERCIALIZATION, CULTURE AND LIVELIHOODS

Intsia bijuga still plays an important role in local culture, since most major traditional functions continue to require the presentation of artefacts such as the *tanoa* (kava bowl), *lali* (slit gong) and canoes. In addition, a commercial woodcarving industry has sprung up since the 1960s alongside the tourism industry. Fiji's cash economy is based on a relatively narrow base, principally tourism, sugar cane production, the garment industry and marine products. Tourism, Fiji's fastest-growing and largest foreign revenue earner, is worth FJ$496 million per year (19.2 per cent of GDP), twice the value of the sugar industry (FJ$222 million per year, 8.5 per cent of GDP) and higher than the garment industry (FJ$313.9 million per year, 12.3 per cent of GDP) (Narayan and Prasad, 2002), followed by commercial timber (2.5 per cent of GDP) (ITTO, 2004). In 2004, more than half a million tourists visited Fiji (Fiji Islands Bureau of Statistics, 2007), many of them buying woodcarvings, no doubt encouraged by a 20 per cent devaluation of the Fijian dollar in 2005. Over 100 handicraft businesses are located in major tourism areas such as Nadi, Sigatoka and Suva, including both large business premises and small stalls at markets frequented by tourists. In addition, woodcarvings, most of *Intsia bijuga*, are exported through middlemen and freight companies.

In Fiji, woodcarving, most of it using *Intsia bijuga*, has an ambivalent relationship with tourism. On one hand, the tourism industry can be partially blamed for the demise of *Intsia bijuga* due to the increased demand by foreign visitors for cultural carvings. On the other hand, tourist demand encourages carvers to capitalize on their traditional skills and knowledge, maintaining art forms that would have otherwise been lost, such as the production of war clubs, headrests and divining plates, few of which are used by communities any longer.

Woodcarving is the main source of income on both Kabara and Fulaga in the Lau Group, as well as for many families who have relocated from these islands to urban settlements such as Suva. On Kabara, which has a resident population of just 482 people in four coastal villages – Naikeleaga, Tokalau, Lomati and Udu – 96 per cent of households depended in 2003 on carving the wood of the *Intsia bijuga* tree as their main or only source of cash income. Other income sources nominated by villagers during our study in 2003 were copra (37 per cent), sea cucumbers (18 per cent), paid employment (7 per cent), small business (4 per cent) and other handicrafts (3 per cent). Since 2003, household income from copra and sea cucumbers has gradually fallen due to poor domestic prices and disruptions to the inter-island shipping service. Although subsistence farming (cassava, sweet potatoes, coconuts) and fishing are important, cash income from carving is crucial to people's livelihoods, as cash is needed for boats, boat motors and fuel, fishing equipment and school fees. Alternative sources of income such as the sale of trochus shells and sea cucumbers are poorly developed, and copra prices are low, so these are less lucrative than woodcarving.

TECHNOLOGY AND CHANGE

Since the 18th century, changing tools and technology have had an impact on the harvesting and trade of *Intsia*. African blacksmiths have been smelting iron and producing iron tools for millennia, but woodcarvers in West New Britain, Papua New Guinea, first obtained iron tools as recently as 1895, a century after they were available in Fiji. This was particularly significant because *Intsia bijuga* wood has a fine grain, rarely cracks and takes a fine polish, but is difficult to carve. Before the arrival of metal tools in the Pacific in the 18th and 19th centuries, stone adzes and shell scrapers were used by carvers in a slow process.

The availability of metal blades for adzes on Kabara increased the speed at which wood could be carved, but the process was transformed even more drastically by chainsaws, which are used now to fell the *Intsia* trees and to rough-saw and shape sections of logs for *tanoa* (kava bowls). In the past, it took a month of one person's time to make a single *tanoa*. Today, the same bowl can be carved in a day. As carvers seek to match the income attained in urban areas, the result is a cycle of rapid, unmanaged harvesting of *Intsia*. New production processes and power tools have made production quicker, but they also mean that new strategies for resource conservation and management are needed, with new laws and policies.

RESOURCE DEPLETION, COLLAPSE OF LIVELIHOODS AND THREATS TO A CULTURALLY IMPORTANT SPECIES

The result of heavy dependence on *Intsia bijuga* trees for income generation, primarily through the kava bowl trade (Figure 9.2D), has resulted in an unsustainable level of extraction. Furthermore, unprocessed *Intsia bijuga* for urban enterprises is derived mainly from islands far from where products are sold. Consequently, *Intsia bijuga* harvests from Kabara support not only local groups, but also carvers in urban centres. This puts added pressure on the limited resource and threatens the basis of a wide range of groups' cash income.

Notes: A. Kava bowls from *Intsia bijuga* are essential for the preparation of kava (from roots of *Piper methysticum*).
B. Large kava bowls, crucial for ceremonial use, have to be made from large, old *Intsia bijuga* trees.
C. Commercial production and the use of chainsaws result in high levels of wood wastage.
D. Local and tourist demand for kava bowls is high: all these kava bowls stacked in a storeroom in Suva are from Kabara island.

Source: A. B. Cunningham.

Figure 9.2 *Cultural value and production of kava bowls* (tanoa)

An inventory of existing *Intsia bijuga* stocks on Kabara has demonstrated that remaining natural stands are limited to the centre of the island (8 per cent of the island total area) in limestone landscapes that are difficult to access. In addition, harvested areas throughout the island show poor regeneration in sample plots and the standing stock suitable for future woodcarving activities is depleted, portending a total collapse of the island's carving industry within 10–15 years. Anecdotes from carvers in neighbouring Fulaga also indicate that *Intsia bijuga* trees of the size required for carving are heavily depleted in their forests. Generally, on both islands, large quantities of wood have been wasted as a result of only prime sections being removed for making *tanoa* (Figure 9.2C).

The diminishing status and possible loss of genetic diversity of many Pacific native trees has been highlighted in numerous regional reports (Thaman et al, 2004). *Intsia bijuga* is among the top ten priority species for immediate conservation and management attention due to its presence in ecologically sensitive ecosystems such as littoral forests and mangroves in the Pacific. The rapid depletion of *Intsia bijuga* is not limited to Fiji, but is widespread throughout the Asia-Pacific region. The biology of *Intsia bijuga* makes it inherently vulnerable to increased harvest pressure: the species does not disperse well, is extremely slow-growing (maturing at more than 75 years) and is extremely scattered in distribution (Atherton and Martel, 1998).

Furthermore, in Fiji there is a glaring lack of conservation and research priority placed on Fiji's native trees, with more emphasis being given to the introduced and more commercially viable species of pine and mahogany. Though highly lucrative as commercial timber, these two species do not have such a diverse range of uses and or cultural importance as *Intsia bijuga*, which, aside from its timber and woodcarving uses, is traditionally valued as a medicine, source of dyes and part of Fijian culture, folklore and symbolism.

It is clear that customary law and controls are no longer sufficient to protect resources that are under powerful new commercial pressure from the tourist trade and subject to significantly more intense harvest pressure due to advances in technology. Forests on islands such as Kabara are valuable for biodiversity conservation and to the local community, but are remote from State control. For this reason, the conservation and sustainable use of these forests has to involve local communities, in the context of national forest policy development.

NATIONAL AND INTERNATIONAL POLICY CONTEXT

The main areas of governmental policy that affect the harvest, management and trade of *Intsia bijuga* are land tenure and resource rights, forest policy and environmental law.

Land tenure and resource rights

Landownership systems in Fiji, as elsewhere, are central to the ways in which forests are managed. There are three basic categories of landownership in Fiji. The first is land under customary tenure ('native lands'), constituting 82.5 per cent of total land area. This land is administered by the Native Lands Trust Board (NLTB), which is chaired by the Minister for Fijian Affairs, with members nominated by the Great Council of Chiefs, plus one or two government representatives. The second category is Crown or State land (9.5 per cent), which includes national parks and reserves, and the third is freehold land (8 per cent). While the proportion of land under customary tenure in Fiji is high at 82.5 per cent, it is higher in Papua New Guinea (97 per cent) (Lakau, 1991). However, it is much lower on Pacific islands such as Hawaii, where colonization has resulted in the opposite situation, with most land in private ownership and the remainder owned either by the State of Hawaii or the US federal government (Banner, 2005).

In Fiji, 82.5 per cent of native lands are communally owned, but this proportion includes land specifically demarcated by boundaries according to clan or *mataqali* groups. This landownership regime resulted from the British colonial government in the 1880s forbidding the sale of land to colonial settlers, in recognition of the crucial link between culture and customary landownership. Fiji is consequently one of the few countries in the world where the interests of its indigenous population are protected through widespread customary land tenure. Native land cannot be sold, but it can be leased through the NLTB, set up in 1940 to act as a trustee for traditional landowners and to ensure the equitable sharing of benefits derived through leasing customary land.

Forest policy

The original legislation governing Fiji's forest resources was the Forest Act (1953), which resulted from Fiji's first national forest policy, developed and endorsed by the colonial legislative assembly in 1950. Although the Forest Act guaranteed Fijian customary rights for forest product extraction for both traditional and subsistence needs, the Act was geared towards timber and did not mention NTFPs. In addition, it did not provide for the protection of native forests, nor require that reforestation following logging should use native species (rather than introduced species such as pines or mahogany). In 1960, a government inquiry into the utilization of natural forest resources identified limitations in the 1953 Forest Act as arising in part from the complex nature of landownership and consequent difficulties in developing a coherent policy that promoted the sustainable use and conservation of local forests.

In the late 1980s, the Fiji government initiated another review of its forestry sector, which highlighted the need for more sustainable forest management practices, and recommended the repeal of the 1953 Forest Act, which was subsequently replaced by the Forest Decree of 1992. The decree reaffirmed sections of the original Forest Act, such as customary rights to forest resource access and utilization, also extending its scope to include non-timber forest products. In particular, the Forest Decree of 1992 established a permitting process under which licences are required for the extraction of NTFPs.

In addition to addressing NTFPs, the 1992 decree instituted a mandatory requirement for management plans prior to the issuance of a logging licence, aimed at controlling illegal logging and making logging operations more sustainable. Although the 1992 decree included penalties for unsustainable and destructive logging operations, its technical and personnel capacity for implementation was limited. Despite good intentions, the 1992 decree, like the 1953 Forest Act, continues to emphasize the commercial value of timber extraction over ecosystem health and long-term sustainability. Alongside this 'fiscal forestry', a further weakness is that the requirements for reforestation following logging (section 14) are vague and, as with the 1953 Act, do not require the planting of native species. The economic value of the large areas of Fiji planted with South American mahogany (*Swietenia mahogani*) and pine (*Pinus caribaea*) plantations is high, and this has caused significant environmental problems in Fiji, resulting in large areas of the landscape being planted with introduced, exotic species to the detriment of ecosystem functions and of native species with high cultural value. *Intsia bijuga* is a good example, as this species was heavily exploited on the main Fijian islands for commercial timber.

In 2003, the Fiji government's Forestry Department began to review and reformulate a new national forest policy. This was in response to changing national perspectives on the role of forests and Fiji's international commitments, such as the Convention on Biological Diversity and discussions arising out of the United Nations Forum on Forests. The shift in emphasis away from a timber-centred focus towards conservation and the sustainable use of forests was significant. The Strategic Forestry Plan, 2002–2005, which supplements Fiji's National Forestry Action Plan, for example, has four aims:

- to provide an appropriate institutional and physical infrastructure to support the development of the forestry sector;
- to ensure the sustainable development and management of forest resources;
- to promote community-owned and managed forest-processing and value-adding facilities based on indigenous forests and community-owned plantations; and
- to promote the production and export of value-added timber products (ITTO, 2005).

In 2007 the Forestry Department completed its extensive national consultation and in November of the same year the cabinet endorsed and passed the new national forest policy. Unlike the 1953 Act and the 1992 decree, the new policy explicitly emphasizes the productivity of the forestry sector in the context of sustainability, reiterating the need for effective ecosystem management, the protection of biodiversity, the reforestation of logged areas with native trees and the engagement and capacity building of indigenous landowners to sustainably manage and maximize benefits from their forests.

The policy also encourages the development and sustainable management of NTFPs, advocating greater government investment in non-timber forest enterprises through product development and marketing to enable local communities to realize their full potential and enhance livelihoods and commercial opportunities. Whether sufficient funds will be made available and the capacity built to implement these policy innovations remains to be seen.

International and national endangered species environmental law and policy

In 1997, Fiji became a signatory to the Convention on International Trade in Endangered Species of Wild Fauna and Flora (CITES). Part of its commitment under the convention was to establish a Fiji Islands CITES Management Authority to regulate and control CITES-listed species.

Partly as a response to these obligations, the government enacted the Endangered and Protected Species Act in 2002. Although *Intsia bijuga* is not listed under CITES in any of its three appendices, Fiji included *Intsia bijuga* under Schedule 2 of the Act, which lists all species indigenous to the Fiji Islands not already included under CITES, but considered threatened locally. Those who contravene this part of the Act are liable upon conviction for a first offence to a fine of FJ$20,000, and for any subsequent offence to a fine of FJ$100,000 or imprisonment for five years. This only applies, however, to the export of species included in Schedule 2 without a permit. For domestic commercial trade in *Intsia bijuga*, the 2002 Act requires, under section 23, all persons involved to register with the management authority. If this is not done, traders potentially face a fine of FJ$5,000 or two years' imprisonment.

Although the Endangered and Protected Species Act attempts to provide some control and regulation of the *Intsia bijuga* trade within Fiji, the majority of those involved in the handicrafts industry are still ignorant of this requirement. To date there has been minimal effort to ensure that the local trade in *Intsia bijuga* is monitored or that traders in *Intsia bijuga* and its products comply with the requirements of the Act. Despite the good intentions outlined in this Act, fulfilling those intentions will require higher budget allocations and a commitment to building the technical skills and capacity needed for its implementation. Given the great distance of Kabara from the nation's capital, Suva, the implementation of the Act requires not only capacity building at the level of the State, but also technical support and training at the local community level for resource management planning and monitoring.

COMMUNITY MANAGEMENT OF INDIGENOUS FOREST: SOCIAL LEARNING FOR IMPLEMENTATION

In Fiji, indigenous communal groups have tenure over 89 per cent of the unexploited forests. Commercial logging on land under customary control, or native land, is only done after consent from the *mataqali* and the NLTB. Timber-cutting rights are generally negotiated between concessionaires or licensees and the NLTB, which authorizes the Forestry Department to issue logging licences and to administer concession agreements.

This process works on the main islands (Viti Levu and Vanua Levu), which are close to Suva, where the NLTB sits, but for several reasons the situation on Kabara is different, requiring support for decentralized forest management. First, distance and communication difficulties isolate Kabara from government administration. The island is 250km by sea from Suva, visited once a week at most by a single wooden steamer (the *Kabara*) built on the island. Inter-island communications have improved since the first telephones were installed in 2004, but even today relatively few people have telephones. Second, although *Intsia bijuga* trees are felled commercially, this is not done by concessionaires or licensees, but by community members. Third, tenure over the forest area where *Intsia bijuga* occurs on Kabara is complex and poorly defined.

In the past, two main factors influenced how many *Intsia bijuga* trees were felled for canoe construction. The first was the long and laborious process of tree felling and carving. The second was the ceremonial process required under customary law, which involved gifts of a sperm whale's tooth to local leaders and the main carpenter constructing the sailing canoe (Banack and Cox, 1987). Today this situation has changed. No whales' teeth are required as gifts before *Intsia bijuga* trees are felled for kava bowl production, and they are produced rapidly with the new technology of chain saws and metal adzes (as explained under 'Technology and change' above).

The only practical solution is a process of consultation, training and local capacity building for a community-based management plan for *Intsia bijuga*. This is the process WWF Fiji have been involved in on Kabara since 2004, starting with meetings in each village on why a forest inventory was required, the approach that needed to be taken and who from the local community would work as field assistants to do the inventory. At the completion of analysis of data gathered during field surveys (forest inventory, socioeconomic survey and market survey), presentations were held to inform people of the findings of the survey.

Based on a forest inventory sampling, only 8 per cent of existing forest on the island now supports *Intsia bijuga* populations. Current volumes of *Intsia bijuga* cut annually indicate that all of the large (i.e. 30cm or more in diameter at breast height) trees still standing will have been felled by 2010. The reaction was one of amazement at how much wood wastage took place and how little uncut forest remained. Local people discussed these research results and outlined their thoughts on how they would like to ensure that *Intsia bijuga* did not disappear from Kabara's forests. The major recommendations arising out of the study and presented to the community were:

- to develop and begin implementing an effective community forest management plan;
- to initiate a community reforestation programme in order to restock *Intsia bijuga* in depleted areas;
- to source and develop alternative income options that would alleviate the existing heavy dependence on *vesi* for income;
- to improve the woodcarving skills of community members to reduce *Intsia bijuga* wood wastage and utilize *Intsia bijuga* offcuts for other wood products; and
- to develop a community marketing strategy and outlet for their products.

Since 2005 the Kabara community, WWF Fiji, the Forestry Department and VASS NZAID have been working together to implement these recommendations. A draft community forest management plan has been developed to manage the remaining forest areas on the island and other forest resources. The community have demarcated and endorsed four community forest reserves on the island, covering an estimated 5km², to conserve the remaining *Intsia bijuga* trees there and to restock the areas through a reforestation programme. In 2005 the community established a forest nursery where community youth were trained to collect *Intsia bijuga* seeds and grow seedlings. As of July 2007, roughly 3000 *Intsia bijuga* seedlings had been planted in four of the community forest reserve areas, covering a total area of 10.4ha on the island, and the Kabara community intends to expand this process over the next five years.

The situation on Kabara can best be described as a localized examplification of the dependency theory. *Intsia bijuga* is the mainstay of the local economy, and because there has been no investment in income source diversification on the island, that essentially makes the whole community vulnerable to sporadic external change. The fact that the community do not have control over the market for their products and are not active in negotiations about the price of their commodities exacerbates the problem.

A further concern highlighted by the study is the growing dependence of the community on food imported from the mainland. This problem is compounded by the fact that much of the land area on Kabara is limestone and not suitable for cultivation. However, as a measure to reduce the community's growing dependence on imported goods – and on harvesting *Intsia bijuga* to generate the income needed to obtain those goods – an extension of the project carried out on the island has been to build, through training, the community's capacity for effective integrated land-use management and improved food security. This exercise, supported by the Agriculture Department, has aimed at crop diversification and boosting meat production on the island – the latter by way of chicken coops, managed by the local women's group, that now not only cater for local meat demand, but also enable income diversification.

With further support from the Forestry Department, community members have been trained and certified in the use of woodcarving lathes as part of the effort to begin utilizing the *Intsia bijuga* offcuts strewn throughout the island in vast amounts. Capacity-building training has included diversifying local skills beyond the traditional types of carving to the manufacture of ornaments and the creation of modern artworks. As a complementary exercise, the community women have received training to develop their handicraft skills, another effort to enhance alternative income sources for the community.

As of 2008, the Kabara community are in the process of opening their own woodcarving workshop in Suva, Fiji's major urban centre, and beginning to assess markets for their products. The idea behind this initiative is that with better control and management of their *Intsia bijuga* resource and the production of wood products, and with effective negotiation with buyers and markets, they will be able to leverage better prices. This, coupled with the development of other income opportunities in the long term, should alleviate the pressure on *Intsia bijuga* and sustain industry and craft practices on the island.

CONCLUSION

Without immediate efforts by the government and by communities on Kabara to regulate and improve the management of their *Intsia bijuga* stock, it is likely that the species will suffer a loss of genetic diversity and a diminished role in sensitive ecosystems, which will threaten local livelihoods and the economic future of the island. The projected exhaustion of usable *Intsia bijuga* on the island in the near future would mean a loss of household income for the majority of families. In addition, for millennia the people of Kabara and Fulaga have produced the wood artefacts used in major ceremonial and traditional gatherings in the Province of Lau, and the depletion of *Intsia bijuga* would prevent them from fulfilling their traditional obligations and role effectively. Another possibility is that the loss of *Intsia bijuga* would further erode the community's cultural heritage as younger generations lose traditional skills handed down from father to son.

It is clear that customary law and controls are no longer sufficient, in the face of powerful new commercial pressures from the tourist trade, to protect a species that is slow-growing, occurs in low densities and is subject to significantly increased harvesting intensity due to advances in technology.

Although focused on commercial timber, the aims of the Strategic Forestry Plan (2002–2005) apply equally well to a commercially carved species like *Intsia bijuga*.

Policy-making is not enough, however. What is needed are practical steps to deal with a common need – the 'implementation crisis' that besets conservation at a global level, as described by Knight et al (2006), who recommend five steps in the implementation process:

- links to an appropriate conceptual framework;
- attention to social learning and action research;
- stakeholder collaboration;
- the development of an implementation strategy; and
- links with land-use planning.

In the Kabara case, this would include a planning process for tourism enterprises. Since 2004, through collaboration between the local community and the WWF South Pacific Programme, these steps have been followed. There is no short cut to success, however. *Intsia bijuga* is a slow-growing tree, the stocks of which have been 'mined' rather than managed for a long time. Long-term success depends not only on the implementation of community-based management for *Intsia bijuga*, but also on diversifying the local economy to reduce dependence on this species as the island's major source of income.

REFERENCES

Atherton, J. and Martel, F. (1998) *Sustainable Management of Ifilele in Uafato Conservation Area, Samoa*, South Pacific Regional Environment Programme, Apia, Samoa

Banack, S. A. and Cox, P. A. (1987) 'Ethnobotany of ocean-going canoes in Lau, Fiji', *Economic Botany*, vol 41, pp148–162

Banner, S. (2005) 'Preparing to be colonized: Land tenure and legal strategy in nineteenth-century Hawaii', *Law and Society Review*, vol 39, pp273–314

Cambie, R. C. and Ash, J. (1994) *Fijian Medicinal Plants*, Commonwealth Scientific and Industrial Research Organisation, Canberra

Davies, F. and Bulicokocoko, S. (1960) *Fijian Idioms, Colloquialisms and Customs*, Education Department, Suva, Fiji

Fiji Islands Bureau of Statistics (2007) Statistical Agency for the Fiji Government, www.statsfiji.gov.fj, accessed 20 October 2008

Haddon, A. C. and Hornell, J. (1936) *Canoes of Oceania. P.294: Rotuma and Tongan Dependencies*, Bishop Museum, Honolulu, Hawai'i

Hooper, S. (1982) 'A Study in the Valuables in the Chiefdom of Lau, Fiji', PhD Thesis, University of Cambridge, Cambridge

ITTO (2004) *Achieving the ITTO Objective 2000 and Sustainable Forest Management in the Republic of the Fiji Islands*, report of the Diagnostic Mission, presented at 37th session of the International Tropical Timber Council, December, International Tropical Timber Organization, Yokohama

ITTO (2005) 'Fiji', in ITTO, *Status of Tropical Forest Management*, International Tropical Timber Organization, Yokohama, pp135–141, www.itto.or.jp/live/Live_Server/1238/Fiji.e.pdf, accessed 17 September 2008

Knight, A. T., Cowling, R. and Campbell, B. M. (2006) 'An operational model for implementing conservation action', *Conservation Biology*, vol 20, pp408–419

Lakau, A. A. L.(1991) *State Acquisition of Customary Land for Public Purposes in Papua New Guinea*, Department of Surveying and Land Studies, Papua New Guinea University of Technology, Lae, Papua New Guinea

Narayan, P. K. and Prasad, B. C. (2002) 'Fiji's sugar, tourism and garment industries: A survey of performance, problems and potentials', *Fijian Studies*, vol 1, no 1, Fijian Institute of Applied Studies, Suva, Fiji

Thaman, R. R., Thomson, L. A. J., DeMeo, R., Areki, R. and Elevitch, C. R. (2004) *Intsia bijuga* (vesi), in C. R. Elevitch (ed) *Species Profiles for Pacific Island Agroforestry*, Permanent Agriculture Resources, Honolulu, HI, www.traditionaltree.org, accessed 17 September 2008

Thompson, L. (1940) *Southern Lau, Fiji: An Ethnography*, Bishop Museum, Honolulu, HI

One Eye on the Forest, One Eye on the Market: Multi-tiered Regulation of Matsutake Harvesting, Conservation and Trade in North-western Yunnan Province

Nicholas K. Menzies and Chun Li

INTRODUCTION

During the early 1980s, as China joined the mainstream of international trade networks, the residents of villages in the mountains of north-western Yunnan province found that there was a market in Japan for a mushroom that had previously been of little interest to Chinese consumers. Official statistics for the export of *song rong* 松茸 – the Chinese name for what is known in Japan and internationally as 'matsutake' – show exports to Japan from Yunnan province alone rising from less than 20 tonnes in 1985 to more than 1200 tonnes in 2001, and an estimated 1420 tonnes in 2005 (Leique Research Institute, 2002, p11; Li, 2006). Japan also imports large quantities of matsutake from Sichuan province, Tibet and Jilin province, as well as from North and South Korea, Bhutan, central Mexico, western Canada and the north-western States of the USA. In Yunnan, the dramatic increase in harvest levels has raised concerns about the environmental impacts of possible overharvesting, while unpredictable price fluctuations in a monopsonistic market affect the livelihoods of harvesters and a chain of intermediaries stretching from remote mountain villages to the international airport in the provincial capital of Kunming.

This chapter describes how a multi-tiered system of regulation has emerged to promote the sustainable production of matsutake in the absence of a declared national or provincial policy governing NTFPs. The harvesting and export of matsutake in China are now regulated formally, as it is a protected species under national legislation, implemented by different levels of government administration, and more informally under village rules defining rights of access, monitoring procedures and marketing. This chapter argues that in the case of matsutake in Yunnan, this system has allowed each level of governance to exert control over the aspects of harvesting and marketing that it is best equipped to manage, and that

Source: Nick Menzies.

Figure 10.1 *Jiedi: Matsutake buyers and sellers at the official village matsutake market waiting for the signal to begin trading*

the outcome is likely to be better, in terms of both conservation and local livelihoods, than any attempt to formulate an all-encompassing policy framework under any one level of government.

A BRIEF HISTORY OF MATSUTAKE

'Matsutake' is a Japanese name commonly used to refer to a number of edible species of the genus *Tricholoma*, which associates symbiotically with conifers in some places and with oak or mixed pine and oak forests in others. The genus is found in Asia, the Americas, north Africa and Europe. The most widespread species in Asia is *Tricholoma matsutake*, and closely related species (Matsushita et al., 2005). *Tricholoma matsutake* is a brownish-white mushroom with a cap 5–15cm in diameter when open, although connoisseurs prefer to eat it while the cap is still closed. Its flesh is white, with a fragrance frequently described as 'spicy' and 'aromatic'.

Tricholoma matsutake is one of many edible fungi known as ectomycorrhizal fungi. It forms a sheath of filaments known as hyphae around the new roots of the host tree. The hyphae grow into the roots, forming a network that facilitates the symbiotic exchange of water, minerals, carbon and other essential nutrients between the fungus and its host. The edible mushroom itself is the fruiting body of the fungus, which emerges from the mycelium, a spreading mass of hyphae extending away from the host tree in the duff, or litter layer on the forest floor. The mycelium is sensitive to changes in the condition of the duff layer and to temperature and humidity in particular (Yun et al, 1997; Yang, 2004). The regeneration and sustainable production of matsutake are therefore highly sensitive to disturbance during harvesting, but not enough is yet known to determine the ideal conditions under which mycelium will be present, so active management to increase production is still experimental, and efforts at domestication and cultivation have not been successful (Hosford et al, 1997; Liu et al, 1999). The only viable source of matsutake continues to be harvesting from the forest.

In Japan, matsutake is particularly appreciated for its reputed medicinal and anticarcinogenic properties, and it has a long history as a gift signifying respect. The earliest known reference to matsutake is in a poem written in AD 759. In more recent literature, many popular, bawdy short stories have drawn their inspiration from its phallic shape (Ohara, 1994, cited in Hosford et al, 1997, p5). Once restricted to the imperial court, matsutake has become a prized delicacy served as the heat of summer gives way to the cooler days of autumn, between late August and early October. In the Japanese market, discriminating buyers select matsutake for fragrance, freshness and colour – the whiter the better – with a preference for a large, closed form.

In China, matsutake also has a long, if less exalted, history. Matsutake first appears as a nutritional supplement in a herbal compendium published under the Zheng Zong emperor of the Song Dynasty (AD 1082–1094). In ethnic Tibetan areas of what is now south-western Sichuan province, as well as in Tibet itself, families offered gifts of dried matsutake as a valued gift on special occasions. Early 20th-century records show an annual average of ten tonnes of matsutake traded between 1909 and 1912 in the markets of Kangding in what is now the Ganzi Tibetan Autonomous Prefecture in Sichuan province (Leique Research Institute, 2002, p10). In China as a whole, however, matsutake had little commercial value in the past. It was so insignificant that it did not even feature until very recently in the otherwise apparently inexhaustible repertoire of ingredients in Chinese cuisine.

Japan has always been, and still is, the world's major consumer of matsutake. Until the mid-20th century, domestic supply satisfied demand. With increasing prosperity, the Japanese market for luxury foods has expanded. At the same time, matsutake production has declined with successive outbreaks of a devastating blight affecting pine forests. A further factor influencing production has been a change in the composition of community forests – one of the most important forms of ownership in Japan – as management has shifted away from harvesting for fuelwood and charcoal, favouring hardwoods and limiting pine regeneration (Hosford et al, 1997, p6). By the early 1980s, there was a dramatic increase in the level of imports, with China being one of the main sources. Some 60–70 per cent of imports from China came from Yunnan, with an annual estimated value of US$30-50 million (Liu, 2003, p1).

The metamorphosis of matsutake from a relatively obscure medicinal fungus of some ceremonial value among Tibetans to a luxury commodity exported to one of the world's most demanding markets has had important consequences for rural livelihoods and conservation. At the same time, in a political economy described as a 'socialist

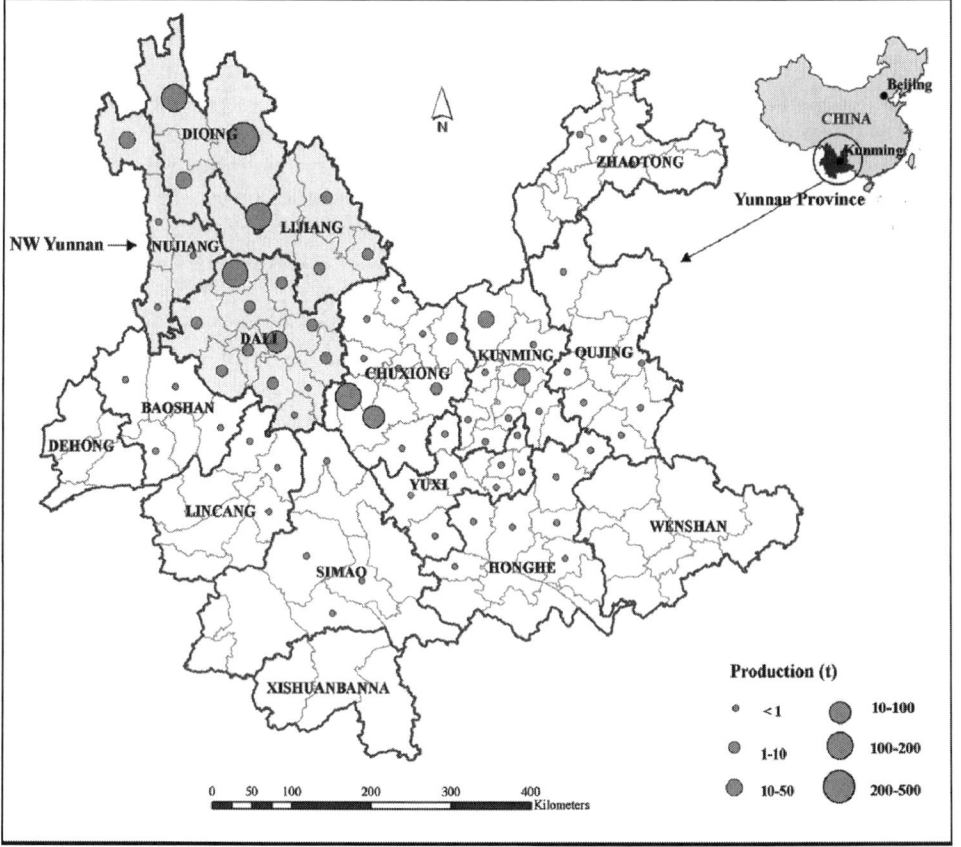

Source: Map produced by Ethnoecology Research Group, Kunming Institute of Botany, Chinese Academy of Sciences, 2006.

Figure 10.2 *Distribution of matsutake production in Yunnan (data based on year 2005)*

market economy', where the state still plays an important role in regulating access to and the functioning of markets, there is some official discomfort at the speed with which entrepreneurial traders have created a marketing chain that successfully moves matsutake from the forest to the supermarkets of Tokyo and Osaka in less than 24 hours. There is widespread agreement that some form of regulation is important to promote sustainable use, to secure rural livelihoods and to protect fragile montane forest ecosystems. It has proved challenging, however, to determine what kind of regulatory systems would address these multiple concerns and where responsibility should lie for implementation and monitoring.

MATSUTAKE IN YUNNAN

The distribution of matsutake inevitably depends on the distribution of its host species, usually in pine or mixed pine and oak forests. In south-west China, it is found in the south-eastern Hengduan mountains, which mark the boundary between Yunnan province and Tibet. Within Yunnan itself, the main areas of matsutake habitat are in the remaining forests of the central plateau and in the Diqing Tibetan Autonomous Prefecture in the mountainous north-west. Smaller populations of matsutake are also found in the south and south-east of the province, where temperature and precipitation regimes are favourable (Leique Research Institute, 2002, p13). The mountains of south-western China are one of the 34 'biodiversity hotspots' listed by Conservation International as globally significant centres of biodiversity, and they are also culturally diverse, home to many ethnic minorities including Tibetans, Yi and Naxi.[1]

Following a pattern common to many mountain communities around the world, villagers cultivate cereals, usually wheat or barley, in the valley bottoms and graze livestock near the village or on higher-elevation summer pastures. Forests and the harvesting of medicinal herbs and other NTFPs have long been important sources of cash income supplementing agricultural production. Large-scale timber harvesting took place in north-western Yunnan from the late 1970s to the late 1990s. For much of this period, however, most forestland was officially under state ownership, and the state only granted concessions to timber companies, which were owned at the time by agencies of the government or the military. The companies employed able-bodied villagers on a daily basis as loggers, or on longer-term contracts as drivers and equipment operators, but all income from the timber sales went to the company, with no royalties or fees accruing to the village itself. Forest grazing and NTFPs were of more direct importance to farmers' livelihoods than timber.

In 1998, alarmed by disastrous floods in the densely populated Yangtze basin, the central government imposed a ban on logging in the upper Yangtze and its tributaries and introduced an ambitious programme known as the Natural Forest Protection Programme (NFPP). The NFPP withdrew all sloping land from agricultural production, offering free grain for up to eight years as compensation for lost production. During that time, farmers were expected to allow forest to regenerate, to plant trees or orchards, or to convert land to pasture as alternative sources of income.[2] With the end of timber harvesting, NTFPs became even more central than they had been before to the livelihoods of communities in forested areas. In Shangri-la county, in villages with mixed land-use systems including agriculture and livestock, income from matsutake represented as much as 25–30 per cent of farmers' incomes in 2000 (Li et al, 2004). In a survey of villages surrounding the Baimaxueshan Nature Reserve in the Diqing prefecture, WWF (Worldwide Fund for Nature) China found that shortly after the logging ban, some 95 per cent of villagers' income came from the sale of NTFPs, with a total of 80 per cent coming from the sale of matsutake alone (Chen, 2001). For these villagers, it was fortunate that the enforcement of the logging ban and the implementation of the NFPP coincided with rising prices for matsutake due to increased demand from Japan. The two related policies, conceived as a strategy to protect and regenerate forest resources, with little or no reference to NTFPs, had the unintended, if predictable, consequence of increasing harvesting pressures on NTFPs, including matsutake.

The sale of matsutake has become one of the mainstays of household livelihoods in the forested areas where it is found. It is, however, a precarious source of income. The availability of matsutake from several different parts of the world gives Japanese buyers the advantage of a monopsonistic market, driving prices downwards, with wide and dramatic fluctuations depending on shifting sources of supply, in a way that is almost incomprehensible to sellers. In Yunnan, local governments, deprived of the revenue they had been earning from taxes on timber, were quick to impose a host of taxes and fees on harvesters, middlemen and traders in the county matsutake market. The subsequent cancellation of many of these taxes and fees has been a welcome relief, but the apparent ease with which they can be imposed and revoked adds to the uncertainty faced by both harvesters and traders (He, 2004, pp7–8). Harvesters' weak position in an uncertain market is a further incentive to seek short-term benefits from intensified harvesting, further adding to the pressure on the resource.

As north-western Yunnan is a recognized centre of biodiversity, large areas containing much of the remaining forest have been placed under some form of protected status over the past 20 years, including national parks, national- and provincial-level conservation areas, and one World Heritage Site. Many communities now find themselves within or adjacent to a protected area, with restrictions on their customary access to and uses of the resource. There are, for example, 117 settlements within or on the boundaries of the Baimaxueshan Nature Reserve in Deqin county, with a total population of 15,000 (WWF, nd). The rapid increase in matsutake harvesting over the past 10–15 years has raised concerns about sustainability and possible impacts on biodiversity in the fragile, high-elevation forest ecosystems where the mushrooms are collected. As a consequence, government agencies and non-governmental organizations (NGOs) with an interest in environmental issues have highlighted the need for programmes to ensure the sustainable harvesting of matsutake as an essential component of conservation strategies in north-western Yunnan. In 1999, at almost the same time as the enforcement of the logging ban, matsutake was listed as a 'second-class protected species' under China's 'Regulations concerning the protection and management of wild plants', under which exports are subject to quotas and certification from the CITES (Convention on International Trade in Endangered Species of Wild Fauna and Flora) Authority of China. The implementation of these regulations and their place in the larger picture of NTFP policies in Yunnan will be examined later in this chapter.

There has been extensive research on the distribution and ecology of matsutake, and it is possible to infer from harvesters' reports and from market surveys that there has been a steady decrease in the quality, size and availability of matsutake in the forests over the past 25 years (Yang, 2004). It is interesting to note, however, that there has been little if any ecological research to date that has studied the environmental impacts of matsutake harvesting on forest ecosystems. Nevertheless, the formal regulatory framework governing the management and harvesting of NTFP has come to be driven by concerns for conservation focused on individual species, rather than a more inclusive set of policies. At the same time, Chinese chapters of international conservation NGOs such as the WWF, The Nature Conservancy, and Flora and Fauna International have found that incorporating the conservation and sustainable harvesting of an emblematic species, matsutake, into their programmes allows them to work with both government agencies and communities in pursuing their own larger vision of protecting the biodiversity of this region.

GOVERNMENT REGULATION OF FORESTS AND FOREST PRODUCTS – NATIONAL AND PROVINCIAL LEVELS

An analysis of government policies for NTFPs must begin with a discussion of forest tenure systems, which have closely followed the fortunes of land tenure since the establishment of the People's Republic of China in 1949. In a step-by-step process beginning in the early 1950s, rural land – and forests – came under collective owner- ship, leading in the 1960s to the 'people's communes'. The commune combined rural administration and agricultural production in one institution, which collectively owned all means of production. Some forests came under communal ownership and management, while most of the productive forests became state-owned and state- managed, sometimes with disastrous results when mass campaigns for forced indus- trialization or for the expansion of agriculture consumed unsustainable volumes of timber, seriously depleting and degrading forest resources (Menzies and Peluso, 1992; Shapiro, 2001).

The decollectivization of agriculture extended to forests in 1983. Forests are now divided into three major categories, each with its own set of rights and responsibilities:

- State-owned forest (*guoyou lin* 国有林) is under the jurisdiction of the State Forestry Administration and its local offices extending to the township level. Management is subject to national planning based on a system of harvest quotas – although since the logging ban, there has been no officially sanctioned harvesting on state- owned forests in Yunnan.
- Collective forests (*jiti lin* 集体林) are an institutional legacy of the communes, and are owned by loosely defined 'village/township collective economic organizations or other entities or individuals who are engaged in cultivation, protection and utilization of forest resources on rural collective land' (Miao et al, 2004). In prac- tice, the 'collective' usually means the local village government. The collective manages the forest itself within constraints established by the local forestry bureau, which in turn follows national planning guidelines. The degree of local autonomy in management decisions varies widely and depends to a great extent on the rela- tions between bureau staff and local leaders.[3] Most villages, however, have control over access to and utilization of NTFPs in the forest and often formulate and enforce village regulations adapted to local conditions and social norms. Arguably, it is these local rules that are the most important mechanism governing NTFPs in China, particularly where there is a risk of overharvesting a valuable species such as matsutake. The next section examines and analyses how villages have structured their rules to address what they perceive to be some of the most important issues surrounding matsutake harvesting in their collective forests.

- The third category, freehold forests (*ziliu shan* 自留山), is allocated to individual households, although the collective retains ownership. They tend to be the more degraded areas of village forest and are usually held under some form of contractual arrangement with the collective, but in some cases the household has sole responsibility for management. Freehold forests may be a family's main source of firewood, leaf litter (used as bedding for livestock, producing valuable manure) and other products – although matsutake would rarely be found in what is often degraded or regenerating forest close to the village.

The State Forestry Administration has direct control over the management of state-owned forests, and less directly over collective forests through the application of harvest quotas, requirements for regeneration, and reforestation or afforestation programmes, which set quotas for the area of trees to be planted in any administrative unit, including the lowest level of government, the villages under the jurisdiction of a village committee. Currently, national policies and legislation are directed to maintaining or extending forest cover, regulating timber harvest and trade, and the conservation of endangered species. There are no specific policies concerning NTFPs, unless they are officially listed as endangered.

When listed, a species falls under the 'Regulations concerning the protection and management of wild plants' issued by the State Council in September 1996, which went into effect in January 1997. The regulations mandated the State Forestry Administration and the Ministry of Agriculture to jointly formulate the list of protected species and measures for their protection. There is a ban on the collection of plants listed as 'First Class Protected Species' unless for the purpose of scientific research, with a permit issued by the two agencies. The collection of 'Second Class Protected Species' is allowed with the approval of the provincial forestry department and agricultural department. The responsibility for enforcing the protection of listed species falls on the government forestry and agriculture agencies from the national level to the county level, two levels above village committees. The export of both first-class and second-class protected species is only allowed with a permit from the national-level State Forestry Bureau, and with a further permit issued by the provincial office of the CITES Authority for China. Finally, the provincial department of foreign trade supervises all commercial and fiscal aspects of exports, including foods and plants.

In 1999 matsutake was listed as a second-class protected plant. While there may be no national policies specifically targeting NTFPs, matsutake itself is at the centre of a web of regulations intended to protect it from overharvesting and to ensure its regeneration. In conformity with the State Council regulations, Yunnan province has enacted measures governing the harvest, sale and export of matsutake. The CITES office, for example, has ruled that only matsutake greater than 5cm in length may be exported. The provincial departments of foreign trade and forestry have introduced licensing procedures which limit the number of companies allowed to export matsutake and require proof of prior compliance with regulations governing trade in matsutake, a minimum level of capitalization and the ability to meet standards of health and hygiene for exported food products. As of 2006, some 20 companies had received export licences. The intent of this process is to avoid unrestricted competition, which the authorities fear may lead to attempts to bypass regulations and to make short-term profits at the expense of sustainable production (Li, 2006). In practice, Japanese importers and their representatives in Yunnan have complained that the system has favoured companies owned by or closely associated with provincial and local governments that have taken advantage of their official connections to avoid inspections and to sell shipments of lower-quality produce (Liu, 2003, p4).

In Yunnan, the provincial government has chosen to regulate export procedures directly, delegating the formulation and implementation of detailed rules and policies for conservation and marketing primarily to prefectures and counties, the two levels immediately below the province in China's administrative hierarchy. Since 2001, however, in order to maintain minimum coherent standards, the province has issued an annual report on the status of matsutake harvesting and trade, drawing attention to common problems as they emerge. Past reports have enumerated an array of issues including the sustainability of harvesting practices; contractual arrangements for harvesting in matsutake forests; pesticide residues on marketed mushrooms; the adulteration of bulk shipments using metallic objects such as nails to increase their weight; collusion in price-fixing; and calls for a more equitable distribution of profits to benefit rural communities (Li, 2006). It is left to local governments, however, to find ways to correct these problems.

COUNTY GOVERNMENT REGULATION

Historically, counties in China have been the centre of marketing networks linking dispersed smallholder producers to wholesalers for further distribution to other cities and regions. They continue to play the same role in the marketing chain for matsutake, with county markets being the main point of contact between harvesters, middlemen and the companies that ship the mushrooms directly to Kunming, the provincial capital, from where they are flown directly to Japan (He, 2004). County markets, therefore, are the point at which local governments can most easily monitor compliance with conservation-oriented regulations governing the size and quality of harvested matsutake, and where they can levy the taxes and fees that have become so important for their budgets since the implementation of the logging ban.

Many counties took the initiative to control the market for matsutake several years before its listing as a protected species. In 1995, for example, Shangri-la county (then known as Zhongdian county) established the Integrated Management Office of the Matsutake Market to direct and supervise trade in matsutake and other edible fungi. The office restricted all trade in matsutake and edible fungi for shipment out of the county to one designated market in the county town – a large courtyard surrounded by stalls rented by traders. Traders must register with the county commercial bureau. They must undertake to comply with county standards for the size and quality of mushrooms, and they must pay taxes and fees as determined by the county. The market is only open after 9pm, giving harvesters and middlemen time to travel from their villages at the end of the day's picking. The county appoints inspectors to patrol the market, monitoring sales, checking receipts and collecting the relevant taxes and fees.

Annual reports issued by the Yunnan provincial government in 2001 and 2002 urged prefectures and counties to strengthen measures for the conservation of matsutake through the consolidation of control over both production and marketing. Shangri-la county dissolved the previous Integrated Management Office, replacing it with a more inclusive and stronger County Leading Group for the Conservation and Development of Matsutake and other Mushroom Resources. The Leading Group is under the direction of the Communist Party's Disciplinary Commission, giving it considerable authority, with representatives of the county bureaux of finance, trade, tax, commerce, public security and forestry.

The role of the Leading Group is to implement the 'detailed tasks of management' as listed in an announcement, dated 24 June 2002, prominently displayed in public locations in the county town, including at the entrance to the matsutake market.[4] The announcement has 14 articles of which 12 (articles 1–4 and 5–14) cover the role of the Leading Group, procedures for registration, traders' obligations and fees, controls on the transport of matsutake out of the county, monitoring requirements and fines. Just two of the articles concern harvesting and packaging procedures at the village level. Article 4 sets minimum standards, based on CITES regulations, for harvesting techniques and sizes to promote conservation and sustainable management. Harvesters may not dig up or uproot mushrooms, and they may only pick mushrooms larger than 6cm in diameter. Article 5 seeks to build Shangri-la's reputation for quality produce and prohibits harvesters and middlemen from adding metal objects and lead when selling packaged lots of matsutake, imposing a large fine on offenders and the withdrawal of their trading permits. The Leading Group in Shangri-la county has chosen to concentrate its efforts where they can be most effective – supervision and monitoring at the point where matsutake enters the wholesale market.

More recently, local governments have begun to complement direct controls with a mix of target-driven incentives and sanctions. In 2002, Diqing prefecture (which includes Shangri-la county) initiated annual meetings with representatives of county governments at which each county is assigned targets for compliance with harvesting standards, such as a limit on the percentage of mushrooms below the permitted size found for sale. At the end of the year, those that have achieved the target receive a reward of 20,000 yuan (approximately US$2000). Counties that fail to meet their targets are publicly criticized in the local media and are not allowed to be considered for citation as 'advanced units', a significant sanction in a system where promotion and advancement depend on such a designation. Households and companies specializing in matsutake trading now require a permit and have to pay a bond at the beginning of the year. At the end of the harvesting season, the bond is returned if they have not violated any of the rules governing the size of mushrooms sold and harvesting techniques. If they have violated the rules, they lose the bond and may lose their permit (Huang, 2006).

Target-driven policy often diverts scarce resources from other important activities to concentrate them on the narrow task for which a target has been assigned. In light of the questionable merits of the target-driven approach to policy being applied to the counties in Diqing prefecture, it is encouraging to note that the prefecture has also called for action to codify the various rules and regulations and enact prefecture-level legislation on the conservation of and trade in matsutake. Perhaps of even greater importance is the call for villages to formulate and implement their own rules adapted to local conditions, which the prefecture would recognize as having force of law.

VILLAGE RULES AND LOCAL REGULATION

The management of forests and forest resources by village communities has a long recorded history in China. Published geographical gazetteers from as early as the 16th century document forest management systems that can still be observed today in some places. Stone stelae erected more than two centuries ago can still be found with inscriptions recording decisions taken by communities to protect their forest to prevent springs from drying up or to ensure a supply of large timber when needed for the maintenance of temples or schools (Menzies, 1987). Recorded or not, these village rules and regulations, known as *xianggui minyue* 乡规民约 (literally, 'village rules and popular compacts'), have played and continue to play a vital role in village life and in the local management of natural resources.

Matsutake became a valuable commodity in the 1980s, just as the allocation of forests between the state, communities and households was taking place. At a time when national policies defining rights of tenure and usufruct were still in a state of flux, disputes between individuals and between villages over rights of access to forests and the precise demarcation of boundaries were common, leading at times to violence (Yeh, 2000). In the light of their new responsibilities for managing collective forests and monitoring the household use of freehold forest, many villages in matsutake-producing counties have devised *xianggui minyue* – codes of conduct and rules governing access to the forest, monitoring harvesting and, in some cases, controlling local markets. The rules usually spell out sanctions against offenders, enforced by village authorities or the harvesters themselves.

In the diverse geographic and cultural setting of north-western Yunnan, the ability of communities to formulate rules adapted to their own context is perhaps the most significant, if unstated, aspect of the national policy framework for forest management affecting NTFPs. Inevitably, the relevance of the rules and the effectiveness with which they are enforced depend on the social capital on which the community can draw internally, on the strength of village leadership and on the degree of support they receive from government and state agencies. Nevertheless, in recognizing village regulations as having the force of law, the Diqing prefectural government is acknowledging the force and legitimacy of local regulation and monitoring. In the discussion of village regulations that follows, we have chosen to use examples from frequent field visits in Diqing prefecture, but it is important to acknowledge that communities in other matsutake-producing prefectures and counties have adopted similar regulations.

Setting boundaries and controlling access

One of the essential conditions for the effective management of common pool or common property resources is the demarcation of boundaries, both spatially and socially in terms of who may and who may not use the resource (Ostrom, 1988). Most villages allow residents to harvest matsutake anywhere within the collective forest, although some may rotate harvesting between different sections of the forest from one year to the next. Given the uneven distribution of matsutake within a forest, attempts to allocate individual household plots within a village forest have not been successful.

The village of Shusong, adjacent to the Baimaxueshan Nature Reserve, had the opportunity to test both approaches before deciding against household plots. Shusong residents are allowed to harvest matsutake within the reserve, but they are prohibited from allocating household 'patches' for harvesting. In the collective forest, the village did try for several years to allocate plots but found the system to be unsatisfactory and abandoned it. It is worth noting, too, that Shusong, like many other communities, has different rules for different NTFPs. Residents may harvest matsutake and the caterpillar fungus (*Cordyceps sinensis* – a medicinal product found at high elevations) anywhere within the collective forest. Each household, however, has its own plots in the same forest for gathering leaf litter or firewood. Local regulation has proved to be an important mechanism favouring adaptive management, allowing the village to make adjustments or to craft new rules as it learns from experience and responds to emerging problems and new opportunities.

Boundaries are rarely uncontested. Geographical boundaries between villages are not always clear, nor are the social boundaries marking who is or is not a member of the 'community'. Disputes over village boundaries are frequent in a region that has experienced dramatic changes in administrative and property systems over the past 50 years. Where they have had recourse to the legal system, villages have not always accepted the verdict of the courts. Some have taken matters violently into their own hands: in a dispute between the village of Da Huotang in Shangri-la county and its neighbours, some villagers even set booby traps to stake their claim. More recently, village committees, which are now elected bodies representing the residents of several 'natural villages', have become more active in mediating disputes, with the capacity to enforce their decisions. In Shusong, for example, the village committee held a meeting in 2002 with all the village heads under its jurisdiction and with the management office of the nature reserve to reach agreement on the boundaries between them.

Social boundaries that determine who has access to the resource can also be difficult to draw. Local regulations are more likely than national, provincial or even county policies to navigate the complexities of family ties and kinship, custom and social norms that define a 'community'. A few villages in Diqing prefecture may occasionally invite 'outsiders' into their forest, or allow them to pay a daily or seasonal fee. Most villages, however, restrict to residents the right to harvest matsutake – i.e. people living in homes within the village boundaries. Some villages have even devised quite elaborate procedures to ensure that no 'outsiders' enter their collective forest. In Naidu village in Shangri-la county, harvesters must gather at 6am in a designated place. At 6.30am, the village head blows a whistle as a signal and everyone is allowed to head into the forest. They must work in groups of six – which ensures that everyone knows exactly who is in the forest. No one is allowed to enter the forest later in the day. A village leader explained: 'This is so that we can check on who is going. We can make sure that there are no children who are missing school and there are no outsiders in the group.'

Codes of conduct: How and when to harvest

To a matsutake harvester, 'conservation' is a set of practices that ensure a continued harvest and source of income year after year. It has a different meaning from the vision of conservation as the protection of biodiversity, which drives government policy and mobilizes international NGOs. The goals of conservation as articulated in village rules and in government policy converge, however, and both are served by the common restrictions they impose on the tools and techniques for harvesting. The rules are based on CITES recommendations, which appear in slightly different formulations in article 4 of the announcement posted at the Shangri-la market and in village regulations throughout the Diqing prefecture, as well as in other parts of Yunnan. The People's Government of Luoji township presents the rules and principles as the Five Forbiddens and Five Promotes, using the catchy, rhythmic form of many political slogans:

The five forbiddens:

1   It is forbidden to pick or to process young matsutake smaller than 6cm.
2   It is forbidden to pick or to process matsutake with an open cap or when they are overripe.
3   It is forbidden to dig up, injure or destroy matsutake when picking matsutake.
4   It is forbidden to sell young matsutake or overripe matsutake.
5   It is forbidden to act violently when buying and selling matsutake, in order to keep the market orderly.

The five promotes:

1   Promote a civilized way of picking matsutake, and work together to protect the environment and ecology of matsutake.
2   Promote everybody's involvement in preventing any activities that harm the resource.
3   Promote protection of the hills by the whole population and work for sustainable use.
4   Promote everybody's conscious awareness of conservation to escape poverty faster.
5   Promote the civilized habits of mutual assistance and mutual respect to advance social progress.[5]

In some villages, there may be further elaboration on the rules in response to specific conditions or incidents that have occurred. In 2002, market prices dropped dramatically when Japanese importers claimed to have found traces of pesticide in a shipment of matsutake from Yunnan. Believing that the contamination of the affected shipment could have been due to residue on the plastic recycled from used pesticide bags in which some of the mushrooms were wrapped, villages were quick to ban the use of plastic and to require the use of readily available organic materials such as azalea or moss. While the province has taken measures since then to improve testing for residue at the point of export, villages were able to act immediately to prevent a recurrence of the problem: the village of Naidu introduced a rule prohibiting plastic before the end of the 2002 harvest season.

The matsutake harvesting season in Yunnan extends from the end of May to the middle of November, depending on factors such as location, forest cover and elevation. Following one of the basic principles of sustainable utilization of a resource, most villages declare an open and a closed season for harvest. Local knowledge of the forest environment allows village leaders to time the season to correspond to what they know are indicators of the best conditions for matsutake growth. In Naidu, temperature and humidity begin to favour matsutake growth in July. Two village leaders visit the forest every day, and when they find mushrooms larger than 5cm in diameter, they declare the harvest season open. The season ends on a fixed day in mid-September when the barley harvest starts.

It is not unusual, too, to find further rules that require a 'rest' period after several days of harvest to allow for regeneration and to avoid overharvesting. Open and closed seasons and 'rest' days are important conservation measures, but their effectiveness is, in fact, probably enhanced by the absence of province-wide regulation, which would require an unworkably complex zoning system to cope with the diversity of environments in Yunnan.

Monitoring and sanctions

Rules are only as good as the level of enforcement and the penalties for breaking them. The 1998 logging ban was a drastic measure taken after years of destructive timber extraction, much of it by state-owned companies, despite a national policy of sustained yield with the backing of legislation, and the full apparatus of the police powers of the state. The capacity to enforce compliance with policy is not a function of the capacity to mobilize resources and to apply force. It requires insitu monitoring coupled with transparency in applying prompt but fair sanctions against offenders. Proximity to the resource and a community interest in maintaining matsutake productivity in the collective forest make it likely – though by no means inevitable – that local monitoring and sanctions will be more effective than a reliance on policing by staff from government agencies such as the State Forestry Bureau or the Public Security Bureau (police).

The first step in monitoring compliance with village rules is to ensure that unauthorized outsiders do not get access to the forest. Procedures such as Naidu's early morning start, and the clear definitions used in other villages of who is and who is not a member of the community for purposes of matsutake picking, eliminate any possible uncertainty about whether a harvester is authorized to be in the forest and whether she or he can claim ignorance of the rules. Most villages then reinforce this filtering process with mechanisms to ensure that harvesters only use the approved tools and techniques. Naidu requires pickers to work in groups of at least six as a form of mutual supervision. Many villages organize regular and frequent patrols. The village of Gulongpu, adjacent to the Baimaxueshan Reserve, assigns ten people to patrol duty from every 50 households involved in matsutake harvesting. Each harvester pays a daily fee to cover the salaries of the patrols. Any funds left over at the end of the season are used for community development activities such as road improvement or school maintenance.

The most common forms of sanctions are the levying of fines and the confiscation of mushrooms in the offender's possession. Informally, villagers admit that they may also beat up outsiders found in the forest, and there is rich anecdotal evidence to confirm that arguments are often decided by violence. Many of these outbursts can be traced to disputes between neighbouring villages over the boundaries of the collective forest – an issue discussed earlier in this section and which has yet to be fully resolved.

Members of a community will not willingly accept and conform with an intrusive regulatory environment unless they are convinced that it applies impartially to all and that its application is public and transparent. A critical element in building legitimacy is the way funds are handled. If there is any doubt about the use of fines, for example, harvesters may treat regulations as an imposition and just another way for unaccountable officials to get rich. Some villages, but not all, allow residents free access to the accounts. Some villages ensure that the use of matsutake-related funds is not discretionary and can only be for specified and very public purposes. Gulongpu uses funds left over after paying patrol salaries for visible public works. Naidu opens its accounts to public scrutiny and specifies that funds can only be used to pay a forest guard and to organize an annual community meeting with associated festivities and entertainment. It is surprising, though, that many communities have not put such confidence-building measures in place, possibly putting at risk the long-term legitimacy and viability of the local regulatory system.

Marketing

While the most important markets in terms of the volume and value of transactions are in the counties, most villages have some kind of point of sale at which middlemen purchase mushrooms directly from the harvesters and take them to the county market. In Shusong, sellers congregate at the end of the afternoon along both sides of the main road leading to Shangri-la county, while several kilometres down the same road at the entrance to the town of Benzilan, buyers have filled a narrow alley with stalls where they wait for sellers to come down from the forests at the end of the day. Naidu and the other hamlets that form the Jidi village committee have built an enclosed square of stalls, which they have designated as the only place where trade in matsutake can take place. Confining transactions to a designated location is, of course, the principal mechanism for the exchange of goods and information about supply, demand and prices. In the case of a regulated good such as matsutake, it also facilitates monitoring compliance with rules about size and quality.

The rationale for local control of marketing locations and procedures is the same as it is in county-level markets. If there is demand for a product, someone will be willing to sell it. To ensure that there is no incentive to harvest undersize mushrooms, for example, it is not enough to prohibit the harvest of mushrooms smaller than 6cm in length; it is also essential to prevent traders from buying them. With just one approved venue for sales, it is feasible for the village leaders to fine not just harvesters but also any trader found to be violating local regulations. At the county level, the Leading Group is focused on quality control for the export market and on conservation. Villages, by contrast, are acting to safeguard their livelihoods by protecting an important resource from overexploitation. For both, however, some degree of market regulation is a necessary part of policy implementation.

DISCUSSION: MULTI-TIERED POLICIES FOR NTFP MANAGEMENT

According to a Chinese proverb, 'the mountains are high and the emperor is a long way off' (*shan gao huangdi yuan* 山高皇帝远). The phrase neatly captures the dilemma of policy formulation in a country the size of China, and even within one province with the physical and cultural diversity of Yunnan. National legislation and policy are crude instruments with which to manage natural resources and biodiversity.

In the case of NTFPs, or in the case of a single NTFP such as matsutake, management has many goals that may be difficult to articulate in one policy. For a product that is valued in domestic or international markets, policy goals may be oriented towards conservation, or to foreign trade, or to both. There may be concerns that overharvesting will endanger the survival of the species or its habitat. Trade policies might seek to complement conservation goals by regulating trade in the species. They may also seek to ensure the quality of exported produce in the face of market competition from other countries. The human and financial resource constraints of any particular level of government or state agency limit its capacity to realize these objectives. Beyond the broadest principles framing national goals for the conservation of biodiversity and an orderly export market, central government is poorly equipped to regulate in situ resource utilization and marketing. This is particularly true in China, where so much forest is under collective, not state, ownership. Regional or local levels of administration – provincial and county governments in the case of China – are more able to formulate appropriate regulations and to monitor compliance at points where the commodity enters marketing chains. The Yunnan provincial government can track shipments as they move into the Japanese market through Kunming airport. County governments can supervise and police transactions in the county markets, where matsutake moves from the producer or harvester to the wholesale trader.

Government control at key points in the marketing chain is unlikely to have much impact in the forests and mountains where harvesters pick matsutake. Even undersized mushrooms can easily pass through informal networks to find a market in processing and canning plants in the neighbouring province of Sichuan. Government at any level does not have the resources or the detailed local knowledge needed to regulate and monitor the widely dispersed harvesting of matsutake by individuals or small groups. Villages in the matsutake-producing areas of Yunnan have developed local codes of conduct, including 'social fencing', restricting who has rights of access to the forests, and procedures for monitoring, enforcement and the punishment of violations of the rules. Having participated in formulating the regulations, harvesters are likely to enforce them more effectively than outsiders such as more detached and distant government agencies.

The case of matsutake in China shows how a multi-tiered set of institutional actors has emerged to form a strategy that is probably more effective than one all-encompassing policy governing NTFPs. The process exemplifies what is referred to as the 'muddling through' approach to policy formation rather than the 'rational decision-making model'. Practice has created a de facto policy framework based on a hierarchy of responsibilities. This has had the advantage that each level of the hierarchy, from the central government to the village, is responsible for what it is best equipped to do, even if each level may have a different perspective on the need for regulation.

The multi-tiered system of management and regulation distributes responsibilities among the actors best able to meet them. It is not without its problems, however, which are often a function of uncertainties in the political, legal and physical environment in which they operate.

Village regulations may draw clear social and physical boundaries to mark out their turf. As described above, though, this does not mean that there are no conflicts that can turn violent. The clarity with which communities delineate boundaries and rights of access is often compromised by fuzziness in the legal demarcation of boundaries and unclear property rights over collective forests. The process of decollectivization and the allocation of forests that took place in 1983 created new boundaries between new categories of property. The work teams involved in the process did not usually put any official markers in place, and it is difficult to locate the records, documents and maps that were used. Yeh (2000) reports a conversation with a forestry official in Kunming who 'confirmed that only the older generation of officials who had participated in boundary-drawing work teams would be privy to that knowledge, which is not shared with local people, and which is kept by individuals rather than being institutionalized'. Even where the boundaries of village collective land are not in dispute, its exact ownership and the rights attached to ownership are not clear, due to the ambiguous legal definition of 'collective'. Since community-level regulation of matsutake harvesting hinges on clear, acknowledged territorial and social boundaries defining the territory over which the village has jurisdiction, its effectiveness is weakened when the boundaries are contested – which, ironically, places the success of local management back in the national arena of broader legislation and policy concerning tenure property rights.

Village committees and the hamlets under their jurisdiction have direct, if contested, control over their collective forests and, to some extent, over the freehold forests allocated to residents. As noted above, however, villages do not have any authority over the extensive tracts of forestland in north-western Yunnan dedicated to the conservation of biodiversity as 'national parks' or 'nature reserves'.[6] In most, if not all, of these recently gazetted protected areas, the introduction of restrictions on land use and access to forest resources in the interests of conservation has had a profound impact on the livelihoods of communities adjacent to or within their boundaries. For these communities, the multi-tiered structure of policy-making is interrupted by a gap between the national level and the village. The management of protected areas follows nationwide rules, which do not, in principle, acknowledge local regulations and practices. Over the past 15 years, the Chinese chapters of international conservation NGOs and, increasingly, domestic NGOs have been able to cross administrative lines to bridge the gap between national policies and local institutions. In north-western Yunnan, NGOs including WWF China and The

Nature Conservancy have played an important role in adapting stringent national regulations to make matsutake harvesting in affected villages compatible with the national policies that now shape their landscape. WWF China, for example, has worked with the Baimaxueshan Nature Reserve and neighbouring village committees to develop local rules to meet the requirements of reserve management.

Government and NGO programmes continue to emphasize conservation and the importance of educating the public about the life cycle of the matsutake and recommended sustainable harvesting techniques, even though there appears to be quite extensive indigenous knowledge about matsutake habitat and ecology. A video CD about matsutake jointly produced by the Kunming CITES office, WWF China and other concerned agencies includes a three-minute description by a Tibetan village leader of local knowledge about the relationship between matsutake regeneration, matsutake size and dead or decomposing logs, and other detailed information his community had used in developing their local rules for harvesters (Kunming Division of CITES Administration Authority, China, et al, 2006). Teaching villagers how to harvest matsutake sustainably may be less important than finding ways to assist them in enforcing the rules they have formulated.

The involvement of international NGOs and the emphasis on regulating the minimum size of mushrooms and on harvesting techniques that is seen in all the regulations, from national to community levels, reflect the priority formal policies place on conservation. The fear that overharvesting might be endangering the survival of *Tricholoma matsutake* is understandable. Quantities and sizes of matsutake offered for sale in the market, though, may reflect changing responses to changing market conditions as much as the dynamics of matsutake populations and distribution in the forest. There appears to be surprisingly little data based on ecological research, either to indicate the direction and dynamics of changes in matsutake populations or to demonstrate what impacts harvesting might have on forest ecosystems and what action might be needed to mitigate them. For the long term, there may be value in sharper questioning of what 'the problem' is and thus what the focus of present and future management efforts should be.

CONCLUSION

This chapter describes a situation in which there is no stated policy governing the management of NTFPs. National forest policy in China is concerned with the forest as a source of timber and of environmental services, not with the huge number of NTFPs – sometimes known in Chinese as 'secondary products (*fu chanpin* 副产品)' – gathered or harvested there. The example of one commercially valuable NTFP, matsutake, shows how the absence of a policy for a category of products has allowed the elaboration of a multi-tiered regulatory environment addressing its dual status as an exported commodity and a possibly threatened species. The composite policy framework that has emerged for matsutake is more likely than a single all-encompassing national policy to be effective in promoting both conservation and sustainable harvesting to benefit rural livelihoods.

What is perhaps unique about the Chinese case is the importance accorded to local village systems of regulation. Village rules and codes of conduct are internally generated, usually in a transparent and participatory manner. They are not codified in law, but the higher levels of government acknowledge them and recognize their legitimacy, allowing for local initiative and adaptation to the diverse ecological and social conditions of Yunnan. Within a community, peer pressure or social fencing is a more effective monitor and a more powerful enforcer of village-level codes of conduct than the stretched resources of the lowest branches of a state agency implementing regulations handed down from remote, higher levels of government.

In an uncertain market, the effectiveness of the present multi-tiered system of regulation is, of course, vulnerable to unexpected changes in prices, buyers' suspicions about quality control, and competition from suppliers in other countries. From the perspective of biodiversity conservation, there is still only minimal research on the environmental impacts of matsutake harvesting. Government and NGO programmes continue to recommend tools and techniques for ecologically sound harvesting, accompanied by calls for intensive programmes of public education, whose efficacy in the past has been questionable and which may be misplaced in the light of evidence that harvesters and marketers are, in fact, well aware of practices that either promote or threaten sustainable production. There are also some indications that higher levels of government such as Diqing prefecture are pressing for more rigid, top-down, target-oriented control over harvesting and marketing, which would run counter to what are arguably the most effective adaptive characteristics of the present framework. An exclusive focus on control and sanctions leaves no place for incentives and explorations of how a more transparent and informed market might promote improved, sustainable practices.

In the management of matsutake and other NTFPs, there is room to move now to complement regulation and policing with a more intensive effort to build human capital: the capacity to manage and monitor and adapt to changing circumstances.

NOTES

1   See the description of the 'Mountains of Southwest China' at www.biodiversityhotspots. org/xp/Hotspots/china/.

2   For critical studies of the impacts of the NFPP both on rural economies and on the environment, see Xu et al (2002). For a study on the impacts of the logging ban on timber imports from Myanmar, see Kahrl et al (2005).

3   For a study of the complexities of decision-making and management in the collective forests of four villages in Yunnan province, see Weyerhaeuser et al (2006).

4   The information about the Leading Group and its responsibilities is transcribed from an announcement posted in the Shangri-la county matsutake market and photographed by one of the authors in August 2002.

5   Transcribed in August 2002 from an announcement posted at the Da Huotang village office in Luoji township.

6   National parks are directly under the administration of the National Parks Agency of the Ministry of Construction in Beijing. Most nature reserves are administered by the State

Forestry Administration and its provincial or county offices. Both categories are dedicated to the protection of significant natural and cultural values, but the management of national parks tends to be more oriented to public access and recreation.

REFERENCES

Chen, Y. (2001) *The WWF Integrated Conservation and Development Project with Communities Adjacent to the National Level Baimaxueshan Nature Reserve (Sustainable Utilization of Matsutake Sub-project)*, project report for WWF China, Kunming Office, Kunming, China (in Chinese)

He, J. (2004) *Globalized Forest Products: Commodification of Matsutake Mushroom in Tibetan Villages, Yunnan Province, Southwest China*, paper presented at 12th Conference of the International Association for the Study of the Commons, Oaxaca, Mexico, www.dlc.dlib.indiana.edu/archive/00001409/, accessed 19 June

Hosford, D., Pilz, D., Molina, R. and Amaranthus, M. (1997) *Ecology and Management of the Commercially Harvested American Matsutake Mushroom*, General Technical Report PNW-GTR-412, Pacific Northwest Research Station, Corvallis, OR

Huang, H. (2006) *An Outline of Matsutake Trade and Management in Diqing Prefecture*, Report for Kunming Division of CITES Administration Authority, China, Kunming (in Chinese)

Kahrl, F., Weyerhaeuser, H. and Su, Y. (2005) *An Overview of the Market Chain for China's Timber Product Imports from Myanmar*, Forest Trends, Washington, DC

Kunming Division of CITES Administration Authority, China, WWF China, Yunnan Province International Chamber of Commerce Matsutake Chapter, Kunming Educational TV (2006) *The Little Matsutake, a Star that Leads to Prosperity*, Kunming Division of CITES Administration Authority, China, Kunming (video CD, in Chinese)

Leique Research Institute (2002) *A Study of the Present Situation of the Matsutake Resource and Trade in Yunnan Province*, report prepared for the Kunming Division of CITES Administration Authority China, Leique Research Institute, Kunming (in Chinese)

Li, C. (2006) Personal communication by e-mail, 6 March and 11 June

Li, C., Chen, Y., Huang, H., Zheng, Z., Song, D. and Mao, W. (2004) 'Report of a Study of Matsutake Production in Shangri-la County', unpublished research report for the Kunming Division of CITES Administration Authority, China, Kunming (in Chinese)

Liu, P., Wang, X., Sun, P. and Yang, X. (1999) 'Notes on the resource of the matsutake group and their reasonable utilization as well as effective conservation in China', *Journal of Natural Resources*, vol 14, no 3, pp245–252 (in Chinese)

Liu, Z. (2003) 'Comments on the Current Situation and Difficulties Experienced by Matsutake from Yunnan in the Japanese Market', unpublished report for Kunming Division of CITES Administration Authority China, Kunming (in Chinese)

Matsushita, N., Kikuchi, K., Sasaki, Y., Gerin-Laguette, A., Lapeyrie, F., Vaario, L., Intini, M. and Suzuki, K. (2005) 'Genetic relationship of *Tricholoma matsutake* and *T. nauseosum* from the northern hemisphere based on analyses of ribosomal DNA spacer regions', *Mycoscience*, vol 46, no 2, pp90–96

Menzies, N. (1987) 'A survey of customary law and community control over trees in China', in L. Fortmann and J. Bruce (eds) *Whose Trees?* Westview Press, Boulder, CO

Menzies, N. and Peluso, N. (1992) 'Rights of access to upland forest resources in southwest China', *Journal of World Forest Resource Management*, vol 6, no 1, pp1–20

Miao, G., Zhang, K. and Zhao, S. (2004) *An Overview on Current Status and Development of Collective Forests in China*, draft paper presented at the Workshop on China Forest Products Trade, Industry, and Livelihoods: Asia-Pacific Supplying Countries and China, Beijing

Ostrom, L. (1998) *Self-Governance and Forest Resources*, plenary presentation at the International CBNRM Workshop, 10–14 May, World Bank, Washington, DC

Shapiro, J. (2001) *Mao's War Against Nature: Politics and the Environment in Revolutionary China*, Cambridge University Press, Cambridge

Weyerhaeuser, H., Kahrl, F. and Su, Y. (2006) 'Ensuring a future for collective forestry in China's southwest: Adding human and social capital to policy reforms', *Forest Policy and Economics*, vol 8, no 4, pp375–385

WWF China (nd) *WWF China Yunnan Baimaxueshan Conservation and Development Project*, World Wide Fund for Nature China, Yunnan Office, Kunming

Xu, J., Katsigris, E. and White, T. A. (eds) (2002) *Implementing the Natural Forest Program and the Sloping Land Conversion Program: Lessons and Policy Recommendations*, China Forestry Publishing House, Beijing

Yang, X. (2004) 'Modelling the Spatial Distribution of *Tricholoma Matsutake*', unpublished MSc thesis, International Institute for Geo-Information Science and Earth Observation, Enschede, Netherlands

Yeh, E. (2000) 'Forest claims, conflicts and commodification: The political ecology of Tibetan mushroom-harvesting villages in Yunnan province, China', *The China Quarterly*, no 161, March, pp212–226

Yun, W., Hall, I. R. and Evans, L. A. (1997) 'Ectomycorrhizal fungi with edible fruiting bodies. 1. *Tricholoma matsutake* and related fungi', *Economic Botany*, vol 51, pp311–327

Managing Floral Greens in a Globalized Economy: Resource Tenure, Labour Relations and Immigration Policy in the Pacific Northwest, USA

Rebecca J. McLain and Kathryn Lynch[1]

INTRODUCTION

The coniferous forests in the Pacific Northwest of the United States have long offered rural residents a variety of products needed to support forest-based livelihoods – floral greens, moss, edible berries, fruits and mushrooms, to name just a few. Some of these products are harvested on a very small scale, primarily for household consumption or as gifts for others. Others, such as floral greens and wild mushrooms, are harvested in large quantities and traded globally.

The term 'floral greens' is used in the floral industry when referring to the stems, branches and leaves of plants used for decorative purposes. These plant materials provide the background for flower arrangements and are made into wreaths and garlands for weddings, funerals and festivals. In Europe, Canada and the United States, decorations made from coniferous trees figure prominently in Christmas festivities, driving a strong seasonal market in evergreen boughs. Floral greens harvested in the Pacific Northwest and southern British Columbia include leafy branches from shrubs, such as salal and evergreen huckleberry; coniferous boughs, such as Douglas fir, noble fir, white pine and western red cedar; the leaves of plants, such as beargrass; and numerous types of mosses and ferns.

Floral greens are an important component of a global floriculture industry that has a total annual value of over US$8 billion (Draffan, 2006), and historically the temperate rain forests in the western United States and Canada have been an important source of raw materials for the industry. The Pacific Northwest floral greens industry draws on supplies from south-east Alaska to northern California.

However, most processing takes place in brush sheds located on the south-eastern corner of the Olympic Peninsula in the State of Washington (Fitzpatrick, 1997; Cocksedge,

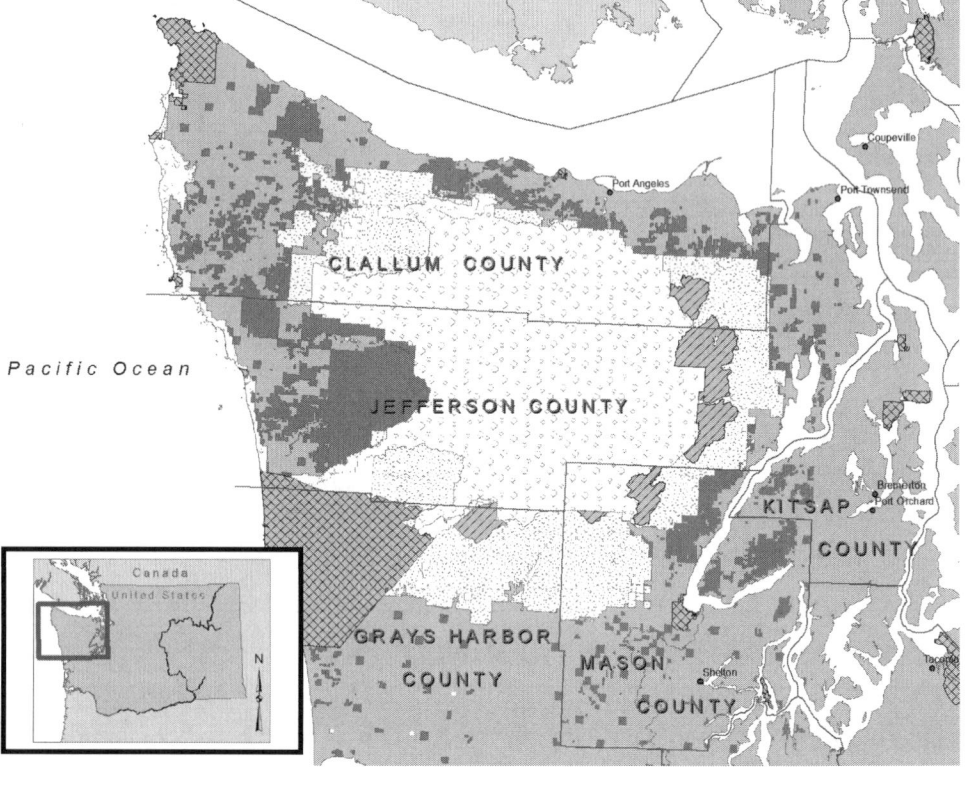

Source: Map created by Jamie Hebert using map layers obtained from Washington State Department of Natural Resources (www.fortress.wa.gov/dnr/app1/dataweb/dmmatrix.html)).

Figure 11.1 *Map of the Pacific Northwest*

2003; Spreyer, 2004). Once processed, roughly 90 per cent of the material is exported to floral markets in Europe (Draffan, 2006).

This chapter examines the transformation of the floral greens industry on the Olympic Peninsula during the past two decades. We look at how the increasingly globalized market for floral greens has changed power relations within the industry, as well as how labour relationships and the land tenure system have changed over time. We pay specific attention to the tensions that arose over labour policy as the workforce became dominated by immigrant workers from Mexico and Central America.

METHODS

Our analysis draws on data gathered from fieldwork and archival analyses conducted in 1993–1994 and 2002 (Kantor, 1994; Robinson, 1994; Lynch and McLain, 2003), with follow-up work in 2007. This longitudinal design provided for the development of a more holistic understanding of the shifting political, economic and social context of the evolving floral greens industry. Phase I of the study, which took place in 1993 and 1994, included interviews with 24 key informants in western Washington and participant observation of scoping meetings and hearings held by the State of Washington in preparation for revising the state's Specialized Forest Products Act, Revised Code of Washington, Title 76, Chapter 48 (RCW 76.48). Phase II, which was conducted in 2002, involved interviews with key informants familiar with the floral greens industry regarding the changes that had occurred in western Washington's floral greens sector between 1994 and 2002.[2]

During Phase I, the research team participated in four NTFP stakeholder meetings sponsored by the Washington State Department of Natural Resources (DNR), a state legislative hearing on proposed revisions of RCW 76.48 and several meetings sponsored by forest extension organizations to discuss industry issues and concerns. The research team interviewed 24 NTFP stakeholders, including pickers, buyers, land managers and extension agents. In Phase II, the research team interviewed 37 NTFP

Source: Kathryn Lynch.

Figure 11.2 *Weighing salal on the Olympic Peninsula*

stakeholders, including pickers, buyers, land managers, local and state government officials, researchers, law enforcement agents and non-profit organization employees. We conducted a literature review and follow-up telephone interviews with key informants in 2007.

OLYMPIC PENINSULA

Located in the north-western corner of Washington State, the Olympic Peninsula's temperate rain forest provides the habitat for many species used as raw materials in the floral greens industry. Chief among these are branches from salal, evergreen huckleberry and swordfern, as well as boughs from several coniferous tree species (noble fir, western red cedar, Douglas fir). The Peninsula's proximity to Seattle, the largest city in the Pacific Northwest, gives floral greens wholesalers easy access to national and global markets.

Landownership on the Olympic Peninsula is a mix of federal, state, tribal and private holdings. Slightly more than half of the Peninsula is controlled by public land management agencies, including the National Park Service (373,000ha), the US Forest Service (252,000ha) and the DNR (261,913ha). A dozen wood-fibre products corporations own most of the 810,723ha of privately held land. The Olympic National Park and Olympic National Forest both occupy large contiguous blocks of forest in the high and middle elevations at the Peninsula's centre. On the Peninsula's periphery, where the region's most productive timberlands are located, private, state and federal lands are distributed as intermingled parcels rather than as large single-ownership blocks. With the exception of the Olympic National Park, where all commercial harvesting is prohibited, most large landholders allow some commercial floral greens harvesting on their lands.

EVOLUTION OF THE OLYMPIC PENINSULA'S FLORAL GREENS INDUSTRY: 1920–1990[3]

Markets and labour

From the early 1920s to the early 1970s, the Pacific Northwest floral greens industry was made up of a few large wholesale companies and numerous smaller, independent buying firms. Most wholesalers based their operations in the south-eastern corner of the Olympic Peninsula. Initially large and small companies alike shipped their product to domestic markets in cities such as New York, Chicago, San Francisco and Los Angeles (Heckman, 1951). The floral greens workforce consisted of native-born citizens or recent immigrants of Euro-American descent living in the region's rural communities and small towns. Brushpicking, the local term for floral greens harvesting, was one component of a multifaceted, seasonally based set of livelihood strategies for both men and women (Heckmann, 1951). Whether part-time or full-time, Euro-American brushpickers were typically self-employed workers rather than employees of the numerous locally based floral greens companies known as brush sheds (Heckmann, 1951).

Beginning in the 1960s, European demand for floral greens expanded as the continent recovered from the devastation of World War II. By the late 1980s, Northwest-based floral greens suppliers were shipping most of their product to Europe, where demand for floral products was much higher. During this period, pickers and buyers alike became accustomed to doing business in an environment of increasing prices.

The shift to an export-oriented market coincided with a change in the floral greens workforce, as South-East Asian refugees from war-torn Cambodia, Laos and Vietnam sought income-generating opportunities in the Pacific Northwest during the late 1970s and early 1980s (Spreyer, 2004). The commercial harvesting of floral greens offered a way for refugees, many of whom spoke limited English, to earn income in a society where they had few job skills (Hansis, 1996). Most of these new immigrants saw floral greens harvesting as a stepping stone to other economic opportunities, a way to earn and save money while laying the groundwork for a shift into better-paying employment (Hansis, 1996; Spreyer, 2004).

In the late 1980s, the ethnic composition of the floral greens workforce changed again when the two largest wholesalers for floral greens and evergreen boughs in the region (one of them being Continental Floral Greens) brought in crews of Latino migrant workers from eastern Washington to fill a temporary gap in locally available labour (Fitzpatrick, 1997). Rather than returning to farm work once the wreath season finished, some of these workers stayed on the Olympic Peninsula to harvest salal, eventually displacing both Euro-American and South-East Asian harvesters (Fitzpatrick, 1997).[4]

Unlike their Euro-American and Asian counterparts, who typically were either citizens or had permanent residency status, many Latino harvesters lacked legal authorization to reside or work in the United States, making them vulnerable to abusive employment practices. This was not unique to the floral greens sector: abusive labour practices increased in all economic sectors, with high numbers of undocumented immigrants, after the passage of the federal Immigration Reform and Control Act of 1986 (Durand et al, 1999). The 1986 immigration law created an opportunity for more than two million Mexicans living or working in the United States without authorization to acquire permanent residency or citizenship status. However, the law also made it illegal for employers to knowingly hire or recruit immigrants lacking authorization to work in the United States (Durand et al, 1999). An unintended consequence of this law was the development of a thriving trade in counterfeit identity documents, which made it very difficult to prove that employers were knowingly hiring unauthorized immigrants.

The 1986 law also provided funding to intensify patrols along the Mexican border, creating an incentive for undocumented immigrants to remain in the United States rather than return home periodically, as many had in the past (Durand et al, 1999). In the long run, the 1986 law also had another unintended effect: it encouraged employers who relied heavily on undocumented workers to obtain labourers through subcontractors rather than by hiring employees in-house, accelerating the trend toward the use of contingent labour and contributing to the decline in wages and labour conditions at the low-paid end of the labour market (Durand et al, 1999).

These immigration policy reforms and the entry of Latino harvesters into the floral greens workforce took place at a time when the export market for Pacific Northwest floral greens was beginning to shift from a seller's to a buyer's market (Spreyer, 2004). In the late 1980s, cultivated floral greens from Florida, New Zealand and Central America entered the European market in large quantities, reducing the demand for wild harvested products and resulting in lower prices for Pacific Northwest floral greens. Speaking at a meeting sponsored by the DNR in 1993 (DNR, 1993), a brush shed operator described the difficulties of trying to sell wild products in this new market environment.

> *Well, you see part of the problem is that the evergreen business is a wild product and it's competing against different floral products that are farmed. So they're getting sprays to take care of the fungus and sprays to take care of bugs. And you know, the florist, when she gets that has got a first quality product, whereas something that's wild, if it's top quality wild product then it sells, but if it's junk, then they may not buy for a while.*

The influx of cultivated floral greens coincided with a drop in the value of the Canadian dollar relative to the US dollar, making Canada's wild harvested floral greens more attractive to European importers. The larger Olympic Peninsula wholesalers had the capacity to obtain raw materials from distant sources. For them, the drop in the value of the Canadian dollar was an opportunity to make more money by sourcing product from Canada. However, smaller companies with limited ability to pull in resources from other areas struggled to stay afloat.

A worldwide trend toward trade liberalization compounded the difficulties of the small and medium-sized brush sheds in the early 1990s. In particular, the impending North American Free Trade Agreement (NAFTA), which went into effect in 1994, fuelled the feelings of crisis among operators of small and medium-sized buying companies. NAFTA reduced trade barriers between the United States, Canada and Mexico, making Canadian floral greens even cheaper, and further limiting the ability of companies dependent on locally sourced products to compete with the larger sheds. NAFTA also had the unanticipated consequence of increasing the flow of immigrants into the United States. One of the selling points of NAFTA was that it would encourage American investment in factories in Mexico, create more jobs for Mexicans in their home country and decrease the numbers of undocumented immigrants entering the United States (Bratsberg, 1995). However, many of the factory jobs never materialized and heavy farm subsidies in the United States undermined Mexican agricultural prices, particularly for the country's two staples, maize and beans (Martin, 2005). The fall in prices for small producers contributed to an increase in both authorized and unauthorized immigration from Mexico to the United States from 1994 (Martin, 2005).

Floral greens tenure system

The changing market environment was accompanied by changes in the floral greens tenure system. Prior to the 1950s, most private landowners on the Olympic Peninsula, as well as the DNR and the US Forest Service, allowed pickers to harvest floral greens on their lands at will. However, as early as the 1920s, some of the floral greens companies that owned large parcels of land required harvesters to pay a small set fee or a portion of the harvested product, known as a 'stumpage' fee, to pick on their lands (Fitzpatrick, 1997; Spreyer, 2004). After World War II, growing demand for floral greens prompted a few of the larger private timber landowners and the DNR to develop similar systems for highly productive floral greens sites (Spreyer, 2004). However, de facto open access conditions continued on less productive sites, including all the land administered by the Olympic National Forest, most of the land managed by the state and most privately held land.

As the number of pickers increased during the 1980s, the DNR, US Forest Service and a number of large-scale private landholders either set up or expanded their existing formal systems for allocating access to floral greens. Three major forms of formalized access existed for floral greens on the Olympic Peninsula in the early 1990s (Spreyer, 2004).

- On lands administered by the US Forest Service, harvesters were required to obtain short-term (two to three weeks) inexpensive permits granting non-exclusive non-transferable access rights to the permit holder. Harvesters could not transfer their harvesting rights to others under these permits.
- On large-scale private holdings, harvesters negotiated with the landholder for annual or semi-annual short-term leases[5] granting the holder exclusive access to harvest in a designated area. Under the terms of these agreements, harvesters could transfer their harvesting rights to one or more other harvesters by drawing up a document known as a 'true copy'.
- On highly productive floral greens sites administered by the DNR, harvesters could acquire through public auctions multiple-year renewable leases that granted the holder exclusive access to a designated area. Under these lease agreements the holder could transfer harvesting rights to one or more other persons by drawing up a true copy.

Most harvesters could afford to buy short-term permits. A few harvesters had the means to acquire short- or long-term leases covering relatively small areas. They generally harvested the floral greens on their leases themselves, although some sublet their harvest rights to other harvesters. The large-area, long-term leases, however, required an upfront investment beyond the means of most harvesters and smaller brush shed operators. Medium- and large-scale brush shed operators acquired these leases and then sublet their harvesting rights to harvesters. The expectation was that the harvesters would sell the products they took off the land to the shed holding the lease. In other cases, the shed operators sublet their harvesting rights to brush bosses, who organized crews of harvesters to do the work on the ground. Again, the leaseholder expected the brush boss and his crew members to sell their harvested materials to his shed. However, by the early 1990s this arrangement led to tense debates that questioned both the true nature of the employer-employee relationship and how to enforce these tenure systems to protect both the natural resources and the harvesters. The following sections explore these complex issues in more depth.

Law enforcement

Washington State's legal mechanism for enforcing brush permits and leases is the Specialized Forest Products Act (RCW 76.48). The Washington State Legislature passed this law in 1967 to reduce thefts of cedar, an exceptionally valuable timber species, and Christmas trees. Over the years, the Act was amended to include other products, including floral greens. In 1992, RCW 76.48 included the following provisions.

- Special forest products harvesters must have written permission from the landowner to harvest on his land. The landowner must use a printed form provided by the DNR.
- Harvesters must validate their permits at the county sheriff's office. Sheriffs can ask the permit holder to show identification documents before validating the permit. A validated permit authorizes the holder to harvest, possess and transport the products included on the permit.
- Harvesters or buyers wishing to transport special forest products within the State of Washington must have documentation of authorization to possess or transport special forest products. This can take the form of a permit validated by the sheriff, a true copy of such a permit, a bill of lading or a sales invoice.

The penalties for failure to obtain and possess a special forest products permit included confiscation of the products and the possibility of a fine of up to $1000 and/or up to one year's confinement in jail. However, enforcing these provisions on the Olympic Peninsula in the early 1990s was a difficult task because of the region's steep, heavily forested terrain, extensive network of all-weather logging roads, and understaffed and underfunded county law enforcement departments. The risk for a floral greens poacher of being caught was small, and, for all practical purposes, open access conditions prevailed. Accusations of brush poaching, price-fixing, the hiring of undocumented immigrants, the flouting of labour laws and unsustainable harvesting abounded. By 1993, pickers, buyers and landowners were all calling on the State of Washington to help resolve the tensions.

DEFINING THE FLORAL GREENS POLICY PROBLEM: POACHING AND UNAUTHORIZED IMMIGRATION

In fall 1993, the DNR sought to resolve the tensions in the floral greens industry through a public meeting that brought together representatives of the large sheds, small independent shed operators, harvesters and state land managers. The transcripts from this meeting (DNR, 1993) show how the participants defined Washington's floral greens policy problem at that time, as well as how they framed the solutions to that problem.

Participants in the floral greens meeting identified a range of difficulties facing their industry. Many of these problems were linked to market conditions over which the State of Washington had little influence, such as low prices on the European export market and stiff competition from suppliers in other countries. However, one issue, the upswing in the incidence of floral greens poaching since the late 1980s, surfaced repeatedly as an area needing state intervention. A representative from one of the largest brush sheds in the region described the economic impact of floral greens poaching on a brush lease held by his company.

> *We estimate that we lose between 20 and 25 per cent of our crop every year off of that [leased parcel] to illegal harvest. It's a chequerboard affair where we have land throughout the county. It's not in a huge block. We can't patrol it on a daily basis and ensure that we don't have people out there.*

Jill, a picker with a one-year renewable lease on a much smaller parcel of private land, explained how poachers had affected her livelihood.

> *I've been employed in harvesting the floral evergreens from the forests in the state of Washington for close to 60 years. And the last four years is a complete turnaround from anything that I ever knew before. It has become a very stressful situation. There are so many people encroaching on the lands that I have a personal permit on that I have no time to work for policing my lands.*

Some shed owners argued that poachers brought inferior quality product to the sheds, resulting in lower prices overall for Pacific Northwest floral greens. Scott, owner of a medium-size brush shed on the eastern Peninsula, said that, 'With the product being stolen; the quality level … has really come down. So in essence we're killing some markets out there.' Jill drew a connection between the high level of poaching, poor harvesting techniques and the declining quantities of brush on her lease.

> *They [brush poachers] pick everything that's in front of them. Brush that might have been shoulder high on me, when they get through with it, it's knee high. It's going to be years before that comes back to anything that's any good.*

Stan, a shed owner and brush leaseholder, reiterated this theme, noting: 'Not only are they taking our crop this year, but they're taking next year's crop in the process too.'

However, not everyone agreed that floral greens were being overharvested.[6] Nick, a brush shed owner with a large lease on the Tahuya State Forest said,

> *There's so much salal available out there to buy that it's unreal. And every year I talk to little side plants that we buy from, and pickers, and I never hear nothing about over-harvesting except from people that have small pieces of ground.*

Many of the participants linked increases in the incidence of poaching to the increase in undocumented Latino harvesters. Peter, the owner of a small buying operation on the west side of the Peninsula, described how the industry had changed in his area.

> *So we're seeing a shift in illegal aliens coming up here, and being brought up and being sent out by someone who is a crew boss, or whatever you want to call him, out there in the field.*

A representative from Hiawatha, the region's largest brush shed, attributed over-harvesting on remote areas of his company's brush lease to undocumented Latino poachers.

> *And in a lot of those [remote] areas, those people get into them and they feel that it's a safe area because they're not being harassed and they'll spend a lot of time in the same area and go pick over and over in the same area with this security feeling that they're not going to be caught.*

Others, such as Nick, did not agree that poaching by undocumented Latino harvesters was a widespread problem.

> *I've never had a problem about theft, other than a few Americans jumping in on my ground … So it's not the immigrants that I've seen any problems with.*

Another shed owner concurred, declaring that the Latinos who picked brush on his leasehold not only had permits for the land they picked on, but also had legal residency status.

Although participants differed in their views about who was to blame for brush poaching, most agreed that it was widespread and that the weak enforcement of resource management laws and regulations by the state and federal government contributed to its persistence. Peter described what happened in his area when law enforcement officers caught people transporting brush without a permit.

> *There's no enforcement. When they do catch people, their hands are slapped, 'Don't do it again.' And they know that they'll be right back there next week, picking the stuff.*

Several participants interjected that when they caught poachers on their brush leases, law officers were reluctant to come on site and charge the offenders. A DNR administrator responded to these criticisms.

> *We want to rely on [county] law enforcement, but then in a real practical sense, their hands are tied, they have neither the resources, nor is this high enough on their priority list for them to take large amounts of time to deal with it.*

He added that the DNR also lacked the resources to assist with regulatory enforcement.

> *DNR has some regulatory functions but most of them are funded from the general fund and you know the general fund is being reduced dramatically by the legislature. So we want to do these things. It's just that as a matter of reality, we don't have the resources to be able to do that.*

Many brush shed operators agreed that buyers needed to take on some responsibility for enforcement, but most were pessimistic about buyers being able to accomplish much alone. As one shed owner pointed out, even if some buyers joined together and refused to buy suspect brush, 'There's the guy down the road that will buy it, and the guy in the garage who will buy it.'

Through this meeting and others held in the early 1990s, Washington's floral greens policy problem became defined first as a poaching problem, due to the inability of the State of Washington to enforce the existing special forest products law, and second as a consequence of the presence of immigrant harvesters lacking authorization to reside or work in the United States.

TAKING STEPS TO RESOLVE THE TENSION

In 1995, the state legislature took steps to remedy the enforcement problem by adding a section to RCW 76.48 requiring buyers of special forest products to record the harvesting permit number, the permit holder's name and the type and amount of products harvested for each transaction. Buyers had to keep these records for one year. In addition, wholesale buyers were required to include the licence plate number of the vehicle transporting the products and the seller's permit number on their bills of sale. These new provisions were designed to force the floral greens sheds to buy products only from harvesters with valid permits and to make it easier for law enforcement officers to track down or verify suspected floral greens poachers.

At the same time, the DNR modified its floral greens leasing programme. On its most productive ground, the DNR began gradually consolidating its floral greens leases as they came up for renewal, combining small and medium-sized parcels into a much smaller number of large parcels. On less productive sites, the DNR set up a system of short-term exclusive-use permits to allocate access to floral greens harvesters.

Aside from providing harvesters with additional access to state lands, the new permit system also generated revenues for the state. Neighbouring landholders followed suit: several large private landholders adopted permit systems similar to the DNR's new system in the mid-1990s, and in 2001 the Olympic National Forest began an open-bid lease system for floral greens in some areas.

Addressing the undocumented immigrant harvester problem was much more complicated, in part because immigration policy is set and enforced by the federal government and not amenable to state-level policy solutions. In an early version of the proposed revisions to RCW 76.48, several members of the state legislature sought to include a provision requiring county sheriffs to check harvesters' identification documents when validating harvest permits, a provision designed to identify and weed out undocumented immigrants. However, two politically powerful groups – the owners of the larger brush sheds, who needed a source of cheap labour to remain competitive in the European market, and the county sheriffs, who feared they would be saddled with an unfunded enforcement mandate – objected to the provision and the legislature removed it from the final version of the revised law.

In 1995, federal, state, tribal and county law enforcement agents and security officers from several large private timber companies sought to address the poaching and immigration issues simultaneously by sharing information and coordinating enforcement efforts on the ground. In 1999, the group organized a 'special emphasis' operation in which law enforcement agents from 20 organizations set up control stations along the major brush transportation routes on the west end of the Olympic Peninsula to apprehend brush poachers and undocumented immigrants. Of the 114 harvesters they caught picking without permits, 76 were undocumented immigrants and were deported. Although brush poaching declined for a short time following the operation, by 2002 it had returned to previous levels. The special emphasis operation and subsequent smaller periodic controls by the federal Immigration and Customs Enforcement agency (Loose, 2005) have not significantly reduced the numbers of undocumented immigrants working in the floral greens industry.

REDEFINING THE FLORAL GREENS PROBLEM AS A LABOUR RELATIONS ISSUE

The county sheriffs responsible for enforcing the new provisions of RCW 76.48 lacked the means to do so since the legislature opted not to allocate funds for enforcement. The DNR suffered from chronic budget and personnel shortages through the 1990s, and thus was unable to enforce its lease and permit programmes. Efforts to reduce the numbers of unauthorized immigrants harvesting floral greens were equally unsuccessful. By the end of the 1990s, it was clear that neither the changes to RCW 76.48 nor DNR's lease consolidations nor the interagency special emphasis operations had reduced floral greens poaching.

As the shortcomings of these policy solutions became apparent, independent harvesters and the smaller brush sheds began redefining the floral greens policy problem as a failure of the state to enforce labour regulations (Hansis, 2002). This issue surfaced in 1999, when several forest labour contractors filed a complaint with the Washington State Department of Labor and Industries (DLI) about illegal labour practices in the floral greens industry (Jefferson Center, 2002). The issue at stake was whether the buying shed operators that sublet harvest sites to pickers met the definition of employers under state labour laws.[7] According to Washington State labour regulations (RCW 51.08.195), floral greens harvesters who call themselves independent contractors are self-employed contractors only if they have a choice in where they can sell their product. If they are obliged to sell their product to the brush shed operator who holds the lease on the land where they pick, the law considers them shed employees. This legal distinction matters because employers must comply with federal and state laws regulating basic work conditions for employees (see Box 11.1), restrictions that do not apply to goods and services obtained through independent contractors. For example, employees are entitled to receive the state minimum wage and must be paid time and a half when they work more than 40 hours a week. Employers must contribute to the state's workers' compensation insurance fund, which provides funds for medical treatment when employees are injured on the job, in terms of the Federal Insurance Contributions Act, which funds the nation's system of old-age, disability and hospital insurance. Additionally, under the 1986 Immigration Reform and Control Act (8 USC 1101), employers must attest that they have examined documents such as

Box 11.1 Federal and Washington State labour law requirements

Federal law	Enforcing agency	Key provisions
Fair Labor Standards Act (29 USC 201-209)	US Department of Labor	Employees must be paid at least US$5.85 per hour (2007). Employees are entitled to be paid time and a half for hours worked over 40 in a week.
Migrant and Seasonal Agricultural Worker Protection Act (29 USC 20)	US Department of Labor	Farm labour contractors must: • register with the US Department of Labor; • pay workers their wages when they are due; • keep records of workers' hours, rates of pay and earnings; • provide workers with itemized, written statements of earnings; and • ensure that vehicles used to transport workers are properly insured, are operated by licensed drivers and meet federal and state safety standards.
Federal Insurance Contributions Act (26 USC 21)	US Internal Revenue Service	Employees and their employers must contribute to Social Security and Medicare, federal programmes that provide benefits to retirees, disabled persons and the children of deceased workers. In 2007, the required tax was 12.4% of the employee's gross salary for Social Security and 2.9% for Medicare. An employer must pay half of both taxes for each employee. The employee pays the other half.

Washington State law	Enforcing agency	Key provisions
Minimum Wage Act (RCW 49.49.46)	Department of Labor and Industries	Employees must be paid at least US$7.93 per hour (2007). Employees are entitled to be paid time-and-a-half for hours worked over 40 in a week.
Industrial Insurance Act (RCW 51)	Department of Labor and Industries	Employers must pay into the state's industrial insurance programme, which provides benefits to workers who are injured in the course of their employment or develop an occupational disease as a result of their required work activities. The amount paid per employee is based on the risk associated with the employee's job category.
Farm Labor Contractor Act (RCW 19.19.30)	Department of Labor and Industries	A farm labour contractor is defined as 'an individual, firm, partnership, association, corporation or government agency that, for a fee, recruits, solicits, employs, supplies, transports, or hires agricultural workers'. The state includes forestry and reforestation workers in the category of agricultural workers.

Washington State law	Enforcing agency	Key provisions
		Farm labour contractors must:
		• be licensed with the state as a master business;
		• be licensed with the state as a farm labour contractor;
		• post a bond annually with the state to ensure the payment of wages to workers;
		• keep wage, hour and earnings records for crew members;
		• provide workers with a written itemization of their earnings, hours worked and rates of pay;
		• pay all required taxes; and
		• use only properly insured vehicles to transport workers.
Employment Security Act (RCW 50)	Employment Security Department	Employers must contribute to the state unemployment insurance fund, which provides partial income replacement on a temporary basis for workers who become unemployed through no fault of their own.

birth certificates, social security cards or immigration documents proving that their employees have authorization to work in the United States.

The question of whether harvesters are independent contractors or brush shed employees is doubly complicated for Latino harvesters, many of whom participate in the *raitero* system, a transport system that gives harvesters access to the Olympic Peninsula's forests. Most Euro-American and South-East Asian harvesters, as well as Latino harvesters who have resided in the United States for many years, own vehicles. However, few newly arrived Latino harvesters have their own transport. Instead, they rely on a system borrowed from migrant farm worker culture. Van owners, called *raiteros*, convey harvesters to and from harvesting sites. *Raiteros* often obtain subleases for floral greens harvesting sites from the larger brush sheds. At the end of the day, the *raitero* picks up the harvesters along with the brush they've gathered and takes them to the buying shed holding the lease to sell their product. The *raitero* charges the harvester a small set fee for his transportation services (usually between US$5 and US$10) and, in some cases, also collects a percentage of the value of the day's harvest.

In 2001, owners of the six largest brush sheds petitioned the Washington State court for a declaratory judgment that floral and brush wholesalers were not 'employers' for industrial insurance purposes. In their complaint, the companies denied that an employer–employee relation existed between them and pickers they sublet harvesting rights to, since the permits they issued stated that brush harvesters were free to sell their product wherever they chose (Box 11.2). Pickers and labour rights advocates, however, argued that in practice harvesters with these permits had to sell their brush to the shed issuing the permit, making them employees.

The Mason County Superior Court judge, who heard the case in 2003, laid out five criteria that sheds had to meet in order *not* to be classified as employers (Box 11.3) (DLI, 2005). If a shed failed to meet all the criteria, it would be considered to have an employer–employee relationship with pickers.

Box 11.2 Excerpt from a permit for harvesting on a brush shed lease

This permit is sold to the harvester for the purposes of harvesting evergreen products from designated locations stated on the permit. The harvester has the choice and right to sell evergreen products to any buyer he or she selects. The harvester is NOT an employee of the permit seller, and therefore agrees to pay all necessary payroll taxes or government imposed charges due the Internal Revenue and State Department of Revenue for the business and occupation tax or other obligations to said department, the Washington State Department of Labor and Industries, and the Washington State Department of Employment Security. (Personal communication, anonymous, 2002)

However, no harvesters were willing to testify in court that they did not have a choice of where to sell their product, for fear that the sheds would refuse to issue them with harvesting permits in the future, and the court found that in the specific case in question, the floral greens companies did not meet the criteria to be employers (DLI, 2005).

In the wake of the Mason County Superior Court decision, the DLI began an audit programme to assess compliance among sheds and *raiteros* with forest labour contractor laws, as well as labour laws governing workplace safety (DLI, 2005). At the same time, department employees began an educational campaign to inform pickers of their rights as workers, and brush sheds and *raiteros* of their responsibilities as employers or farm labour contractors (DLI, 2005). Department managers view the floral greens audit programme as a test case that they will eventually be able to use to show that employee–employer relationships exist in a number of other industries (shellfish, construction and reforestation) where the use of independent contractor arrangements has become widespread.

Box 11.3 Five-step checklist for determining whether a brush shed is an employer

A brush shed that

- sells permits to a vendor-picker;
- does not require the vendor picker to sell the product back to the company;
- does not control the vendor-picker's work;
- is not in the picking business, but rather is in the buying and packing business; and
- requires a vendor-picker to be solely responsible for his or her own taxes and complying with all other business regulations

is not an employer.

Based on audits conducted from 2003 onwards, the DLI has found that most sheds are employers of both pickers and *raiteros* under the criteria laid out in the 2003 court case (Holt, 2007). The audits indicate that in cases where pickers have access to floral greens on a sublease through a shed, 95 per cent of the pickers sell all their product to the shed holding the lease. In contrast, pickers who have permits or leases directly through landholders typically sell their products to several sheds. The DLI interprets these results as evidence that pickers who have access to harvesting sites through shed subleases (or through a *raitero* holding a shed sublease) do not exercise control over where they sell their product, and thus the state considers them to be shed employees rather than independent contractors.

Most of the small and medium-sized sheds have complied with the audit findings, including paying compensation for previously unpaid workers as well as for current employees (Holt, 2007; Jenkins, 2007). Some of these audited companies, though not all, are currently reporting worker hours. However, a coalition of 15 brush sheds, including all the region's largest sheds with their subsidiary sheds, has appealed against the audit findings.

The DNR's favouring of large-scale exclusive-use leases to allocate floral greens harvesting rights has exacerbated conflicts within the industry by concentrating control over the most productive floral greens sites among a handful of large, highly capitalized brush shed owners who have the financial means to acquire those leases. The concentration of leases in the hands of a relatively small number of companies limits the harvesters' options for obtaining legal access to resources. The lack of legal access, coupled with inadequate law enforcement capacity, creates an environment conducive to poaching. At the same time, because most harvesters must sell their products to the sheds whose owners hold the leases that they work on, they are forced to accept whatever price those sheds offer.

COMPARISONS WITH THE WILD MUSHROOM INDUSTRY IN THE PACIFIC NORTHWEST

The labour status of wild mushroom pickers in the Pacific Northwest, many of whom are South-East Asian or Euro-American, provides an interesting contrast with the floral greens industry.[8] Euro-American mushroom pickers participate in the harvest either as self-employed individuals or in self-employed groups of two to four (McLain, 2000; Jones, 2002). South-East Asian matsutake harvesters typically work in self-employed extended family groups (Richards and Creasey, 1996).

In 1989, the year in which Continental Floral Greens brought Latino workers into its processing operations, a field buyer for the region's largest wild mushroom company tried a similar tactic with South-East Asian immigrants during the matsutake mushroom harvest. The buyer set the workers up in a field camp and offered to show them how to harvest the matsutake on condition that they sold their mushrooms to him. However, the experiment failed when crew members realized that other buyers would pay them quite a bit more and that they would be better off pooling their money to purchase vehicles and work for themselves. The *raitero* system has also failed to take root in the wild mushroom industry.

Several key differences between the wild mushroom and floral greens industries offer a plausible explanation for the fact that floral greens workers have the characteristics of de facto employees, while wild mushroom workers have those of self-employed contractors (Box 11.4).

First, the two industries differ greatly in how control over resource access is structured. In the floral greens sector, most harvesters do not obtain access to harvesting sites directly from the land manager but instead pay either a *raitero* or a brush shed operator for permission to harvest. As leases have become concentrated in the hands of a few brush shed operators, pickers have become increasingly dependent on those sheds for access to harvesting sites and exercise less choice over where they sell their products. In contrast, most mushroom harvesters obtain access to mushroom patches by buying relatively inexpensive non-exclusive permits directly from the land manager. Because the wild mushroom buying companies do not control picker access to mushroom harvesting sites, they cannot use the threat of denying future access to mushroom patches as leverage to force pickers to sell to them. Wild mushroom pickers thus have more choice over where they sell their product.

Second, the legal status of the harvesters differs greatly between the two industries. In the wild mushroom industry, many harvesters have either US citizenship or legal permission to reside and work in the United States. In contrast, many floral greens harvesters lack permission to live and work in the United States. They dare not

Box 11.4 Comparison of floral greens and wild mushroom industries

Variable	Floral greens industry	Wild mushroom industry
Markets	Almost exclusively an export market, with most product shipped to European floral markets	A strong domestic market as well as a strong export market oriented to Europe for chanterelles, boletes, and morels, and to Japan for matsutake
Access to gathering sites	Most pickers obtain harvesting permits from brush bosses or brush sheds. Typically these are permits for harvesting on lands that the brush sheds have leased from the landowners.	Most pickers obtain harvesting permits directly from landowners.
Access to transport	Many pickers lack their own means of transport and rely on *raiteros* to reach harvesting sites.	Most pickers have their own means of transport.
Dominant citizenship or residency status	A significant number of pickers are immigrants who lack authorization to reside or work in the United States.	Many pickers are either US citizens or legal residents with permission to work in the United States.
Dominant form of labour relations	Pickers are nominally self-employed contractors, but in practice wage labour conditions prevail.	Most pickers are self-employed individuals or families.

complain about their working conditions for fear that the shed operators or *raiteros* might report them to the immigration service, leading to their arrest and deportation.

Third, the two industries differ in the way in which access to their key markets is distributed. Virtually all of the floral greens harvested in the Pacific Northwest are exported to Europe, and, since the early 1990s, a handful of floral greens companies have controlled that market (Spreyer, 2004). In contrast, at least 30 per cent of the region's wild mushrooms are sold on the domestic market, and the export market is split between Europe for chanterelles, morels and boletes, and Japan for matsutake (Schlosser and Blatner, 1995). Smaller companies are able to compete effectively in the domestic market, particularly if they focus on high-end niche customers. Large companies in the wild mushroom industry exercise much less control over markets than do companies in the floral greens sector.

CONCLUSION

During the 1993 floral greens meeting (DNR, 1993), a high-level DNR administrator voiced concern that globalization might well bring with it fundamental changes that neither the state nor the floral greens industry were prepared for.

> *We're going into NAFTA. Once you open the borders, what do you have then? Maybe we're wasting our time even talking about it if that's the case.... It's getting so complicated and there's so many multiple layers that we can't even identify what the real issues are, let alone how to resolve them.*

Events of the past 14 years have shown that his concern was justified. When brush poaching became widespread in the Pacific Northwest in the early 1990s, state land managers concluded that the main issue was inadequate enforcement and that the solution was to consolidate numerous small brush leases into a smaller number of larger leases which would be easier to police. Later, as the shortcomings of this strategy became apparent, a coalition of law enforcement and regulatory agencies sought to reduce poaching by implementing an intense multi-agency control operation aimed simultaneously at apprehending harvesters without harvesting permits and undocumented immigrants.

This approach also proved ineffective, and by the early 2000s, efforts to address the brush poaching issue shifted to expanding the state's capacity to enforce labour laws governing the distinction between employees and independent contractors. Enforcing these labour laws on a wide scale is unlikely to be possible, however, as long as large numbers of undocumented immigrants continue to enter the United States. And the stream of undocumented immigrants is unlikely to cease as long as multilateral international trade policies, such as NAFTA and the Central American Free Trade Agreement, undermine the economic livelihoods of rural Mexicans, Guatemalans, Salvadoreans and Nicaraguans, and put pressure on them to look elsewhere for work.

In sum, crafting policies likely to reduce brush poaching in the Pacific Northwest requires land managers and natural resource policy-makers to understand the dynamics of today's global economy, including the changes taking place in historical relationships between resource access, market access, labour relations, labour force migration, trade relations and ecological conditions.

ACKNOWLEDGEMENTS

We dedicate this chapter to Beverly A. Brown, who was instrumental in clarifying the links between floral greens management and labour and immigration policies in the Pacific Northwest region. We thank Leilan Greer for her work as editor and Eric T. Jones for his input as technical reviewer. This work draws in part on research funded through the US Forest Service, Pacific Northwest Research Station, and the US Environmental Protection Agency's Science To Achieve Results (STAR) graduate fellowship programme.

NOTES

1   Based on research funded partly through USDA-Forest Service Agreement PNW 02-CA-11261975–128.
2   Pseudonyms are used throughout this chapter to protect the identities of the interviewees.
3   See Spreyer (2004) for an in-depth description of the evolution of the Pacific Northwest's floral greens industry.
4   The term 'Latino' obscures the national diversity within this segment of the harvester population, which includes US citizens, Mexicans, Guatemalans, Salvadoreans and Nicaraguans. It also obscures linguistic and cultural differences: a sizeable number are native speakers of various Indian languages, whose Spanish may be limited.
5   Technically these agreements were permits, but the people who used them called them leases.
6   Foresters have tried unsuccessfully for decades to eliminate salal, which competes with tree seedlings, from the forests of the Pacific Northwest. It is thus unlikely that even extremely intense commercial harvesting significantly affects the long-term viability of salal populations. Two recent ecological studies of the impacts of commercial harvesting on salal had mixed results (Ballard, 2004; Cocksedge and Titus, 2006). Ballard (2004) found that light harvest treatments had a negative impact on new shoot growth, but new shoot growth after heavy harvest treatments was nearly the same as on shrubs that were not harvested. Cocksedge and Titus (2006) found that new growth was greater in harvested plots than in undisturbed plots. Both studies were short-term (three years or less), covered very small geographical areas, and are of limited applicability to the region as a whole. Widespread forest land conversion to residential and industrial uses is a much greater threat to salal populations than commercial harvesting.
7   A report issued by the Jefferson Center (2002) provides an overview of the labour policy debate taking place in the floral greens industry on the Olympic Peninsula.
8   Latinos participate in the commercial wild mushroom harvest, but make up a substantially smaller portion of the (visible) mushroom picker population than is the case with floral greens.

REFERENCES

Ballard, H. (2004) 'Impacts of harvesting salal (*Gaultheria shallon*) on the Olympic Peninsula, Washington: Harvester knowledge, science, and participation', PhD dissertation, University of California, Berkeley, CA

Bratsberg, B. (1995) 'Legal versus illegal U.S. immigration and source country characteristics', *Southern Economic Journal*, vol 61, no 3, pp715–727

Cocksedge, W. (2003) 'Social and ecological aspects of the commercial harvest of the floral greenery, salal (*Gaultheria shallon* Pursh; Ericaceae)', master's thesis, University of Victoria, BC, Canada

Cocksedge, W. and Titus, B. D. (2006) 'Short-term response of salal (*Gaultheria shallon* Pursh) to commercial harvesting for floral greenery', *Agroforestry Systems,* vol 68, no 2, pp103–111

Department of Labor and Industries (DLI) Washington State (2005) *Harvesting Washington's Brush: Monitoring Compliance with Labor Laws in the Floral Greens Industry*, Washington State Department of Labor and Industries, Olympia, WA

Department of Natural Resources (DNR) Washington State (1993) Floral Brush Issues Meeting, Olympia, WA, 21 September. Unpublished meeting transcript on file with authors

Draffan, G. (2006) *Report on the Floral Greens Industry,* The Evergreen State College Labor Center, Olympia, WA, www.endgame.org/floral.pdf, accessed 24 August 2007

Durand, J., Massey, D. S. and Parrado, E. A. (1999) 'The new era of Mexican migration to the United States', *Journal of American History*, vol 86, no 2, pp518–536

Fitzpatrick, S. (1997) 'Nontimber forest products: Using local knowledge', master's thesis, Evergreen State College, Olympia, WA

Hansis, R. (1996) 'The harvesting of special forest products by Latinos and Southeast Asians in the Pacific Northwest: Preliminary observations', *Society and Natural Resources,* vol 9, pp611–615

Hansis, R. (2002) 'Workers in the woods: Confronting rapid change', in E. T. Jones, R. J. McLain and J. F. Weigand (eds) *Nontimber Forest Products in the United States*, University of Kansas Press, Lawrence, KS, pp52–56

Heckman, H. (1951) 'The happy brush pickers of the high Cascades', *Saturday Evening Post,* 6 October, pp36–38, 103, 105

Holt, M. (2007) Personal communication, 13 September

Jefferson Center (2002) 'Are forest brush harvesters employees? The legal battle over nontimber forest products workers heats up in Washington State', report on the current status of the debate and its relationship to national issues in nontimber forest products, *Bulletin 4*, Jefferson Center, Wolf Creek, OR

Jenkins, A. (2007) 'Washington State battling to regulate brush pickers', *OPB News*, 23 August, Oregon Public Broadcasting, www.news.opb.org/article/washington-state-battling-regulate-brush-pickers, accessed 29 August 2007

Jones, E. T. (2002) 'The political ecology of wild mushroom harvester stewardship in the Pacific Northwest', PhD dissertation, University of Massachusetts, Amherst, MA

Kantor, S. (1994) 'Local knowledge and policy development: Special forest products in coastal Washington', master's thesis, University of Washington, Seattle, WA

Loose, S. (2005) 'Justice in the brush industry?' *Jefferson Center News*, vol 4, no 2, pp4–6

Lynch, K. A. and McLain, R. J. (2003) *Access, Labor, and Wild Floral Greens Management in Western Washington's Forests*, General Technical Report no PNW-GTR-585, US Department of Agriculture, Forest Service, Pacific Northwest Research Station, Portland, OR

Martin, P. (2005) 'NAFTA and Mexico–US migration: Policy options in 2004', *Law and Business Review of the Americas*, vol 11, no 3/4, pp361–385

McLain, R. J. (2000) 'Controlling the forest understory: Wild mushroom politics in central Oregon', PhD dissertation, University of Washington, Seattle, WA

Personal communication (2002) (source wishes to remain anonymous) April, copy of permit on file with authors

Richards, R. T. and Creasy, M. (1996) 'Ethnic diversity, resource values, and ecosystem management: Matsutake mushroom harvesting in the Klamath bioregion', *Society and Natural Resources*, vol 9, pp359–374

Robinson, C. (1994) 'Multiple perspectives: Rules governing special forest products management in coastal Washington', master's thesis, University of Washington, Seattle, WA

Schlosser, W. E. and Blatner, K. A. (1995) 'The wild edible mushroom industry of Washington, Oregon and Idaho: A 1992 survey', *Journal of Forestry*, vol 93, no 3, pp31–36

Spreyer, K. K. (2004) 'Tales from the understory: Labor, resource control, and identity in western Washington's floral greens industry', PhD dissertation, University of California, Berkeley, CA

LEGISLATIVE REFERENCES

United States, 8 USC 1101 Note, Immigration Reform and Control Act as amended

United States, 26 USC 21, Federal Insurance Contributions Act as amended

United States, 29 USC 20, Migrant and Seasonal Agricultural Worker Protection Act as amended

United States, 29 USC 201–219, Federal Fair Labor Standards Act as amended

Washington, Revised Code of, Title 19, Chapter 19.30 (RCW 19.30) Farm Labor Contractors Act as amended

Washington, Revised Code of, Title 49, Chapter 49, Section 46 (RCW 49.49.46) Minimum Wage Act as amended

Washington, Revised Code of, Title 50 (RCW 50) Washington State Employment Security Act as amended

Washington, Revised Code of, Title 51 (RCW 51) Washington State Industrial Insurance Act as amended

Washington, Revised Code of, Title 76, Chapter 48 (RCW 76.48) Specialized Forest Products Act as amended

APPENDIX

Table 11.1A *Common and scientific names of plants mentioned in the text*

Common name	Scientific name
Beargrass	*Xerophyllum tenax* (Pursh) Nutt.
Bolete	*Boletus edulis* Bull. Fr.
Chanterelle	*Cantharellus* spp.
Douglas fir	*Pseudotsuga menziesii* (Mirbel) Franco
Evergreen huckleberry	*Vaccinium ovatum* (Pursh)
Matsutake	*Tricholoma magnivelare*
Morel	*Morchella* spp.
Noble fir	*Abies procera* Rehd.
Salal	*Gaultheria shallon* Pursh
Sword fern	*Polystichum munitum* (Kaulfuss.) K. Presl.
Western red cedar	*Thuja plicata* Donn ex D. Don
Western white pine	*Pinus monticola* Dougl. ex D. Don

NTFP Policy, Access to Markets and Labour Issues in Finland: Impacts of Regionalization and Globalization on the Wild Berry Industry

Rebecca T. Richards and Olli Saastamoinen

INTRODUCTION

In this chapter, we address how sociodemographic changes have altered NTFP harvesting practices in Finland in the context of post-Soviet Union and European Union (EU) regionalization. We examine how social forces of regionalization in conjunction with globalization have altered the competitive advantage of the NTFP industry in Finland relative to neighbouring Sweden and Russia. Following Ribot and Peluso's (2003) theory of access, we discuss the strengths and limitations of traditional Finnish NTFP rights-based resource access policy in addressing emerging structural and relational market and labour problems. We conclude with the implications for sustaining and developing equitable NTFP use and trade in Finland.

FINLAND'S FORESTS AND CHANGES IN NTFP HARVESTING

Finland is Europe's most heavily forested country. Geographically, most of Finland is situated at a latitude between 60°N and 70°N, and thus a significant area of the country extends north of the Arctic Circle. Because of the moderating influence of the Gulf Stream, even the northernmost areas are forested. About 20 indigenous tree species occur in Finland, the most common being pine (*Pinus silvestris*), spruce (*Picea abies*) and birch (*Betula pendula* and *B. pubescens*). Typically, two or three tree species dominate a forest (Boreal Forest Website, 2002).

In all, 86 per cent of the total land area in Finland comprises forested land, including treeless tundra and open peatlands, or more than five hectares of forest

Source: Marjut Turtiainen.

Figure 12.1 *Map of Finland*

per Finnish citizen. For most Finns, forests are a part of everyday life. Currently, the National Forest Programme of Finland aims to further develop forest management and protection to provide citizens with as much work and as many sources of livelihood as possible, to maintain healthy, vital and diverse ecosystems, and to ensure spiritual and physical recreational opportunities for the people (Finnish Ministry of Agriculture and Forestry, 2005).

The most commonly collected NTFPs in Finland are forest and peatland wild berries, mushrooms, decorative lichens, medicinal plants, forage, game and reindeer meat and parts (e.g. antlers), and minor woody parts of trees for domestic utensils and handicrafts, as well as birch sap and tar (Lund et al, 1998; Saastamoinen et al, 1998). Of these, wild berries are the most significant NTFPs because of their relative abundance, nutritional benefit and extensive commercialization (Aarne et al, 2005). However, wild mushrooms are also becoming more widely gathered and marketed for export (Moisio, 1999; Aarne et al, 2005).

Box 12.1 The rise of the cep (*Boletus spp.*) export industry in Finland

The most significant wild mushroom species harvested in Finland are chanterelles (*Cantarellus cibarius*), ceps or porcinis (*Boletus edulis* and *B. pinophilus*), and northern milkcaps (*Lactarius trivialis* and *L. utilis*). The annual value of commercially harvested wild mushrooms in Finland has been estimated to be €2.52 million (Aarne et al, 2005). In general, Finnish wild mushrooms have not been exported widely. However, in recent years, ceps have rapidly become the most commercially valuable wild mushroom exported from Finland. In 2003, a very productive wild mushroom year, the cep commercial crop was estimated at about 1 million kg – although in less productive years it has fallen to about 100,000kg (Mäntynen, 2005). This cep export industry has been largely credited to an Italian entrepreneur who settled in eastern Finland and has capitalized on the Italian market through marketing, networking and efficient supply logistics (Pohjois-Karjalan TE-keskus, 2006).

NTFP harvesting in Finland is significant in terms of both traditional household use and market export. However, despite the traditional and commercial importance of NTFPs in Finland, significant sociodemographic changes are affecting the levels of household engagement in NTFP harvesting and use. Rapid urbanization and the subsequent migration of many rural Finns from northern, central and eastern Finland to the southern region of the country have been major factors in reducing household participation in frequent, consumptive wild berry and other NTFP harvesting activities (Saastamoinen et al, 2000; Pouta et al, 2006). Nevertheless, tradition continues to play a strong role in NTFP use because two-thirds of the wild berries and mushrooms gathered from Finnish forests are harvested for recreational picking and home use (Yrjölä, 2002). At the same time, the commercial harvesting of NTFPs continues to be an increasingly significant economic activity,

---

Box 12.2 Finnish participation rates in wild berry harvesting

Finnish national surveys have found that the national household participation rate for harvesting wild berries was 59.5 per cent in 1997 (Saastamoinen et al, 2000) and 55 per cent for residents aged 15–74 from 1998 to 2000 (Pouta et al, 2006). Thus more than half Finland's households and residents pick wild berries, with the estimated average annual volume harvested per household ranging from 25.8kg in 1997 to 22.6kg in 1998. Of the total volume harvested in 1997, 72.7 per cent was picked for household use. The remainder (over one-quarter) was sold, with 4.8 per cent of the households surveyed reporting that they engaged in commercial picking (Saastamoinen et al, 2000).

particularly in the rural areas of northern and eastern Finland (Maaseutupolitiikan yhteistyöryhmä, 2001; Aarne et al, 2005).

Wild berry harvesting in the regional context of Finland, Sweden and Russia

In the forests of Finland and neighbouring Sweden and Russia, wild berries are abundant, although some species in some regions have experienced a decline in productivity due to air pollution and the resulting soil acidification and eutrophication (Statistics Sweden, 2001), peatland drainage (Salo, 1995) and decades of timber harvesting (Chibisov, 1999). Because of the general abundance of wild berries and the relative lack of other fruits and vegetables in an immense region of poor agricultural soil, extensive forests and harsh climate, the collection of wild berries in the boreal north historically has been critical to dietary sustenance and nutrition as well as supplementary household income (Saastamoinen et al, 1998; Panteleeva, 2004). More recently, wild berry harvesting has increasingly contributed to the rural economy as a local industry (Kangas, 2001a; Aarne et al, 2005).

The species of wild berries collected are diverse, but the three most commonly harvested are lingonberry or cowberry (*Vaccinium vitis-idaea*), bilberry (*Vaccinium myrtillus*) and cloudberry (*Rubus chamaemorus*). Wild cranberry species (*Vaccinium* spp.) are widely collected from the peatlands, particularly in eastern Finland and Russia. Quantity varies greatly from year to year and from species to species.[1]

Finland

In Finland, the average annual consumption of wild berries is 8.3kg per person (Maaseutupolitiikan yhteistyöryhmä, 2001). It has been estimated that approximately 35,000–50,000 people in Finland (out of a population of 5.3 million) annually engage in commercial wild berry picking (Aarne et al, 2005). Harvest estimates indicate that

---

Box 12.3 Wild berry yields in Finland, Sweden, and Russia

In Finland in an average year on mineral soils (excluding peatlands and open fells), according to model-based estimates, the lingonberry yield is 243.8 million kg and the bilberry yield is 168.4 million kg (Turtiainen et al, 2005). More detailed estimates have been produced in Sweden, which is perhaps the only nation to have conducted extensive national field inventories of forest berries. From 1975 to 1977, Swedish forests produced 255 million kg of bilberries and 155 million kg of lingonberries (Eriksson et al, 1979). In the Republic of Karelia, the region of Russia bordering Finland, there have been estimates of biological resources based on forest plans of 50 million kg of lingonberries, 70 million kg of bilberries and between 16 million and 18 million kg of cranberries. However, the cranberry crop exceeds 10 million kg only in very productive years (Belonogova and Zaitseva, 1989; Myllynen and Saastamoinen, 1995).

Notes: The *Boletus edulis*-picking boom for sale started in mid-1990s in North Carelia, Finland.
The two Finnish words on the wall say 'berries' and 'mushrooms'.

Source: Olli Saastamoinen.

Figure 12.2 *Mushroom pickers waiting their turn to sell their weekend crop on a Sunday afternoon in 1995 outside 'Tuote ja Vihannes Ky', the oldest firm in the Joensuu region in the wild berry and mushroom business*

40 million kg of Finnish wild berries are gathered annually, and of this harvest 25 per cent is sold commercially (Aarne et al, 2005).

In productive years, the volumes of wild berries harvested are higher. Results from a national survey indicate that the total yield collected by Finnish households was 56.5 million kg in 1997 and 49.7 million kg in 1998. In 1997, 4.8 per cent of Finnish households surveyed were engaged in picking wild berries for sale. These households collected 15.4 million kg for sale, or 27 per cent of the total wild berry harvest reported in the survey (Saastamoinen et al, 2000).[2]

Engagement in commercial wild berry harvesting varies regionally: the rates of households picking wild berries for sale in six municipalities in eastern and northern Finland in 1997 ranged from 8 to 31 per cent (Kangas, 2001b). This variability appears to be directly related to regional differences in the Finnish standard of living. For example, based on annual statistics from the *Finnish Statistical Yearbook of Forestry* (Finnish Forest Research Institute, 2004), we determined important patterns in the regional distribution of income earned in 2003 from commercially harvesting wild berries. Of the total income, 45 per cent was earned in the northern Finnish province of Oulu (which includes Kainuu, the poorest area in Finland). Similarly, 97 per cent of total income from the commercial cloudberry harvest was earned in Lapland, the poorest and northernmost province in Finland. Related to this, 83 per cent of total income from the commercial harvest of wild mushrooms was earned in eastern Finland, which is also one of the poorest regions of the country.

This regional distribution of NTFP income results from the higher absolute and per capita abundance of wild berry and mushroom resources in the most sparsely populated and least developed regions of Finland (Turtiainen et al, 2005). Regional NTFP income differences also result from the higher unemployment rates in these regions as reflected in the 1997 national survey, in which the highest rates of commercial wild berry harvesting were reported in Finnish households consisting of members of active working age who were involuntarily unemployed, households with lower-paid workers and pensioner households (Kangas, 2001b; Saastamoinen et al, 2005).

Sweden

Rates of wild berry harvesting in neighbouring Sweden are not well documented but are reportedly lower than those in Finland. Statistics Sweden estimates that 45 million recreational trips are made annually by Swedes to collect wild berries and mushrooms (Statistics Sweden, 2001). However, despite strong berry-picking traditions, rural-to-urban migration in Sweden reduced the volume of wild berries picked for home consumption by two-thirds between 1977 and 1997 (Lindhagen and Hörnstein, 2000). Approximately 51 million litres (or roughly 35 million kg) of Swedish wild berries was estimated to have been harvested in 1999. Of this harvest, 59 per cent was sold commercially (Statistics Sweden, 2001).

Russia

In neighbouring Russia, household dependence on wild berries is much greater than in Finland or Sweden, particularly in rural areas (Sossinsky, 2002). Since the dissolution of the Soviet state in 1991 and the subsequent instability in the Russian food product industry, potato and bread consumption in Russia has increased, while the dietary intake of foods rich in protein, minerals and vitamins has declined (Panteleeva, 2004). Hence, Russian household consumption of wild berries is critical for nutritional health. In addition, the sale of wild berries provides significant supplementary household income for many Russian rural households, especially since the collapse of the state-supported forest company infrastructure in many Russian villages (Piipponen, 1999). For some Russian households, wild berry sales alone can comprise two-thirds of household annual income (Paneteelva, 2004).

In the part of Russia bordering Finland, the Republic of Karelia,[3] wild berries are reportedly harvested in high volumes, although no republic-wide studies exist. Most of the NTFPs that residents harvest are for personal household use and only wild berries and mushrooms are generally sold (Polevshchikova, 2005a). During the Soviet era prior to 1990, wild berries were bought by state institutions. It has been reported that in the five-year period from 1981 to 1985, 1.76 million kg of wild berries were collected annually in the Republic of Karelia, and in the subsequent period from 1986 to 1990, the annual average harvest for delivery to the state rose to 5.13 million kg (Volkov et al,

2003). However, since the post-Soviet transition in 1991, harvesting statistics have not been maintained (Volkov et al, 2003).

One non-governmental organizational study found that in the Republic of Karelia region of Pudozh, the average annual wild berry harvest approximated 3 million kg and provided roughly 40 thousand roubles (US$1750) of extra income for 1500 people, or 15 per cent of the Pudzoh population (Karvinen et al, 2004). Similarly, Piipponen et al (1999) found that in one Republic of Karelia village in 1997, 80 per cent of village households collected berries, with 42kg harvested on average per household. At the time, most berries were consumed within the household, with only 7 per cent of families reporting that they sold wild berries. However, the true figure may actually be higher, given the general reticence about reporting extra income, even to researchers (Nadezhda Polevshchikova, pers. comm. (interview) 19 November 2005).

Despite the socioeconomic importance of wild berry collection in Russia, only a small proportion of wild berry production is harvested and sold, because in some northern regions of Russia only 10 per cent of the total crop is accessible (Lukin and Gushchin, 1999). In addition, the wild berry industry in Russia is poorly developed because of weak marketing and a lack of modern processing equipment (Polevshchikova, 2005a). Low levels of NTFP commercial development in general can be explained by the low road density, low wages and the low per capita income of NTFP producers (Ruiz-Pérez et al, 2004), factors that are all characteristic of the wild berry industry in Russia. Nevertheless, Russian processing facilities are improving to the extent that in recent years, Swedish and Finnish buyers have found it difficult to purchase sufficient volumes of Russian wild berries at competitive prices. Despite such progress, however, the industry in Russia remains relatively undeveloped (Panteleeva, 2004; Polevshchikova, 2005a, 2005b).

FINLAND'S COMPETITIVE ADVANTAGE IN THE WILD BERRY INDUSTRY

For decades, Finland has exported wild berries to Sweden, Germany and Austria, as well as to other parts of central Europe. Because yields fluctuate greatly from year to year, export amounts also vary, depending on the crop, on price and on amounts already stockpiled in freezers. In central and southern Europe, demand for wild berry raw material by processors has been declining. Since the lingonberry crop failure of 1993, and the immediate subsequent loss of European processors who bought lingonberries, European processor demand has continued to decline due to the greater demand for the sweeter berries that replaced lingonberries as a raw material following the crop failure. Lingonberry exports from Finland and Sweden have also declined because of the growth in cheaper Russian and Chinese lingonberry exports (Maaseutupolitiikan yhteistyöryhmä, 2001).[4] In contrast, bilberry demand in central and southern Europe has remained relatively stable, although price levels are falling because of increased bilberry exports from Eastern Europe (Maaseutupolitiikan yhteistyöryhmä, 2001).

Since the economic transition in Russia, Russian wild berry exporters have entered the EU market, where they can earn greater returns on raw material sales than in the domestic Russian market. In general, wild berry imports into the EU have grown fastest either from countries outside the EU or from the former Soviet states only recently admitted to the EU. Increasing imports from Russia, China, Belarus, Ukraine, Poland and the Baltic states of Estonia, Latvia and Lithuania, as well as increasing imports of cultivated highbush blueberry (*Vaccinium corymbosum*) and cultivated cranberry (*Vaccinium macrocarpon*) from the United States and Canada, have reduced wild berry prices offered by EU processors (Maaseutupolitiikan yhteistyöryhmä, 2001). In part, this price decline is due to tax-free import prices for wild berries from Eastern Europe and China. In addition, North American imported cultivated cranberries and blueberries have reduced the Finnish market share in wild berries, not only in Europe but also in Japan (Maaseutupolitiikan yhteistyöryhmä, 2000).

Domestic wild berry demand has remained fairly stable in Finland, despite the increases in Russian imports during years of good berry crops and the fact that Russian berries ripen earlier than Finnish berries. Companies that clean and freeze wild berries annually buy between 5 and 10 million kg for both domestic consumption and export (Kangas, 2001a). Wild berry imports into Finland have grown the fastest in the frozen berry sector that produces jam, juice and other processed berry products. For example, between 1994 and 1999, Finnish lingonberry imports increased 220 per cent, while exports declined 86 per cent (Maaseutupolitiikan yhteistyöryhmä, 2000). However, the fresh berry market in Finland still relies primarily on domestic wild berry supplies.

Impacts of EU regionalization

Despite the steady demand for wild berries, Finland's entry into the EU in 1995 created numerous adverse changes in the export tax structure for wild berries as well as the fresh wild berry markets in rural areas. For instance, lower domestic price supports for Finnish farmers resulted in a loss of more than a quarter of Finnish farms between 1993 and 2003. The resulting decline in rural consumers was accompanied by the concentration of the Finnish retail trade sector into very large hypermarkets that outcompeted smaller shops, especially after Finland adopted the euro in 2002. This concentration was reflected in a sharp 46 per cent decline in the number of operating village and local shops between 1995 and 2002. As a result, only 20 per cent of Finnish food products are purchased locally (Laine, 2003; Niemi and Ahlstedt, 2005). These changes are reflected in the decline in the number of wild berry buyers belonging to the natural product development organization Arktiset Aromit from 330 in 1994 to only 80 by 2001 (Maaseutupolitiikan yhteistyöryhmä, 2001). This dramatic 76 per cent drop resulted largely from a decline in the number of local rural shops that bought wild berries locally and sold them on at regional level.

NTFP POLICY AND ACCESS TO RESOURCES IN FINLAND, SWEDEN AND RUSSIA

It has been argued that in Europe generally, NTFPs like wild berries are environmental and recreational goods and services (Mantau et al, 2001) that cannot be successfully marketed because they lack rivalry and exclusivity, and therefore demand is too limited for them to maintain a competitive advantage. In this view, wild berry marketability increases as rivalry and exclusion opportunities increase, because producers control scarce resources and are thus more willing to make long-term investments in their enterprises (Janse and Ottitsch, 2005). However, in Finland, Sweden and Russia, wild berries are not scarce but biologically abundant resources (although the availability of a single species may be very low in some years and regions). Furthermore, the unique Nordic policy institution of 'everyman's rights' ensures public access to NTFPs such as wild berries as non-rival and inclusive resources in both Finland and Sweden.

'Everyman's rights' NTFP policy in Finland and Sweden

'Everyman's rights' is the Nordic convention of property rights based on customary laws that allow public access to and use of both NTFPs and services not only on public

---

Box 12.4 Everyman's rights restrictions on cloudberry harvesting in Lapland

A specific Finnish law permits regional restrictions on harvesting wild berries and similar NTFPs in certain areas of Finnish Lapland where NTFPs have particularly significant economic importance. Here the Ministry of Agriculture and Forestry can prohibit anyone other than local residents, when the best interests of local residents so requires, from picking wild berries on state lands. In practice, these restrictions only apply to cloudberry harvesting. The law does not reverse everyman's rights, but in this case limits the harvest to local residents only (Kuusiniemi, 1998). These restrictions have been implemented only rarely, but in principle they provide the means to regulate conflicts between local and non-local harvesters, including foreign pickers.

Theoretically the law allows open access to cloudberry resources on state land to be shifted temporarily to exclusive access within the category of 'club' products; that is, products that are available to only a few on the basis of residence. In the northern municipalities of Lapland, more than 90 per cent of forest land (including open fell and peatland areas) belongs to the state. This is why the law only refers to state forests. Even in these northern areas of Lapland, the private landowner cannot prohibit the picking of cloudberries by others based on this or any other law. However, the 'proximity' principle of this customary law grants the landowner, if resident on site, the moral – and, in most cases, practical – privilege of discouraging others from collecting cloudberries in areas very near their home.

land, but on land that is privately owned.[5] Although widespread private forest owner-ship was introduced in Finland with the allocation of forest land parcels to local farms during the 18th century, these land tenure changes did not affect the public's right to use the forest (Pouta et al, 2006).[6] Everyman's rights have since been somewhat restricted by various Finnish laws (especially by the Finnish criminal code), but have not actually been codified in law, and are instead recognized as the customary law of the right of public use. At present the content of everyman's rights in Finland is described in guidelines disseminated by the Finnish Ministry of the Environment (2006). In both Finland and Sweden, access to land subject to everyman's rights is free to the public, and it is illegal to charge a fee or prohibit entry. Moreover, access does not require the landowner's permission, although everyman's rights are restricted in some areas, such as in national parks and nature reserves, and the use of these areas is regulated by national environmental legislation.

Under everyman's rights, not only recreational use such as hiking and skiing but also NTFP household and commercial use such as wild berry and mushroom picking are freely allowed, excluding some products.[7] In return for public access and use, everyone is expected to refrain from causing damage or disturbances, including disrupting the landowner's privacy. Thus it is forbidden to kill or disturb animals; damage growing trees; collect moss, lichens, herbs or wood; build an open fire; or

---

Box 12.5 Everyman's rights do not apply to lichen harvesting

Although a minor NTFP in Finland, the reindeer lichen genus (*Cladonia*) is a commer-cially significant product for many rural areas. However, lichen and moss are consid-ered part of the actual property and are thus excluded from everyman's rights. In the reindeer husbandry regions of northern Finland, as well as northern Sweden and Norway, reindeer lichen are the staple diet of reindeer and lichen plant communities have declined because of overgrazing. According to the Finnish Reindeer Management Act, reindeer can graze and eat lichen in these areas, even on private forests, where otherwise lichen resources would belong to the forest owner. In other words, only rein-deer have 'everyman's rights' to lichen.

Elsewhere the ownership of lichen resources has economic importance only locally and in the case of a specific species. Large amounts of *Cladonia alpestris* are used for ornamental purposes, especially in making wreaths that are primarily exported to Germany and central Europe (Salo, 1995). The annual lichen harvest is estimated to be approximately 500,000kg, and the value of the harvest in 2002 was €1.5 million (Aarne et al, 2005). The centre of the lichen export trade is in northern Finland, south of the rein-deer management area near Oulu, where the gathering of lichen has been of significant economic importance: for example, in the best years, lichen production and harvesting contribute about a third of the total income of the small island of Hailuoto (Kauppi, 1993). The Finnish taxation laws apply to lichen sold for decorative purposes (see text). Where lichen production and harvesting occur, hikers are warned to avoid damaging produc-tive lichen areas.

drive a motorized vehicle across private land without the landowner's permission (Aarne et al, 2005).

Everyman's rights apply not only to Nordic citizens but also to foreign nationals, with certain exceptions related to local boating, fishing and hunting rights. In addition, the application of everyman's rights to foreign nationals is restricted by immigration laws. In conjunction with everyman's rights, the income derived from gathering wild cones, berries, mushrooms and other NTFPs used for human nourishment or medicinal use has until recently been regarded as tax-free under both Finnish and Swedish tax regulations for citizens and foreign nationals. In Finland, however, the actual end use of the NTFP rather than its status as a raw material governs this exemption, so that income earned from NTFPs gathered for decorations or handicrafts has not been tax-exempted. This exception has caused some confusion among harvesters (Aarne et al, 2005), and a Finnish parliamentary initiative has been introduced to ensure that income derived from harvesting any NTFP is subject to consistent tax regulation.[8] In Finland, harvesters ara liable for a reduced amount of value-added tax (VAT) on income earned from NTFPs if the total annual income from sales exceeds €8500. If annual income exceeds €20,000, the harvesters have to pay the full amount of VAT, which in Finland is 22 per cent (Aarne et al, 2005). These income levels are seldom exceeded.

Post-Soviet NTFP policy in Russia

In contrast to the widespread private forest ownership in Finland and Sweden, almost all Russian forests are federally owned and controlled by state organizations in a hierarchical system with the State Forestry Service of Russia (formerly the Federal Forest Service), under the aegis of the Ministry of Natural Resources in Moscow.

The reform of the Russian Federation's Forest Code, which has been the primary source of legislation and regulation governing forest management and use, including wild berry access to Russian forests, has recently been completed, and the new code came into force on 1 January 2007. Articles 11 and 35 of the new code allow Russian citizens free access to the forests to gather wild berries, mushrooms, nuts, medicinal plants and other NTFPs for their own needs. Determination of what constitutes 'own needs' or personal use will be regulated by laws applying to subjects of the federation; for instance, Republic of Karelia laws will govern NTFP harvesting for personal use in that republic.

However, article 34 of the new Russian Forest Code also defines the gathering of edible forest products and medicinal plants as a permissible entrepreneurial activity and allows such resources to be harvested, stored and transported from the forest. Thus Russian citizens and legal residents can lease forest parcels and put equipment, storage facilities and other temporary structures on the leased parcels for these purposes. The appropriate federal authority will regulate how such commercial harvesting is conducted (Venäjän federaation metsälaki, 2006).

The new Russian Forest Code thus clearly distinguishes between harvesting NTFPs for personal use and their commercial sale.[9] This could lead to restriction of free-use access to NTFPs, as in the case of ginseng (*Panax ginseng*) leases in the Russian Far East (Zakharenkov, 2003). Currently, few if any data exist on the degree to which access to

Russian forest land is held in leases or concessionary agreements for NTFP harvesting purposes, and most if not all Republic of Karelia forest lands have been open for free access to NTFPs by Russian citizens (Polevshchikova, 2005a). It is too early to determine whether the new Russian Forest Code, regional laws and federal regulations will change free access to NTFPs in Russia – access that in practice has been relatively unrestricted by rival, exclusionary land tenure laws and regulations. At present, the greatest barrier to NTFP access in Russia is the weak transportation network, which limits the harvesting of wild berries, as well as their sale and marketing, to geographically limited and often overexploited forest areas near roads, waterways, railway lines and population centres.

NTFP POLICY AND COMPETITIVE ADVANTAGE: REGIONAL CHALLENGES TO WILD BERRY MARKETS AND LABOUR

Access to markets in Finland and Sweden

Both Finland and Sweden entered the EU in 1995.[10] Since then, former import tariffs on goods from non-EU countries have been removed so that competition in both domestic and export markets has increased. Weakened competitiveness has resulted primarily from the higher prices paid to, or expected by, Finnish and Swedish wild berry harvesters compared to the lower prices paid to berry pickers in new Baltic and Eastern European EU member countries. Hence, Finland's and Sweden's competitive advantage has been eroded by the entry into the wild berry market of new European competitors. In addition, wild berry exports from non-EU nations such as China and substitutes in the form of cultivated berries from North America have also increased market barriers for Finland and Sweden. Given non-rival and inclusive access to resources under everyman's rights, therefore, NTFP access policy is a less significant factor in the competitive advantage of the Finnish and Swedish wild berry industries than access to markets.

For wild berry enterprises in both Finland and its competing neighbour states, the marketing of wild berry raw materials and manufactured products is constrained by wide seasonal crop variability and fairly small sector markets. These conditions prevent producers from relying on, and specializing in, any individual wild berry product and discourage long-term enterprise investments in improving processing efficiency and innovation. In addition, many wild berries are sold simultaneously in different local, national and international markets that require different marketing approaches (Lintu, 1998). Thus wild berry producers across the region face high risks in accessing competitively priced resources and overcoming barriers to entering global markets.

Access to labour in Finland and Sweden

Although most forested land in Finland and Sweden is privately owned, most wild berries sold commercially are harvested from the central and northern rural areas, where rapid outmigration over the past decades has depopulated the countryside. Because wild berries are viewed as overabundant forest products in Finland and Sweden and wild berry prices remain low relative to imported alternatives, the Nordic policy of everyman's rights has not been perceived as resulting in overharvesting, but rather linked with the underutilization of wild berries in both countries. Access to labour – that is, the availability of independent pickers – rather than access to resources has become a critical policy concern in strengthening the competitive advantage of the Finnish and Swedish wild berry industries.

Wild berry enterprises in Sweden and more recently in Finland are employing increasing numbers of foreign migrant pickers from the Ukraine, Belarus, Thailand and other non-EU nations to collect wild berries, because of difficulties in finding enough domestic harvesters. Although national studies have not yet been conducted, it is generally thought that many Finnish and Swedish berry pickers believe that harvesting wild berries is not sufficiently lucrative.

Foreign berry pickers migrate to Finland and Sweden because wages in their own countries are much lower and picking wild berries for Finnish and Swedish berry processing companies gives them an opportunity to earn badly needed income relatively quickly. Foreign nationals often borrow money for travel and accommodation so that they can pick wild berries during the summer season. Berry processing companies in turn generally arrange visas and accommodation, often in empty rural schools that are vacant because of the migration to urban areas. In Finland, companies may provide foreign nationals with help with the visa application process, lodging and a motor vehicle for berry picking. However, foreign berry pickers are not considered employed workers if they stay in Finland for less than four months on a tourist visa and as long as they can freely choose to whom to sell their berries. Pickers are free, at least in principle, to sell their berries (*Helsingin Sanomat*, 2006b; Rantanen and Valkonen, 2006). Hence, companies are deemed to 'invite' pickers, rather than contract them.

A Thai berry picker who takes a year to earn 2000 in Thailand can earn the same amount or more in a few months of berry picking in Finland or Sweden – but then he and his companions must harvest large volumes. In 2005, for example, 92 Thai pickers in the northern Finnish community of Savukoski harvested 317,000kg of wild berries, which they sold to the company that had invited them. In the previous year, the same company had bought all its wild berries from local pickers and secured only 17,000kg (*Helsingin Sanomat*, 2006a). Seeing the competitive advantage of employing migrant pickers, the same company contracted 650 Thai pickers in 2006, and they collected well over 1 million kg of wild berries across eastern Finnish Lapland in 2006 (*Helsingin Sanomat*, 2006c).

Notwithstanding the prospect of significant earnings from harvesting wild berries, foreign pickers accept a high degree of risk in borrowing their fare to Finland or Sweden because of the unpredictability of yields. In 2006, the bilberry crop failed and the Ukrainian and Russian bilberry pickers who had travelled to Finnish Lapland in anticipation of earning a substantial income could not even harvest enough bilberries to pay for their return home or for their accommodation, so that local citizens and authorities had to provide emergency services (*Helsingin Sanomat*, 2006b).

Finland and Sweden have to compete for foreign migrant berry pickers because harvesting rates by their own citizens are falling and migration out of berry-producing areas is increasing at a time when berry processors require greater volumes of cheaper berries to compete in the EU and global markets. Until 2006, Finland and Sweden had similar regulations regarding the taxation and employment status of foreign berry pickers. Then the Swedish Ministry of Finance ruled that foreign nationals harvesting wild berries in Sweden for less than six months were to be considered employed by the companies inviting them, and the companies were required to deduct a 25 per cent special income tax from the pickers' wages. This requirement applied whether the berry picker was paid by the kilogram or by the hour (Rantanen and Valkonen, 2006; Skatteverket, 2006). Current Swedish tax regulations thus constitute a new barrier to wild berry market entry that does not exist in neighbouring Finland. According to industry representatives, the new regulations are jeopardizing much of the annual income of the approximately 5000 foreign pickers who annually migrate to the Swedish wild berry harvest (*Skogsbärbranschens Intresseförening*, 2006). Hence, barriers to labour access are currently greater policy concerns for the Finnish and Swedish wild berry industry's competitive advantage than either access to resources or markets.

Access to market and labour in Russia

Finland's and Sweden's closest competitor for wild berry market entry is northern Russia. Aside from the poor Russian transport infrastructure and the lack of processing equipment, the absence of competition among wild berry buyers in local Russian markets means that harvesters receive only spot market prices (Polevshchikova, 2005b).[11] Such prices tend to follow the supply-driven curve typical of other raw material commodities, with wide price fluctuations depending on seasonal timing and productivity.

Other barriers to market entry include the problems that inventive NTFP Russian entrepreneurs have encountered with licensing requirements and health and safety inspections when trying to organize ecotours involving the collection of wild berries and mushrooms (National Parks for Joint Benefits, 2004). These experiences are echoed by those working with the NTFP industry in the Russian Far East, where the three primary factors limiting NTFP business growth have been identified as the lack of business financing, weak management expertise and costly bureaucratic obstacles that often require legal assistance (Warner et al, 2002). Hence, barriers to market entry currently exert a greater braking effect on the competitive advantage of strengthening Russian wild berry and related NTFP enterprises than either access to resources or labour.

IMPLICATIONS AND RECOMMENDATIONS

Notwithstanding variations in yield from year to year, it is estimated that no more than 5–10 per cent of the two most common wild berry species and only 1–3 per cent of the wild mushrooms available in Finnish forests are collected annually (Salo, 1995; Saastamoinen et al, 1998). Every summer the media comments on the underutilization of such forest resources, with calls for their more efficient use for nutrition and health, as well as for income and employment (Saastamoinen et al, 1998).

As demonstrated in the case of wild berry resources, Finland's open-access policy has produced a successful NTFP management system because everyman's rights are founded on the premise that NTFP resources are abundant but the intensity of NTFP utilization is low (Rekola, 1998).[12] Moreover, the very recent trend of attracting foreign nationals by allowing access to resources in Finland and lowering market entry barriers through the same taxation exemptions as those enjoyed by domestic pickers has supported the Finnish wild berry industry during a period in which wild berry prices in Europe have fallen and citizen picker numbers have declined. However, some local criticism of the use of foreign pickers in northern and eastern Finland, where unemployment rates are high and the unemployed or underemployed are subsidized by the state, has prompted a political debate as to whether the scope of everyman's rights should be restricted to Finnish citizens only (Rantanen and Valkonen, 2006).[13]

Policy interventions in Finland to increase NTFP utilization by citizens have included several decades of NTFP training and research. Between 1969 and 1999, about 3000 voluntary extension advisers were educated on NTFP utilization needs such as species identification and the proper handling of products for sale. The training largely concentrated on wild mushrooms because mushrooms have had lower rates of utilization than wild berries. Since 1999, training has been extended to include the processing and marketing of herbs and other products. In Finnish universities and research institutes, NTFP studies have concentrated on biological and nutritional questions relating to wild berries and mushrooms, but product development and marketing have recently received more attention (Saastamoinen, 1999). For example, a leading Finnish cosmetics company has created a popular cream that contains cloudberry seed oil, which is rich in essential fatty acids, carotenoids and phytosterols. The company intends to expand its operations internationally, with the most significant growth anticipated in the Russian and US markets (Virtual Finland, 2005). All these efforts focus on increasing the decreasing participation of Finnish citizens in the NTFP industry rather than directly addressing the politically sensitive phenomenon of NTFP collection by foreign nationals.

In Russia, citizen participation in the collection of NTFPs remains high and the conflicts associated with foreign harvesters in Finland and Sweden are generally absent. However, Russia faces greater market entry barriers than its Nordic neighbours. Increasing Russia's NTFP market access through niche marketing contracts with diverse but cooperating buyers has been viewed as a means of increasing regional and global demand for higher-value-added NTFPs (Perner, 2004). Building niche markets means fewer seasonal fluctuations and more economic activities, and hence more employment and higher income along the NTFP value chain. However, as Perner (2004) notes, at least one of the barriers to NTFP niche marketing is the difficulty that Russian producers face in 'organically certifying' their products, since new NTFP companies in Russia have not been able to implement such certification as required by global market suppliers. The lack of certification policies in Russia poses significant barriers to market entry, while giving both Finland and Sweden a significant competitive advantage, since both countries have met EU certification standards for NTFPs.

Nevertheless, NTFP companies in Finland and Sweden are not spared obstacles to market entry since they tend to be small, diversified and primarily raw material suppliers. These characteristics increase their vulnerability to supply and demand cycles and weaken their competitive advantage in the EU and global markets. Recent policy recommendations for strengthening Finnish NTFP companies' market entry opportunities include focusing company business strategies, capturing unique product niches, finding marketing and supply partners, and building marketing networks (Aarne et al, 2005).

The sustainable development of the Finnish NTFP industry will therefore increasingly require more processed and innovative products to allay the concerns of increasing numbers of EU and global consumers regarding healthy, environmentally friendly and ethical harvesting. The NTFP industry will also need to develop links with nature-based tourism and recreational enterprises in eastern, central and northern Finland in order to sustain the local rural economy (Aarne et al, 2005). In supporting such efforts, Finnish policy-makers will need to ensure that low-income foreign nationals and low-income citizens are given the same protection under everyman's rights as commercial harvesters and wealthier foreign nationals seeking to gather NTFPs as a recreational tourist activity.

CONCLUSION

At present Finland enjoys a competitive advantage in the wild berry industry over Sweden, where recent taxation decisions have decreased the financial incentive for foreign nationals to harvest resources. Finland also maintains a competitive advantage over Russia, where cheap labour costs are offset by barriers to market entry due to evolving certification processes, operational costs and regulatory uncertainty with the new Russian Forest Code. However, Finland's competitive advantage may be temporary and may already have weakened because of its failure to develop policy supports for sectoral and cross-sectoral markets in the long term.

To maintain its advantage, Finland must strengthen its NTFP market access policy, not only in marketing and product innovation, but also in harvesting logistics and efficiency, as well as research and education. In addition, Finland must also scrutinize its NTFP labour policy and overcome its reluctance to enter the debate on the entry of foreign nationals under the institution of everyman's rights. In addressing these issues, Finnish policy-makers should promote the same organizational innovations that have been used to attract and employ hard-working foreign NTFP harvesters in order to find new ways of mobilizing more domestic harvesters, particularly in those regions of Finland characterized by high unemployment rates and abundant NTFP resources.

ACKNOWLEDGEMENTS

Support from the Fulbright Senior Scholar programme, the University of Montana Faculty Exchange programme and Academy of Finland Project 104940 at the University of Joensuu is gratefully acknowledged.

NOTES

1   In addition, sea buckthorn (*Hippophae rhamnoides*) is becoming an increasingly valuable but less abundant NTFP along Finnish and Russian Baltic coasts.

2   At 95 per cent confidence intervals, the estimated limits were 52.3 million and 60.4 million kg in 1997, and when an approximate Bayesian bootstrap method was used to obtain the variance, the 95 per cent confidence limits were considerably larger, at 41.3 million and 71.4 million kg, while the average of 56.6 million kg remained the same (Kangas, 2001b). This demonstrates the uncertainty of estimated harvested NTFP volumes that can be produced even from data obtained through a very large sample ($n = 6849$ households) and a solid response rate (60 per cent).

3   The Republic of Karelia is a federal region of the Russian Federation.

4   In Finland, lingonberry prices paid to pickers fell steadily after 1993 to a record low of 0.65 euro cents per kg in 2005, in conjunction with the sharp decline in Finnish exports and the rapid increase in exports from non-EU countries.

5   'Everyman's rights' (in Finnish *jokamiehen oikeudet* and in Swedish *allemansrätten*) are in fact gender- and age-neutral and apply not only to men but also to women and children.

6   Currently, most forested land (52 per cent) in Finland is owned by private non-industrial land-owners, slightly over one-third (35 per cent) by the state and 8 per cent by the forestry industry or other companies. The remaining 5 per cent of forested land is owned by municipalities, churches and other minor landholder groups (Finnish Forest Research Institue, 2005).

7   Under environmental legislation protecting endangered species, the Finnish regulations governing mushroom harvesting include the listing of species that can be picked for commercial use. In addition, pickers are advised to obtain a special certificate to prove they are familiar with legislation and that they have shown adequate knowledge in identifying mushroom or herb species. Buyers usually require this certificate from their raw material suppliers (Aarne et al, 2005).

8   The initiative refers to *luonnontuotteet* or 'natural products', a Finnish term for NTFP. The initiative has not yet produced new laws.

9   Such a distinction was also the intent of the previous Russian Forest Code, under which forest parcels could also be made available through leases or concessionary agreements, with a fee established for each particular NTFP (Kukuev, 1999; Polevshchikova, 2005a).

10   However, Sweden rejected the euro and membership of the EU Economic and Monetary Union in 2003.

11   In Russia, the wild berry industry is relatively centralized in a major local buyer, and economic value-added activities are few (Polevshchikova 2005b).

12   In some instances, though, everyman's rights to some of Finland's NTFPs, such as the less abundant sea buckthorn, are no longer practicable because of conflicts between collectors, who find harvesting the sea buckthorn berries costly, and landowners, who find losing potential access to sea buckthorn equally if not more costly (Rekola, 1998).

13   Historical precedent for confining everyman's rights to Nordic citizens is found in the case of the Saami. In northern parts of Sweden and Finland, the crown authority did not begin to expand its interests in the vast areas of wilderness until the medieval era and in institutionalizing the laws governing common land boundaries and property rights overlooked the aboriginal population, the reindeer-herding Saami (Sunderberg, 2002).

REFERENCES

Aarne, M., Hänninen, R., Kallio, M., Kärnä, J., Karppinen, H., Ollonqvist, P., Packalen, K., Rimmler, T., Toppinen, A., Kajanus, M., Matilainen, A., Rutanen, J., Kruki, S., Peltoniemi, J. and Saarinen, J. (2005) 'Finland', in L Jáger (ed) *COST E30 Economic Integration of Urban Consumers' Demand and Rural Forestry Production. Forest Sector Entrepreneurship in Europe: Country Studies*, Acta Silvatica & Lignaria Hungarica, Sopron, Hungary, Special Edition, pp171–244

Belonogova, T. V. and Zaitseva, N. L. (1989) *Ekologo-biologitsheskie osobennosti Khozyaistvenno tsennyh rastenye Karelya* [Ecological-biological characteristics of economically valuable plants of Karelia], Karelia, Petrozavodsk, Russia

Boreal Forest Website (2002) *Boreal Forests of the World: Finland: Forests and Forestry*, Lakehead University, ON, www.borealforest.org/world/world_finland.htm, accessed 28 May 2008

Chibisov, G. (1999) 'Impacts of forest management on the yield of berries and mushrooms', in A. Niskanen and N. Demidova (eds) *Research Approaches to Support Non-wood Forest Products Sector Development*, EFI Proceedings no 29, European Forest Institute, Joensuu, Finland, pp101–108

Eriksson, L., Ingelög, T. and Kardell, L. (1979) *Blåbär, lingon, hallon, Förekomst och bärproduktion i Sverige 1974–1977* [Bilberry, lingonberry, raspberry: Occurrence and production in Sweden 1974–1977], Rapport 16, Sveriges lantbruksuniversitet, Avdelning för landskapsvård, Uppsala, Sweden (in Swedish with English summary)

Finnish Forest Research Institute (2004) *Finnish Statistical Yearbook of Forestry*, SVT [Official Statistics of Finland] Agriculture, Forestry and Fishery 2004:45, Vammala, Finland

Finnish Forest Research Institute (2005) *Finnish Statistical Yearbook of Forestry*, SVT [Official Statistics of Finland] Agriculture, Forestry and Fishery 2005:45, Vammala, Finland

Finnish Ministry of Agriculture and Forestry (2005) *Interim Evaluation of the National Forest Programme*, www.mmm.fi/kmo/english/NFP_evaluation_report_ENG.pdf, accessed 27 May 2008

Finnish Ministry of the Environment (2006) *Everyman's Right*, www.ymparisto.fi/default.asp?contentid=49256&lan=EN, accessed 28 May 2008

Helsingin Sanomat (2006a) 'Thai berry-pickers are welcome in Finnish Lapland', Helsinki, Finland, 1 August

Helsingin Sanomat (2006b) 'Foreign berry-pickers put on spot by empty promises and poor crop in Lapland', Helsinki, Finland, 4 August

Helsingin Sanomat (2006c) 'Thai berry-pickers return home after earning year's salary in Lapland', Helsinki, Finland, 4 October

Janse, G. and Ottitsch, A. (2005) 'Factors influencing the role of non-wood forest products and services', *Forest Policy and Economics*, vol 7, pp309–319

Kangas, K. (2001a.) 'Commercial wild berry picking as a source of income in northern and eastern Finland', *Journal of Forest Economics*, vol 7, pp53–68

Kangas, K. (2001b) 'Wild Berry Utilization and Markets in Finland', PhD dissertation, University of Joensuu at Joensuu, Finland

Karvinen, S., Markovsky, A., Rodionov, A., Rogov, A., Sikanen, L. and Tsypuk, A. (2004) *Predstavlenija o lesnom sektore respubliki Karelija* [Notions on forest sector of the Republic of Karelia], Institute of Economy, Karelian Science Centre, Petrozavodsk, Russia

Kauppi, M. (1993) 'The gathering of lichens as a trade', *Aquilo Ser. Botanica Tom*, vol 31, pp89–91

Kukuev, Y. (1999) 'Non-wood forest products in Northwest Russia', in A. Niskanen and N. Demidova (eds) *Research Approaches to Support Non-wood Forest Products Sector Development*, EFI Proceedings no 29, European Forest Institute, Joensuu, Finland, pp53–62

Kuusiniemi, K. (1998) 'Jokamiehen oikeudet [Everyman's rights]', in K. Kuusiniemi, P. Majamaa and P. Vihervuori (eds) *Maa-, vesi- ja ympäristöoikeuden käsikirja* [Land, water and environmental rights handbook], Tietosanoma, Helsinki, Finland

Laine, O. (2003) *Kyläkauppa turvaa maaseudun kylien elinvoimaisuuden* [Village shop ensures rural village vitality], Päivittäistavarakauppa ry, Helsinki, Finland, www.pty.fi/tiedotteet/kylakaupat.htm, accessed 29 November 2005

Lindhagen, A. and Hörnstein, L. (2000) 'Forest recreation in 1977 and 1997 in Sweden: Changes in public preferences and behavior', *Forestry*, vol 73, pp143–153

Lintu, L. (1998) 'Development issues related to the marketing of non-wood forest products', in H. G. Lund, B. Pajari and M. Korhonen (eds) *Sustainable Development of Non-wood Goods and Benefits from Boreal and Cold Temperate Forests*, EFI Proceedings no 23, European Forest Institute, Joensuu, Finland, pp179–187

Lukin, I. and Gushchin.V. (1999) 'Non-wood forest products, their harvested volumes and household consumption in Arkhangelsk region, Russia', in A. Niskanen and N. Demidova (eds) *Research Approaches to Support Non-wood Forest Products Sector Development*, EFI Proceedings no 29, European Forest Institute, Joensuu, Finland, pp81–90

Lund, H. G., Pajari, B. and Korhonen, M. (eds) (1998) *Sustainable Development of Non-wood Goods and Benefits from Boreal and Cold Temperate Forests*, EFI Proceedings no 23, European Forest Institute, Joensuu, Finland

Maaseutupolitiikan yhteistyöryhmä (2000) *Luonnontuotealan nykytilan kuvaus ja kehittämisohjelma vuosille 2000–2006*, Luonnontuotealan teemarhymä, Sisäasiainministeriö [Rural policy collaborative work group (2000) Natural product current outlook and development programme for 2000–2006, Natural product theme group, Ministry of the Interior], Helsinki, Finland

Maaseutupolitiikan yhteistyöryhmä (2001) *Luonnontuotteiden talteenoton ja käyton edistämisohjelma*, Luonnontuotealan teemarhymä, Sisäasiainministeriö [Rural political collaborative work group (2001) Natural product recovery and employment promotional programme, Natural product theme group, Ministry of the Interior], Helsinki, Finland

Mantau, U., Merlo, M., Sekot, W. and Weicker, B. (2001) *Recreational and Environmental Markets for Forest Enterprises: A New Approach Towards Marketability of Public Goods*, CABI Publishing, New York

Mäntynen, L. (2005) 'Tattimiehen tivoli alkoi [The funfair of cep-picking has started]', *Helsingin Sanomat*, Helsinki, Finland, 4 August

Moisio, S. (1999) *Luonnontuotealan seurantakysely 1999* [Natural product follow-up survey 1999], Arktiset Aromit ry, Suomussalmi, Finland

Myllynen, A.-L. and Saastamoinen, O. (1995) 'Karjalan tasavallan metsätalous [Forestry in the Republic of Karelia]', *Silva Carelica*, vol 29, pp1–210

National Parks for Joint Benefits (2004) 'Business planning and consultations', *National Parks for Joint Benefits News*, November 2003–March 2004, www.parksandbenefits.com/news/march2004.htm, accessed 4 October 2005

Niemi, J. and Ahlstedt, J. (eds) (2005) *Finnish Agriculture and Rural Industries 2005: Ten Years in the European Union*, Publication 105a, Agrifood Research Finland, Economic Research (MTTL), Helsinki, Finland

Panteleeva, O. (2004) 'Effective food chain management in the wild berries business of Russian rural economies', paper presented at 14th Annual World Food and Agribusiness

Symposium of the International Food and Agribusiness Management Association (IFAMA), Montreaux, Switzerland, 12 June, www.ifama.org/conferences/2004Conference/Papers/Panteleeva1032.pdf, accessed 29 May 2008

Perner, P. J. (2004) 'The business of Russia's forests', *BISNIS Bulletin*, April, www.bisnis.doc.gov/bisnis/BULLETIN/apr04bull5.htm, accessed 29 May 2008

Piipponen, M. (1999) *Transition in the Forest Sector of the Republic of Karelia*, Interim Report IR-99–070, International Institute for Applied Systems Analysis, Laxenburg, Austria

Piipponen, M., Karkinen, K., Klementjev, J., Oksa, J., Polevshchikova, N., Romanov, G., Saastamoinen, O. and Varis, E. (1999) *Household and Living Conditions in the Forest Village of Matrosy, Russian Karelia*, Faculty of Forestry Research Notes 99, University of Joensuu, Joensuu, Finland

Pohjois-Karjalan TE-keskus (2006) *Pohjois-Karjalan alueellinen maaseutuohjelma 2007–2013* [North-Karelia regional rural program 2007–2013], www.pohjois-karjala.fi/Resource.phx/maakuntaliitto/eu-ohjelmat/elinkeinoseminaari/aineisto.htx, accessed 29 May 2008

Polevshchikova, N. (2005a) 'An overview of the poverty and NWFP situation in the Republic of Karelia, Russia', in C. Lacuna-Richman, M. Turtiainen and A. Bzrszcz (eds) *Non-wood Forest Products and Poverty Mitigation*, Faculty of Forestry Research Notes 166, Joensuu, Finland, University of Joensuu

Polevshchikova, N. (2005b) *Non-wood Forest Products and Social-economic Problems of the Forest Settlements*, paper presented at Second Workshop of the Academy of Finland Project on Poverty Alleviation and Non-Wood Forest Products: A Comparative Study in Tropical and Temperate Forest Contexts, Prague, Czech Republic, 18 November

Pouta, E., Sievänen, T. and Neuvonen, M. (2006) 'Recreational wild berry picking in Finland: Reflection of a rural lifestyle', *Society and Natural Resources*, vol 19, pp285–304

Rantanen, P. and Valkonen, J (2006) *Practice of Wild Berry Picking*, paper presented at 23rd Nordic Sociology Congress, Turku, Finland, 19 August

Rekola, M. (1998) 'The problem of NWFP ownership: Property rights analysis', in H. G. Lund, B. Pajari and M. Korhonen (eds) *Sustainable Development of Non-wood Goods and Benefits from Boreal and Cold Temperate Forests*, EFI Proceedings no 23, European Forest Institute, Joensuu, Finland, pp197–202

Ribot, J. C. and Peluso, N. L. (2003) 'A theory of access', *Rural Sociology*, vol 68, pp153–181

Ruiz-Pérez, M., Belcher, B., Achdiawan, R., Alexiades, M., Aubertin, C., Caballero, J., Campbell, B., Clement, C., Cunningham, T., Fantini, A., de Foresta, H, Fernández, C. G., Gautam, K. H, Martínez, P. H., de Jong, W., Kusters, K., Kutty, M. G., López, C., Fu, M., Alfaro, M. A. M., Nair, T. K. R., Ndoye, O., Ocampo, R., Rai, N., Ricker, M., Schreckenberg, K., Shackleton, S., Shanley, P., Sunderland, T. and Youn, Y. (2004) 'Markets drive the specialization strategies of forest peoples', *Ecology and Society*, vol 9, no 2, art 4, www.ecologyandsociety.org/vol9/iss2/art4/, accessed 29 May 2008

Saastamoinen, O. (1999) 'Forest policies, access rights and non-wood forest products in northern Europe', *Unasylva*, vol 50, no 3, pp20–26

Saastamoinen, O., Kangas, J., Naskali, A. and Salo, K. (1998) 'Non-wood forest products in Finland: Statistics, expert estimates and recent development', in H. G. Lund, B. Pajari and M. Korhonen (eds) *Sustainable Development of Non-wood Goods and Benefits from Boreal and Cold Temperate Forests*, EFI Proceedings no 23, European Forest Institute, Joensuu, Finland

Saastamoinen, O., Kangas, K. and Aho, H. (2000) 'The picking of wild berries in Finland in 1997 and 1998', *Scandinavian Journal of Forest Research*, vol 15, no 6, pp645–650

Saastamoinen, O., Lacuna-Richman, C. and Vaara, M. (2005) 'Is the use of forest berries for poverty mitigation a relevant issue in an affluent society such as Finland?', in C. Lacuna-Richman, M. Turtiainen and A. Bzrszcz (eds) *Non-wood Forest Products and Poverty Mitigation*, Faculty of Forestry Research Notes 166, University of Joensuu, Joensuu, Finland

Salo, K. (1995) 'Non-timber forest products and their utilization', in M. Hytönen (ed) *Multiple-Use Forestry in the Nordic Countries*, Finnish Forest Research Institute, Vantaa, Finland, pp117–156

Skatteverket (2006) *Berry Buyers May Be Employers*, www.skatteverket.se/international/international/berrybuyersmaybeemployers.4.906b37c10bd295ff4880001131.html, accessed 29 May 2008

Skogsbärbranschens Intresseförening (2006) Rädda de svenska skogsbären – kaos hotar i skogen [Save the Swedish forest berries – Chaos threatens in the forest], press release, 6 July

Sossinsky, S. (2002) 'Picking wild berries and other gainful employment', *The Moscow News*, 28 November, www.english.mn.ru/english/printver.php?2002–30–16, accessed 13 November 2005

Statistics Sweden (2001) *Environmental Accounts for Forest: Test of a Proposed Framework for Non-ESA/SNA Functions*, May, w36.scb.se/statistik/MI/MI1202/2000I02/MIFT0105.pdf, accessed 29 May 2008

Sunderberg, K. (2002) 'Nordic common lands and common rights: Some interpretations of Swedish cases and debates', in M. de Moor, L. Shaw-Taylor and P. Warde (eds) *The Management of Common Land in North West Europe, c. 1500–1850*, Brepols Publishers, Turnhout, Belgium, pp173–194

Turtiainen, M., Salo, K. and Saastamoinen, O. (2005) *Satomalleilla lasketut Suomen kangasmetsien alueelliset ja valtakunnalliset mustikka-ja puolukkasadot* [Model-based estimates of regional and national bilberry and lingonberry yields on mineral soils in Finland], Faculty of Forestry Research Notes 167, University of Joensuu, Joensuu, Finland

Venäjän federaation metsälaki (2006) [The forest code of the Russian Federation: unofficial Finnish translation] no 200-FZ, 4 December, www.idanmetsatieto.info/tiedostot/tiedotteet/Laki_metsälain_voimaantulosta.pdf, accessed on 17 June 2008

Virtual Finland (2005) *Beauty Through Arctic Berries*, www.virtual.finland.fi/netcomm/news/showarticle.asp?intNWSAID=34696, accessed 29 May 2008

Volkov, A. L., Krutov, V. E., Kozlov, A. F. and Shiskin, A. E. (eds) (2003) *Lesnyje resursy, lesnoje khozajstvo y lesopromishlennyj komleks Karelii na rubeshe XXI veka* [Forest resources, forestry and forest industries of Karelia at the cusp of the 21st century], Russian Academy of Science Karelian Research Centre, Petrozovodsk, Russia

Warner, R., Simonov, E. and Gibson, D. (2002) *Biodiversity Assessment for Russia*, report submitted to USAID/Russia by Chemonics International Inc., Washington, DC, March, www.biofor.org/documents/Russia%20Final%20Version%20PDF.pdf, accessed 29 May 2008

Yrjölä, T. (2002) *Forest Management Guidelines and Practices in Finland, Sweden and Norway*, European Forest Institute Internal Report No 11, European Forest Institute, Joensuu, Finland

Zakharenkov, A. (2003) 'The priority tasks of optimization of NTFP management and usage in the Russian Far East', *International Forestry Review*, vol 5, pp89–90

Navigating a Way through Regulatory Frameworks for Hoodia Use, Conservation, Trade and Benefit Sharing

Rachel P. Wynberg

INTRODUCTION

The complexities of governing NTFPs are vividly illustrated by a case from southern Africa involving species of *Hoodia*, a succulent plant that has undergone rapid commercialization in the past decade. The case is particularly interesting because of the plant's traditional use to stave off hunger and thirst by the indigenous San peoples, the oldest human inhabitants of Africa (Pappe, 1862; White and Sloane, 1937). Policy frameworks that have evolved to regulate *Hoodia* have thus had to take into consideration both the conservation and trade aspects of *Hoodia* use, as well as the emerging legal arena of 'access and benefit sharing', concerned with the rights of indigenous peoples, and ways in which benefits arising from the commercial use of traditional knowledge and genetic resources should be fairly distributed.

This has been complicated by the fact that both the traditional knowledge that was used in the commercial development of *Hoodia* and the species involved cross national borders, involving the governments of South Africa, Namibia and Botswana, as well as indigenous communities of the San, Nama, Damara and other groups. However, each of the three countries with which *Hoodia* and its knowledge are associated has evolved a distinct regulatory approach to the plant's conservation and use, and to the way in which access and benefit-sharing issues are framed.

A bewildering complexity of policies and laws has consequently emerged in southern African countries to regulate the harvesting, trade and commercial development of *Hoodia*, existing at a convoluted interface between biodiversity conservation; access and benefit sharing; intellectual property rights; and traditional knowledge (Figure 13.2). As this chapter illustrates, the manifold laws that regulate each of these components typically have at best little coherence, or at worst are contradictory. Additionally, they are administered in substantially different ways by a range of government

Photo: David Newton.

Figure 13.1 *Flowering* Hoodia gordonii, *Ceres (Karoo), Western Cape, South Africa*

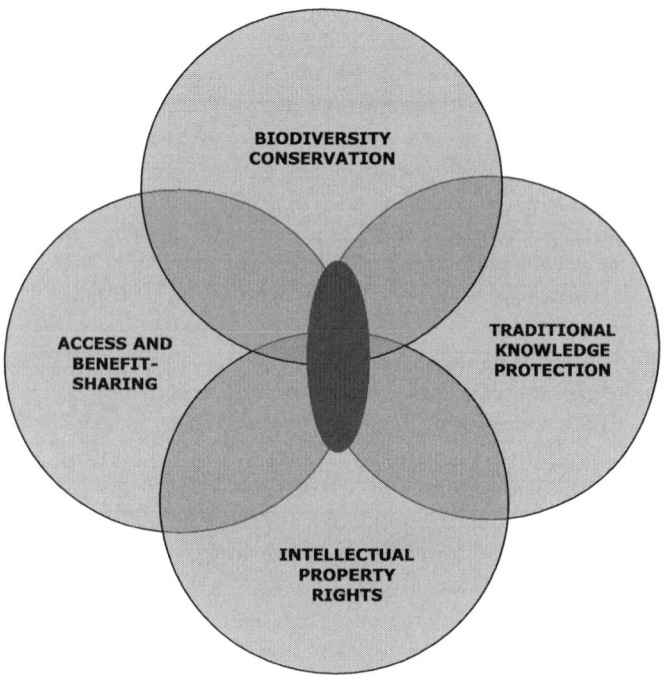

Figure 13.2 *The intersection of regulatory frameworks for* Hoodia

Box 13.1 Overview of the ecology of *Hoodia* spp.

Species of the genera *Hoodia* and related *Trichocaulon* have long been used as thirst quenchers and appetite suppressants (White and Sloane, 1937; Pappe, 1862). Both genera are members of the Apocynaceae family, succulent perennials adept at storing moisture during the long dry spells of their native habitats (CITES, 2004). The unusual flowers are flat and saucer-like in shape and brownish in colour, and form prolifically near the stem tips in summer, when they are often characterized by a distinct carrion smell to attract pollinating flies. The stems are cylindrical, leafless and typically multi-angled, ribbed and spiny. More than 20 species have been recorded from southern Africa, although most commercial attention has focused on *Hoodia gordonii*. Other species of interest for their appetite-suppressing properties are *H. currorii, H. flava, H. lugardii* (now *H. currorii* subsp. *lugardii*), *H. pilifera* (previously *Trichocaulon piliferum*), *H. officinalis* (previously *Trichocaulon officinale*) (White and Sloane, 1937; Van Wyk and Gericke, 2000; patent WO 9846243A2). Vernacular names for the plants include *ghaap* and *!khobab, |goa.-|, |khowa.b, |goai-|, |khoba, |khoba.b|s, |khowab, |goab, otjinove, !nawa#kharab, sekopane* and *seboka*, where !,# and | represent three different click sounds (plosives) (White and Sloane, 1937; Smith, 1966; Malan and Owen-Smith, 1974; Van Wyk and Gericke, 2000; Hargreaves and Turner, 2002; CITES, 2004).

Figure 13.3 illustrates the distribution of *Hoodia gordonii* in the region, and indicates its occurrence in the summer rainfall areas of Angola, Botswana, Namibia

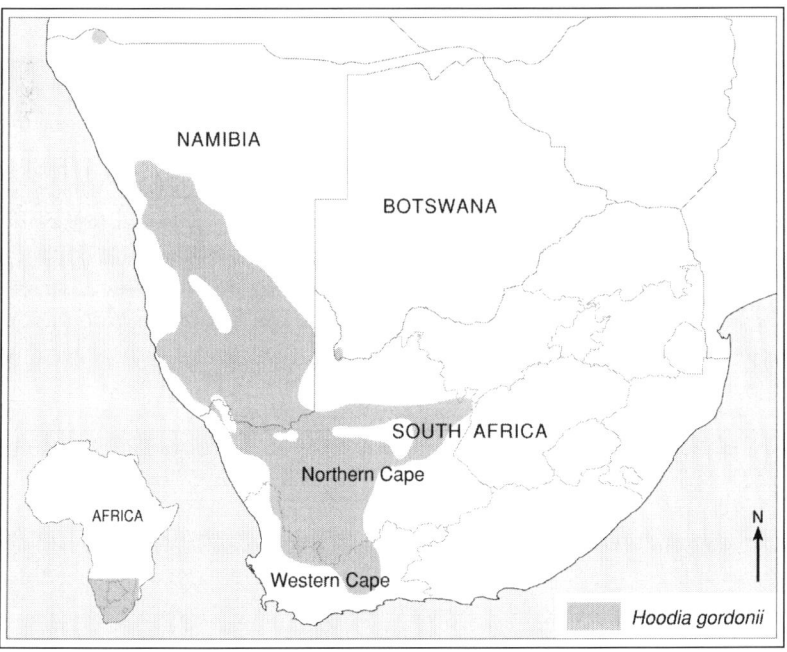

Source: Powell, 2005.

Figure 13.3 *Distribution range of* Hoodia gordonii

and South Africa, as well as winter rainfall areas in Namibia. In general, most *Hoodia* species have patchy distributions, with several occurring in very large populations over vast areas. Little is known about the lifespan, population dynamics and regeneration cycle of *H. gordonii* (Powell, 2005) and this has necessitated a precautionary and reactive management approach from government. Moreover, yields from wild plants appear to be inconsistent, with both reproductive and vegetative productivity being dependent upon environmental events. Rainfall is the chief factor limiting productivity and post-harvest recovery (Wynberg and Newton, 2009). Natural threats to the plants include stem-boring insects, pathogenic infections, rodent infestations and the trampling and grazing of plants by game and other animals. In certain areas destruction of the plant's habitat is a problem, resulting from road construction, mining, agriculture and urban development (Powell, 2005).

institutions with overlapping mandates and unclear roles and responsibilities. This chapter provides a broad analysis of policies with relevance for the conservation, use and trade of *Hoodia* species, exploring the variety of approaches that have been adopted by governments in southern Africa, the effects of these and the constraints faced in ensuring effective implementation.

THE COMMERCIAL DEVELOPMENT OF *HOODIA*

The commercial development of *Hoodia* is a fascinating story that has captured the imagination of policy-makers, academics and community activists alike. In the 1800s and early 1900s, colonial botanists published traditional knowledge about the appetite-suppressing qualities of *Hoodia* (Pappe, 1862; Laidler, 1928; Marloth, 1932). This led to the subsequent inclusion of *Hoodia* in a 1963 project on edible wild plants of the region undertaken by the South African-based Council for Scientific and Industrial Research (CSIR), one of the largest research organizations in Africa.

In 1997, following nine years of confidential development, a patent application was filed in South Africa by the CSIR that included the use of plant extracts and the active constituents of the plant responsible for suppressing appetite and treating obesity. This was done without the consent of the San, the original holders of knowledge about these properties, although the CSIR was eventually pressurized to enter into negotiations with the San and to develop an agreement to share the benefits arising from the commercial development of *Hoodia* (Wynberg, 2004). The CSIR proceeded in 1997 to grant a licence for the further development and commercialization of the patent to Phytopharm, a small UK company that specializes in the development of phytomedicines (Phytopharm, 1997). The agreement granted Phytopharm an exclusive worldwide licence to manufacture and market *Hoodia*-related products and to exploit any other part of the CSIR's intellectual property rights relating to *Hoodia* species. Through a programme dubbed 'P57', Phytopharm developed *Hoodia* to a more advanced stage, leading to a licence and royalty agreement in 1998 with Pfizer, the US-based pharmaceutical giant, for further development and commercialization. However, the 2003 closure of Pfizer's Natureceuticals group led to the later withdrawal of Pfizer from the agreement.

In 2004, Phytopharm granted the consumer giant Unilever PLC an exclusive global licence for *Hoodia gordonii* extracts, with their probable incorporation into existing food brands as a functional weight-loss product for the mass market (Phytopharm, 2004). Under the terms of the agreement, Unilever would buy exclusive rights to the product for an initial £6.5 million, rising to £21 million once it had achieved certain milestones. Phytopharm would also receive an undisclosed royalty on sales of all products containing the extract. Developments included clinical safety trials, manufacturing and the cultivation of some 300ha of *Hoodia* in South Africa and Namibia (Kevin Povey, Unilever research and development programme director, pers. comm., 2007). This situation changed significantly in November 2008, with the announcement by Unilever that it was to abandon plans to develop *Hoodia* as a functional food because of safety and efficacy concerns. Phytopharm is now seeking other partners to further develop *Hoodia* and bring products to market (Phytopharm, 2008).

Much is at stake if a successful product is developed: the global value of functional foods, defined as 'any modified food or food ingredient that may provide a health benefit beyond the traditional nutrients it contains' (Bloch and Thomson, 1995) is estimated at US$65 billion (Phytopharm, 2007), with the overall market value for the dietary control of obesity at over US$3 billion per annum in the US alone (Phytopharm, 2003). The predicted growth potential of functional foods is some 50 per cent from 2005 to 2010, with an accelerating trend towards new products. Potential profits are thus likely to be highly significant, and could result in substantial returns not only for the companies involved, but also for southern African countries and the impoverished San.

The publicity generated by the CSIR-Phytopharm-Unilever agreements, the marketing opportunities offered by the San use of the plant, and the CSIR patent led to a frenzied interest in *Hoodia* among plant traders. By 2002 a parallel market had emerged, based on wild-harvested *Hoodia* that had simply been dried, sliced and exported. By 2005, trade had escalated exponentially – and in many cases illegally – from just a few tons to more than 600 tons of wet, harvested material, sold as ground powder for incorporation into non-patented dietary supplements. In North America in particular, dozens of *Hoodia* products were sold as diet bars, pills, drinks and juice, traded by a myriad of companies freeriding on the publicity and clinical trials of Phytopharm and Unilever. The CSIR patent was focused on the *Hoodia* extract, and nothing prevented other companies from simply selling the raw material for incorporation into herbal supplements. Most products were of dubious authenticity, contained unsubstantiated quantities of *Hoodia* and made unfounded claims, many implying association with the San, who received no benefits. Concerns led to the closer analysis of products by the US Food and Drug Administration (FDA), which revealed many to contain little or no *Hoodia* and to lack adequate evidence of safety (FDA, 2004; Stafford, 2009). The US Federal Trade Commission (FTC) also brought action against spammers sending e-mail messages about *Hoodia* weight-loss products, alleging that the claims made for the products were false and unsubstantiated (FTC, 2007; 2009). In South Africa and Namibia, illegal trade and harvesting of *Hoodia* resulted in a number of prosecutions and arrests; the high prices commanded for the dry product of up to

US$200 per kg had led to the incorporation of the plant into a global underground network of diamonds, drugs and abalone (Wynberg and Chennells, 2009).

POLICY FRAMEWORKS FOR REGULATING THE CONSERVATION, TRADE AND USE OF *HOODIA*

Initial policy responses to *Hoodia* commercialization

The scale of policy interest in *Hoodia* tracked the growing extent of trade in the plant, which was also affected by Unilever's research and development programme, and its initiation of large-scale cultivation projects. However, these policy responses were enfolded within a well-established legal and institutional framework for species conservation in southern African countries. Table 13.1 describes key policies and laws in place relevant to the conservation and use of *Hoodia*, as well as the wider legal and institutional context.

As Table 13.1 shows, at the time of the spike in *Hoodia* trade in 2002 and 2003, most species were already protected to varying extents by nature conservation legislation in South Africa, Namibia and Botswana. In South Africa, the plant was listed as a protected genus in the Northern Cape province, through the Environmental Conservation Ordinance 19 of 1974, and through similar legislation in the Western Cape province, with a permit required from provincial authorities to collect, cultivate, transport or export *Hoodia* spp. Namibia, too, in the Nature Conservation Ordinance of 1975, listed all *Hoodia* species as protected, requiring prior authorization for harvesting and trade. In Botswana, harvesting was controlled by the Agricultural Resources (Conservation) Act and related regulations.

Up until 2002, however, there had been little demand for *Hoodia* and governments thus adopted a passive approach towards its regulation, relying predominantly on existing nature conservation laws. But the escalation in demand necessitated new regulatory approaches. In 2004 concerns about the threats posed to natural populations through unregulated collection led to the inclusion of *Hoodia* spp. in Appendix II of the international Convention on International Trade in Endangered Species of Wild Flora and Fauna (CITES) (CITES, 2004). In response, southern African governments began developing a more tightly regulated permitting system for *Hoodia* use and trade, although this was done differentially both within and between countries.

In South Africa, a different set of approaches evolved between the Northern Cape and Western Cape, the areas in which most *Hoodia* species occur in that country. The initial response from the Northern Cape, which mirrored that of Namibia and Botswana, was to place a moratorium on wild harvesting and trade of any *Hoodia* species. Insufficient information was available about the resource, it was contended, to determine sustainable off-take rates, and therefore a precautionary approach was warranted (Powell, 2005). Moreover, permit applications for harvesting, cultivation and trade of *Hoodia* had increased substantially, along with reports of illegal harvesting, and, because of the difficulties of determining species not in flower, there was a risk of collecting 'look-a-like' but incorrect species (CITES, 2004; Powell, 2005).

Table 13.1 *Key laws and policies pertaining to the use, trade and conservation of* Hoodia *in South Africa, Namibia and Botswana*

South Africa

Policy/law	Relevant provisions/content
Constitution of the Republic of South Africa (Act 108 of 1996)	Conservation and ecological sustainability are given prominence in the Bill of Rights.
Indigenous Knowledge Systems Policy (2004)	An enabling framework to stimulate and strengthen the contribution of indigenous knowledge to social and economic development in South Africa
Policy Framework for the Protection of Indigenous Traditional Knowledge through the Intellectual Property System and the Intellectual Property Laws Amendment Bill (2008)	Provides that the law of trademarks/geographical indications may be able to provide protection of certain names/features associated with traditional knowledge, that a national council consisting of experts on traditional knowledge can advise the Minister of Trade and Industry and the registrar of intellectual property, and that communities may form business enterprises to administer and commercialize their traditional intellectual property.
National Environmental Management Act (107 of 1998)	Gives legal effect to environmental rights in the Constitution; sets in place procedures and mechanisms for cooperative governance; and regulates environmental impact assessments.
National Forests Act (84 of 1998)	Overall purposes include the sustainable use, management and development of forests, the restructuring of state forestry, the protection of certain forests and trees, and the promotion of community forestry. Certain activities may be licensed in state forests, including the collection of biological resources.
National Environmental Management: Biodiversity Act (10 of 2004)	Provides for the management and conservation of biological diversity, the use of indigenous biological resources in a sustainable manner, and the fair and equitable sharing among stakeholders of benefits arising from bioprospecting involving indigenous biological resources. Requires benefit-sharing agreements.
National Environmental Management: Biodiversity Act (10 of 2004): Threatened or Protected Species Regulations	List *Hoodia gordonii* and *Hoodia currorii* as protected species.
National Environmental Management: Biodiversity Act (10 of 2004): Bioprospecting, Access and Benefit-sharing Regulations	Make a distinction between the 'discovery' and 'commercialization' phase of a bioprospecting project. National minister responsible for environment issues permits for bioprospecting and export for bioprospecting purposes. Foreigners may only apply for permits jointly with a South African collaborator. Export must be in the public interest. Benefit-sharing agreements may be refused if there is no provision for enhancing scientific and technical capacity to conserve, use and develop biodiversity or to promote conservation.
Agricultural Pests Act (36 of 1983)	Provides for the prevention and combating of agricultural pests, and regulates the importation of controlled goods, including plants, pathogens and insects. Prohibits any person from importing into South Africa any plant without a permit. The Minister of Agriculture has imposed a number of controls concerning the import of seeds, for example by requiring phytosanitary certificates.

Table 13.1 *Key laws and policies pertaining to the use, trade and conservation of* Hoodia *in South Africa, Namibia and Botswana* (Cont'd)

South Africa

Policy/law	Relevant provisions/content
Patents Act (57 of 1978)	Governs the registration and granting of patents for inventions. Provides for the patenting of microorganisms and microbiological processes but prohibits the patenting of plants and animals. The 2005 amendment requires an applicant for a patent to furnish information on the use of indigenous biological resources or traditional knowledge in an invention.
Customs and Excise Act (91 of 1964)	Provides for the prohibition and control of the importation, export or manufacture of certain goods.
Various Provincial Ordinances and Acts	The provinces have, in all, 28 legal instruments for nature conservation. In general they allow for the establishment and protection of nature reserves, for the conservation of threatened species, and for fishing and hunting. Many of these laws are outdated and the nine provinces are at different stages of phasing out old laws and developing and implementing new ones. *Hoodia* is listed as a protected genus in the Northern Cape, through the Environmental Conservation Ordinance (19 of 1974), and in the Western Cape and Free State provinces through similar legislation, and a permit is required from provincial authorities to collect, cultivate, transport or export *Hoodia* spp.

Namibia

Policy/law	Relevant provisions/content
Namibian Constitution (1990)	Obliges the government to adopt policies aimed at the 'maintenance of ecosystems, essential ecological processes and biological diversity of Namibia and utilization of living natural resources on a sustainable basis for the benefit of all Namibians, both present and future'. Recognizes the existence and importance of customary law, declaring it to be of the same value as common law. Vests ownership of all non-privately owned land and natural resources in the state.
Environmental Management Act (7 of 2007)	The basis for environmental management in Namibia. Aims to prevent and mitigate significant effects on the environment and establishes principles for decision-making. Stipulates that activities including the removal of resources and land-use transformation may require an environmental assessment. Requires benefit sharing. Sets up an advisory council which includes access to genetic resources.
Nature Conservation Ordinance (4 of 1975)	The primary legislation governing nature conservation in Namibia. Sets in place a permitting system for protected species, including *Hoodia* spp., requiring prior authorization for harvesting and trade. Requires a permit for the picking and transport, sale, donation, export and removal of protected plants. Requires the written permission of landowners before any indigenous plant is picked. The 1996 amendment gives rights over wildlife and tourism to communal farmers.
National Agriculture Policy (1995)	Aims to achieve growth in agricultural production and profitability, ensure food security, improve living standards for farmers and farm workers, and promote sustainable use of land and natural resources. Aims to promote diversification of rural livelihoods.

Table 13.1 *Key laws and policies pertaining to the use, trade and conservation of* Hoodia *in South Africa, Namibia and Botswana* (Cont'd)

Namibia

Policy/law	Relevant provisions/content
Traditional Authorities Act (17 of 1995)	Requires traditional authorities to ensure that members of their communities use natural resources sustainably and in a manner that conserves the environment.
Plant Quarantine Act (7 of 2008)	Requires phytosanitary certificates to accompany exports of raw material.

Botswana

Policy/law	Relevant provisions/content
Wildlife Conservation and National Parks Act (28 of 1992)	Governs the use of resources in national parks, protected areas and game reserves, as well as procedures to access biological resources. Implements CITES.
Agricultural Resources (Conservation) Act (2006)	Provides for regulations to control access to biological resources and sets in place permitting requirements for the harvesting, export and trade of *Hoodia*, except when for domestic use. Includes no specifications for benefit sharing and prior informed consent.
Forest Act (38 of 2004)	Protects forests and regulates the use of forest resources.
Tribal Land Act (32 of 2002)	Recognizes that a community collectively owns the land as well as the resources on it, but gives decision-making power over those resources to tribal land boards. Has relevance for the provision of prior informed consent and the negotiation of benefits.

The Western Cape, however, adopted a different approach. A moratorium, they argued, would simply drive the *Hoodia* industry underground and make the trade more difficult to track and manage (Paul Gildenhuys, CapeNature, pers. comm., 2006). Moreover, a comprehensive permitting system already existed to comply with CITES 'non-detriment requirements' (essentially to show that harvesting has been conducted in accordance with sustainability guidelines) and to regulate the export of parts, derivatives or whole plants. Thus a number of permits were issued by CapeNature, the provincial authority in the Western Cape responsible for biodiversity conservation, to traders for the wild harvesting and export of *Hoodia* spp. A number of conditions were attached to the permit, including restrictions on the size of the plant harvested, but, astonishingly, with no specifications of tonnage. This so-called 'open permit' had been used for years by CapeNature, based on a regulatory model developed for the flower industry that had never before presented problems (Kas Hamman and Paul Gildenhuys, CapeNature, pers. comms, 2007). *Hoodia* was different, however, because of the extremely high price that it commanded, encouraging the collection of as much material as possible.

Because of moratoriums elsewhere, the Western Cape was now the only legal point of export in southern Africa for *Hoodia* material, and the open-ended nature of the permit provided the perfect means through which illegally harvested material from the region could be included and legitimately exported under a CITES permit. In 2005, for example, a total of 500 tons was reportedly exported from the Western Cape, far exceeding the estimated amount of plant material available in the province and thus verifying suspicions about the inclusion of material from Namibia and other provinces in South Africa (Gosling, 2006). Over the same period, reports of illegal *Hoodia* harvesting surged in the Northern Cape and Namibia, including stories of microlight aircraft assessing *Hoodia* populations, the nocturnal smuggling of *Hoodia* in boats across the Orange River from Namibia to South Africa using children flashing torch signals, and the hiding of material in animal carcasses (Charles Musiyalike, Ministry of Environment and Tourism, pers. comm., 2007).

Towards proactive policy frameworks

Increasing awareness of these problems, combined with concerns about the quality and safety of material sold as *Hoodia*, and recognition of the need to ensure the sustainability of *Hoodia* supply, led to a rapid response from conservation authorities across the region, along with an attempt to bring greater cohesion and standardization to policies. The Northern Cape lifted restrictions on wild harvesting and, together with the Western Cape, established resource assessment procedures as the basis for determining a quota for each permit, with specific harvesting procedures prescribed. As an interim measure, both the Northern Cape and Western Cape also required anyone harvesting *Hoodia* from the wild to reinvest some of their profit back into the establishment of cultivated *Hoodia* plantations. Restrictions were also now placed on the permissible volumes to collect.

In 2007, however, a decision was taken by both the Northern Cape and Western Cape to stop issuing permits for wild harvested *Hoodia* and all existing farmers were required to cultivate the species if they wished to continue trading it, a situation that still applies today. Moreover, most *Hoodia* growers are now organized to some extent through an organization known as the Southern African *Hoodia* Growers Association (SAHGA). This organization represents the interests of commercial growers of *Hoodia* in South Africa who have agreed to comply with certain standards of best practice, safety, fair trade and benefit sharing, and who wish to supply *Hoodia* as a food or as a dietary supplement as approved by food and drug quality control authorities worldwide.

In Namibia a similar body known as the *Hoodia* Growers Association of Namibia (HOGRAN) has been constituted, and here too there have been incremental efforts to implement a regulatory system for *Hoodia* that both ensures conservation and promotes the development of a viable industry. Initial policy outlawed wild harvesting completely, but now 'salvage' harvesting is permitted of plants that have died through natural circumstances. Such harvesting is only permitted once active cultivation and enrichment planting programmes have been established. Unlike South Africa, where cultivation is predominantly focused on private lands, Namibia has pursued a far greater developmental role for *Hoodia*, actively promoting its cultivation as an economic opportunity for small farmers living on communal lands. This is also the case in Botswana, although there has been little change to *Hoodia* regulation in that country because it contains only small populations of the commercially desirable species.

The commercial development of *Hoodia* and associated controversies also led to greater policy engagement on issues relating to the protection of traditional knowledge and the fair sharing of benefits resulting from its use. The initial acquisition of traditional knowledge about the appetite-suppressing properties of *Hoodia*, without the consent of the San, and the CSIR's subsequent licensing agreement with Phytop-harm to commercially develop a product elicited little, if any, policy response from any southern African government at the time. Only after considerable media attention in 2001 did the CSIR consent to negotiations with the San to develop a benefit-sharing agreement, but this was largely done in a legal vacuum. It was partly the unfolding of these experiences and the high-profile nature of the case that gave impetus to the development of binding laws in South Africa and elsewhere. In South Africa, this was encapsulated by the National Environmental Management: Biodiversity Act (10 of 2004) (Biodiversity Act) and the 2008 promulgation of access and benefit-sharing regulations to give effect to the Act. As described in Table 13.1, this regulatory frame-work addresses for the first time the need for bioprospectors to obtain prior informed consent from custodians of biodiversity and holders of traditional knowledge before initiating any project. It also requires a benefit-sharing agreement to be developed between different stakeholders to ensure that holders of traditional knowledge or custodians of biodiversity are fairly compensated.

Box 13.2 *Hoodia* permit requirements in Namibia

- Anyone wanting to cultivate *Hoodia* must apply for a nursery licence and stipulate from where seed is collected.
- Permits are also required to register a *Hoodia* nursery.
- Under certain conditions 'salvage' harvesting may be allowed of plants that have died through natural circumstances (e.g. animal damage or disease). Applicants need to show that the wild population exceeds 1000 plants and must produce veri-fied distribution and density maps.
- Wild harvesting is only permitted once active cultivation and enrichment planting programmes have been established. The Ministry of Environment and Tourism must be informed of the intention to harvest and the date on which the harvesting will take place, and must be present when harvesting takes place.
- Those wishing to be involved in any manufacture associated with *Hoodia* need to register as a manufacturer and provide proof of registration as a business, proof of experience with medicinal plants, and documentation from the health authorities regarding the operating site.

IMPLEMENTATION CHALLENGES

Hoodia policies for conservation, trade and use have clearly evolved towards being more proactive, flexible and effective. This is well reflected in the positive way in which most *Hoodia* growers and traders across the region have engaged in policy development and compliance, facilitated through the emergence of two organizations, SAHGA and HOGRAN, to represent their interests in South Africa and Namibia respectively. New legal requirements for prior informed consent and benefit sharing also point towards a more proactive policy framework. Nonetheless, a number of constraints still impede successful implementation.

Absence of a comprehensive and integrated regulatory framework

One of the biggest problems is the absence of a comprehensive and integrated regulatory framework for *Hoodia* to address laws and policies acting at different scales, from local through to regional. This is especially pertinent in countries such as South Africa, which has a federal system of government in which there is considerable confusion between national and provincial levels of government over responsibility for managing *Hoodia* species and regulating the associated industry. In part this is because the South African Constitution (Act 108 of 1996) designates most biodiversity functions as areas of concurrent legislative competence, meaning both national and provincial government may take responsibility for species management. This gives provinces some leeway in the way in which they develop and implement policies and laws, provided these are in keeping with national norms and standards. In practice, however, national standards have lagged behind existing provincial laws, with the result that provinces have taken responsibility for CITES implementation and *Hoodia* management, despite a pronouncement that the species is to be managed nationally (Wynberg and Newton, 2009). This incessant yo-yoing of responsibility reflects to a large extent ongoing tensions between those managing the species on the ground, who hold in-depth knowledge of the plant's use and trade patterns, and those attempting to develop and implement a national, coherent policy approach towards *Hoodia*. In Namibia, which has a more centralized government system, *Hoodia* regulation is much simplified by the fact that permits are administered by a single authority, the Ministry of Environment and Tourism, which also provides oversight on all *Hoodia* use and trade, rather than multiple provincial bodies as in South Africa.

Multiple permitting

Across South Africa, Namibia and Botswana, the situation is further complicated by the range of different national government departments involved in regulating discrete aspects of *Hoodia* use and trade. In practice this means that anyone wishing to use or trade *Hoodia* needs multiple permits. Not only are permits required to harvest, grow, manufacture and export *Hoodia*, but also for phytosanitary purposes, and for the ploughing, transformation or rezoning of land. Different authorities administer each of these permits, requiring the applicant to make separate applications to environmental, trade, health and agricultural departments. Individual permits are also required for each trade transaction, and this is considered to be onerous and as acting against the entry of small growers into the system. One way to streamline this could be to introduce a single permit that allows cultivation, harvesting of cultivated material, processing and trade with inspection and renewal on an annual basis (Wynberg and Newton, 2009).

Ineffective monitoring, enforcement and compliance

Such a system would also improve monitoring, enforcement and compliance, which in all three countries are key constraints preventing the effective implementation of the *Hoodia* permitting system. Law enforcement capacity is low, the legal processes are cumbersome and seemingly full of loopholes, and the low penalties do not constitute a sufficient deterrent to transgressors, given the high value of the resource. This is exacerbated by the fact that illegal harvesting typically occurs in remote rural areas, with material quickly transported across borders, especially from Namibia to South Africa. Increasingly, however, governments are collaborating to design joint policies for *Hoodia* management, with steps put in place to join forces more strongly on poaching, trade and the transport of illegally harvested material. This bodes well for future cooperation and suggests a positive environment within which policy resolutions can be found.

Confusing and complex access and benefit-sharing policies

An additional complication is that, as of 1 April 2008, those wishing to trade *Hoodia* in South Africa have been required to obtain bioprospecting and export permits to comply with the regulations promulgated under the Biodiversity Act. To do this they have to develop agreements to share benefits with holders of traditional knowledge and/or custodians of the resource. Although bioprospecting is typically interpreted narrowly to refer to the exploration of biological material for commercially valuable genetic and biochemical properties, the Biodiversity Act defines 'bioprospecting' and 'indigenous biological resources' very widely, with the inference that bioprospecting could be interpreted to go beyond research involving genetic material or biochemical material and include all trade in biological resources, commonly referred to as 'biotrade' (Wynberg and Taylor, 2009).

The wide definitional scope of the Biodiversity Act has significant – albeit unclear and complex – implications for *Hoodia*. For example, *Hoodia* has been developed both as a genetic resource, to be included in patented extracts, and as a herbal medicine, where the raw material is simply dried, cut and incorporated into products. In practice, the use of traditional knowledge for both types of products prescribes the need for a benefit-sharing agreement, but there are clearly overlapping and sometimes artificial boundaries between trade in genetic resources and that in biological organisms. In anticipation of these requirements growers affiliated to the SAHGA have already signed a benefit-sharing agreement with the Working Group of Indigenous Minorities in Southern Africa (WIMSA), which sets out a 'San Levy' requiring SAHGA to pay WIMSA approximately US$3 per kg of dry, processed *Hoodia* to be exported. However, membership of SAHGA is voluntary and the extent to which individual traders will be forced to comply with benefit-sharing requirements under the Biodiversity Act remains to be seen.

While these may seem like unrelated regulatory issues, access and benefit sharing, CITES and the wider trade in species are integrally linked in many ways (INA, 2004; Ruiz and Lapeña, 2007). In fact, the CITES listing for *Hoodia* includes an annotation requiring CITES permits for all parts and derivatives of *Hoodia* species except those bearing a label 'Produced from *Hoodia* spp. material obtained through controlled harvesting and production in collaboration with the CITES Management Authorities of Botswana/Namibia/South Africa under agreement no. BW/NA/ZA xxxx'. The intent behind this annotation was to ensure that countries in which *Hoodia* naturally occurred (the so-called range states) captured the economic benefits accruing from commercialization. Although never implemented, this annotation represented a significant attempt by CITES to link trade and benefit sharing.

In practice, the dovetailing of permitting requirements for each of these various activities has proved extremely difficult, requiring permit applicants to comply with a stream of bureaucratic procedures administered by different authorities. Within government it is also extremely difficult to keep track of such diverse applications. To overcome some of these problems in South Africa, it has been suggested that the access and benefit-sharing permit system and the provincial research permit system be synchronized with current efforts to develop a uniform and coordinated permitting system for CITES, possibly through a single electronic database, which would include information about the application, its status and existing permits granted. Early experiences of implementing benefit-sharing agreements for *Hoodia* suggest that such information could help considerably in determining, for example, the volumes of material traded, and thus the benefits due to traditional knowledge holders. As described in Wynberg and Chennells (2009), the lack of such information has been a major stumbling block preventing compliance with and implementation of existing benefit-sharing agreements relating to *Hoodia* use.

CONCLUSION: CYCLES OF POLICY INTERVENTIONS

The policy interventions that have been made at different stages of the commercialization of *Hoodia* yield broad lessons about the way in which the state and other institutions engage in and respond to the development of a natural product, and changes in its supply and demand. Figure 13.4, which is based on concepts developed by Homma (1992) to characterize the NTFP commercial production cycle, shows how key events – such as the signing of a licence agreement or the upscaling of *Hoodia* cultivation – changed both the nature of extraction and the level of involvement of different institutional players.

In the case of *Hoodia* the role of the state was largely reactive and interventionist in the early stages of commercialization, responding initially to peaked commercial interest and declines in the availability of the resource through policy measures to regulate or restrict use and trade. Thus *Hoodia's* entry into the weight control market in 2001 led to a surge in demand for the raw material that required southern African governments to respond rapidly by introducing a stringent permit system and, in some cases, prohibiting wild harvesting. The international community similarly reacted by including *Hoodia* species in Appendix II of CITES. However, as the resource became better managed and the availability of cultivated material reduced pressure on wild populations, governments responded with a less severe permitting system and the role of the state tapered off. Now that cultivation has been initiated, the function of the state is diminishing to one of monitoring resource use, setting quality and export standards, providing policy support to bolster market opportunities and, in some instances, giving

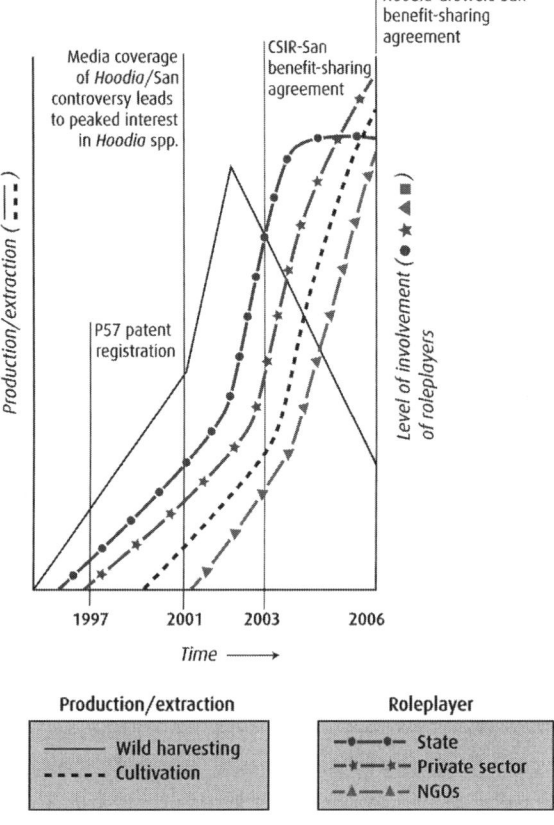

Note: This graph is schematic and based on informed extrapolations.

Figure 13.4 *Production cycle of* Hoodia *spp.*

support to cooperatives. In turn, market requirements for a consistent, high and reliable quality and quantity of cultivated material are leading to the industry adopting a greater self-regulatory role.

The reactive and iterative policy-making that has been described for *Hoodia* has clear drawbacks in its lack of coherence, comprehension and foresight, but it also has its advantages. Many species enter markets that are highly volatile and erratic. Seldom are policy-makers abreast of these developments and able to plan quickly enough or appropriately. In this case the significant changes in *Hoodia* markets, availability and demand clearly necessitated an iterative and flexible approach by government towards permitting and regulation. Reactive policy-making may thus be a vital mechanism to cope with rapidly changing conditions, in this case market and trade fluctuations.

This chapter has described the complexities of regulating a species undergoing rapid commercialization, where information about both the biology of the species and its trade is incomplete and scarce, where several nation states are involved, and where multiple laws apply to regulate harvesting, trade and commercialization. The intractability of traditional knowledge use and benefit sharing adds yet another layer of murkiness to the picture. It is to be expected that the policy outcomes resulting from this situation will be messy.

An important question to ask is: how can policy move forward under such circumstances? Reactive and 'experimental' policy-making provides a partial answer in the short term and for crises, but one that may not ultimately be conducive to ecologically sound or equitable NTFP policies. Deeper consultation with harvesters, processors and traders, drawing on their experiences and insights, provides another important guide for policy-makers. Also vital is for policy-makers to have greater knowledge of and exposure to policies and approaches outside of their traditional sectors. Combined, these imperatives suggest the need for a new approach to NTFP policy-making that takes into account and acknowledges the increasingly complex systems within which NTFPs are regulated, that moves away from positions that are pigeon-holed to specific government departments and sectors, that recognizes the expertise and experience of stakeholders involved in the trade, and that is visionary and bold in achieving integration.

ACKNOWLEDGEMENTS

Many individuals involved in trading or regulating *Hoodia* were interviewed for this research, often giving generously of their time and knowledge. Their contribution is gratefully acknowledged. Particular thanks are due to David Newton of TRAFFIC East and Southern Africa for his insights and contribution towards understanding *Hoodia* policy, and to Fahdelah Hartley for administrative support. I am also grateful to the Wellcome Trust, which partially supported this research as part of a wider project focused on *Hoodia*. Additional financial support was provided by South Africa's National Research Foundation (NRF) although opinions, findings and conclusions or recommendations expressed in this chapter are those of the author and the NRF does not accept any liability for them.

REFERENCES

Bloch, A. and Thomson, C. (1995) 'Position of the American Dietetic Association: Phytochemicals and functional foods', *Journal of the American Dietetic Association*, vol 95, no 4, pp493–496

CITES (2004) 'Amendments to Appendices I and II of CITES', Proposal to 13th Meeting of the Conference of the Parties, Bangkok, Thailand, 2–14 October, www.cites.org/common/cop/13/raw_props/BW-NA-ZA-Hoodia.pdf, accessed 7 April 2006

FDA (2004) '75-Day Premarket Notification of New Dietary Ingredients', US Food and Drug Administration, Washington, DC, www.fda.gov/ohrms/dockets/dockets/95s0316/95s-0316-rpt0218-vol156.pdf

FTC (2007) 'FTC stops international spamming enterprise that sold bogus *Hoodia* and human growth hormone pills', 10 October, Federal Trade Commission, Washington, DC, www.ftc.gov/opa/2007/10/hoodia.shtm, accessed 5 April 2008

FTC (2009) 'FTC charges marketers of "Hoodia" weight loss supplements with deceptive adver-tising', 27 April, Federal Trade Commission, Washington, DC, www.ftc.gov/opa/2009/04/nutraceuticals.shtm, accessed 31 May 2009

Gosling, M. (2006) 'Hoodia "diet plant" under threat from illegal exports', *Cape Times*, 24 November

Hargreaves, B. J. and Turner, Q. (2002) 'Uses and misuses of *Hoodia*', *Asklepios*, vol 86, pp11–16

Homma, A. K. O. (1992) 'The dynamics of extraction in Amazonia: A historical perspective', *Advances in Economic Botany*, vol 9, pp23–31, New York Botanical Garden

INA (2004) *Expert Workshop Promoting CITES-CBD Cooperation and Synergy*, International Academy for Nature Conservation, Isle of Vilm, Germany, 20–24 April, www.cbd.int/cooperation/final-report-CITES CBD_Vilm_Workshop_Report.doc

Laidler, P. W. (1928) 'The magic medicine of the Hottentots', *South African Journal of Science*, vol 25, pp433–437

Malan, J. S. and Owen-Smith, G. L. (1974) 'The ethnobotany of Kaokoland', *Cimbebasia*, series B, vol 2, no 5, p151

Marloth, R. (1932) *The Flora of South Africa*, vol 3, Darter Bros, Cape Town, Wheldon & Wesley, London

Pappe, L. (1862) Silva Capensis: *A Description of South African Forest Trees and Arborescent Shrubs Used for Technical and Economical Purposes*, 2nd edn, Ward & Co., London, pp54–55

Phytopharm (1997) 'Phytopharm to develop natural anti-obesity treatment', press release, 23 June, www.phytopharm.com/news/newsreleases/?page=10&id=1814, accessed 5 April 2008

Phytopharm (2003) *Annual Report and Accounts for the Year Ended 31 August 2003*, Phytopharm plc, Cambridge

Phytopharm (2004) 'Interim results for the period to 29 February 2004', press release, 5 May, www.phytopharm.com/files/news/1588/Interim2004.pdf, accessed 5 April 2008

Phytopharm (2007) 'Phytopharm initiates stage 3 activities of joint development agreement for *Hoodia* extract with Unilever', 24 September, www.phytopharm.com/news/newsreleases/?id=7324, accessed 5 April 2008

Phytopharm (2008) 'Unilever returns rights to *Hoodia* extract', 12 December, www.phytopharm.com/news/newsreleases/?id=16553, accessed 30 March 2009

Powell, E. (2005) 'The medicinal plant *Hoodia gordonii*: a review', internal report no 7, prepared by the Directorate of Nature Conservation Scientific Services, Northern Cape Department of Tourism, Environment and Conservation, South Africa

Ruiz Miller, M. and Lapeña, I. (eds) (2007) *A Moving Target: Genetic Resources and Options for Tracking and Monitoring Their International Flows*, IUCN Environmental Policy and Law Paper No. 67/3, IUCN, Gland, Switzerland

Smith, C. A. (1966) *Common Names of South African Plants*, Botanical Survey Memoir no 35, Department of Agricultural Technical Services, Pretoria

Stafford, L. (2009) 'After another cancelled partnership, the future of *Hoodia* remains unclear', *HerbalEGram*, vol 6, no 2, cms.herbalgram.org/heg/volume6/02%20February%20/Hoodia_Nixed.html, accessed 30 March 2009

Van Wyk, B. E. and Gericke, N. (2000) *People's Plants: A Guide to Useful Plants of Southern Africa*, Briza Publications, Pretoria

White, A. and Sloane, B. L. (1937) *The Stapelieae*, vol III (2nd edn), Pasadena, CA

Wynberg, R. (2004) 'Rhetoric, realism and benefit-sharing: Use of traditional knowledge of *Hoodia* species in the development of an appetite suppressant', *World Journal of Intellectual Property*, vol 6, no 7, pp851–876

Wynberg, R. and Chennells, R. (2009) 'Green diamonds of the South: A review of the San-Hoodia case', in R. Wynberg, R. Chennells and D. Schroeder (eds) *Indigenous Peoples, Consent and Benefit Sharing: Learning from the San-Hoodia Case*, Springer, Berlin, pp89–126

Wynberg, R. and Newton, D. (2010) (forthcoming) *A Policy and Trade Assessment for* Hoodia *spp. in Namibia, Botswana and South Africa*, Environmental Evaluation Unit, University of Cape Town, and TRAFFIC, East Southern Africa

Wynberg, R. and Taylor, M. (2009) 'Finding a path through the ABS maze: Challenges of regulating access and ensuring fair benefit-sharing in South Africa', in E. C. Kamau and G. Winter (eds) *Genetic Resources, Traditional Knowledge and the Law: Solutions for Access and Benefit Sharing*, Earthscan, London, pp203–224

Chapter 14

Laws and Policies Impacting Trade in NTFPs

Alan Pierce and Markus Bürgener

INTRODUCTION

Trade in non-timber forest products (NTFPs) is voluminous, but attempts to quantify it are guesswork at best. Estimating the value and volume of NTFP trade is difficult for a number of reasons, the most significant being that the classification 'NTFP' is not recognized by most government agencies tasked with keeping economic statistics. Vantomme (2003) cites several additional obstacles to gathering NTFP trade data, including:

- NTFPs are often used for subsistence purposes or traded in the informal market sector.
- NTFPs are heterogeneous and are overseen by a variety of government agencies (e.g. departments of forestry, agriculture, horticulture and manufacturing), leading to dispersed and inconsistent reporting.
- NTFP product classification systems are inadequate for tracking NTFPs.

Iqbal (1995, p8) estimated the total world trade in NTFPs at US$11 billion in 1995, but warned that the data were 'indicative only and are to be used with caution'. Table 14.1 presents estimated international trade values for select, high-value NTFPs. Iqbal (1993) notes that international NTFP trade typically flows from developing to developed nations. However, some developing nations import large quantities of NTFPs, particularly medicinal plants. For example, between 1991 and 2003, annual global trade in 'pharmaceutical plants' averaged US$1.2 billion (Lange, 2006). These figures, however, incorporate medicinal plants imported into countries like India and China for processing and re-exportation. Table 14.2 sets out the world's leading importers and exporters of medicinal plants between 1991 and 2003.

In this chapter, we provide an overview of policy and legal issues pertaining to the local, regional and international trade of NTFPs. Readers interested in a more detailed analysis of the topic should consult Iqbal, 1995; Lintu, 1995; Dewees and Scherr,

Table 14.1 *Estimated value of select NTFPs in international trade*

Product	Value (US$)	Origin	Citation
Rattan Various species	6.5 billion (total value of the industry)	Primarily Asia and Africa	ITTO (1997)
Bamboo Various species	2.5 billion (exports, 2000)	Global	Hunter (2003)
Cork *Quercus suber*	1.5 billion	Mediterranean region	Ciesla (2002)
Essential oils Various species	1 billion	Global	Iqbal (1995)
Natural honey	300 million	Global	Iqbal (1995)
Matsutake mushrooms *Tricholoma* spp.	169 million (exports by various countries to Japan, 1996)	Global	Boa (2004)
Pygeum *Prunus africana*	150 million (annual market value)	Africa	Cunningham et al (1997)
Gum arabic *Acacia* spp.	53 million (1990 – Sudan exports only)	Africa, primarily Sudan and Nigeria	Lintu (1995)
Brazil nuts *Bertholletia excelsa*	50 million (export value)	Brazil, Bolivia, Peru	Lintu (1995)
Wild American ginseng *Panax quinquefolius*	25.7 million (1995 US exports)	USA	US Department of Commerce (1995)
Mastic gum *Pistacia lentiscus*	14.4 million	Greece	Moussouris and Regato (2002)

1996; Lange and Schippmann, 1997; Lange, 1998; Kathe et al, 2003b; Lange, 2006; and Bürgener, 2007. Below we discuss how laws and policies affect NTFP harvesters, producers and manufacturers as resources move from forest to local market or store shelf. We conclude with a section on global trade that describes treaties, tariffs and import/export policies applying to internationally traded NTFPs.

LOCAL AND REGIONAL NTFP TRADE

Access to resources

Commerce begins with the procurement of natural resources. Laws and policies governing access to NTFPs are not NTFP trade policies per se, yet they strongly influence NTFP trade. Regulations prescribing the way in which natural resources should be accessed and used often contain provisions describing approved harvest methods, the maximum amount of material to be harvested, the location of harvests and procedures for obtaining access. Access and harvest regulations directly affect gatherers or producers at the beginning of NTFP supply chains and therefore must be considered

Table 14.2 *The 12 leading countries for imports and exports of the commodity group 'pharmaceutical plants' (SITC.3: 292.4 = commodity group HS 1211), average annual trade volumes, 1991–2003*

Country of import	Quantity (tonnes)	Value (US$)	Country of export	Quantity (tonnes)	Value (US$)
Hong Kong	59,950	263,484,200	China	150,600	266,038,500
USA	51,200	139,379,500	Hong Kong	55,000	201,021,200
Japan	46,450	131,031,500	India	40,400	61,665,500
Germany	44,750	104,457,200	Mexico	37,600	14,257,500
Rep. Korea	33,500	49,889,200	Germany	15,100	68,243,200
France	21,800	51,975,000	USA	13,050	104,572,000
China	15,500	41,602,800	Egypt	11,800	13,476,000
Italy	11,950	43,006,600	Bulgaria	10,300	14,355,500
Pakistan	10,650	9,813,800	Chile	9,850	26,352,000
Spain	9,850	27,648,300	Morocco	8,500	13,685,400
UK	7,950	29,551,000	Albania	8,050	11,693,300
Malaysia	7,050	38,685,400	Singapore	7,950	52,620,700
Total	320,550	930,524,400	Total	368,100	847,980,800

Source: Lange (2006), based on COMTRADE database figures.

in any treatment of NTFP trade policies. There are two primary forms of NTFP access and harvesting oversight: governmental, or statutory, and local, including customary law.

Statutory/governmental oversight

States often distinguish between commercial and subsistence harvesting of NTFPs, creating strict rules for the former and lax rules for the latter, even in some cases turning a blind eye to it (see Dyke and Emery, Chapter 5, Richards and Saastamoinen, Chapter 12). Statutory control can take a variety of forms. In some cases, states nationalize trade in important NTFPs by setting prices or harvest quotas, licensing dealers and collecting revenue through fees and taxation. Examples include tendu leaves (*Diospyros melanoxylon*) in India (see Lélé et al, Chapter 3), medicinal plants in Bulgaria (Kathe et al, 2003b) and rattan (see Arquiza et al, Chapter 6). Governments may also control access to state lands, including leasing collection rights to private companies for the harvest of NTFPs, as in the case of Brazil nuts (*Bertholletia excelsa*) in Amazonia (Ortiz, 2002; Pacheco, Chapter 1). Another method of state control involves licensing gatherers, a common approach to regulating the harvest of many wild species worldwide, particularly mushrooms (Boa, 2004, p49).

These forms of controlling access are not mutually exclusive and governments often use a combination of all three approaches. For example, in Washington State in the USA, NTFP harvesters require a state permit to harvest, transport and sell NTFPs, yet if they wish to harvest NTFPs from federal lands, they require an additional permit (see McLain and Lynch, Chapter 11). State control of NTFP trade is most rigid in nationalized schemes and leasing arrangements, and less controlled in the case of licensing systems, mainly owing to the costs of enforcement, the lack of staff to provide oversight and the diffuse nature of wild harvesting. To compensate, some national agencies have resorted to the use of disciplinary power and the threat of surveillance to ensure that gatherers comply with NTFP harvest and trade regulations (McLain, 2000).

Local and customary oversight

In many areas of the globe, access to NTFP resources is still determined by family, clan, tribe or village ties. Under common property resource systems, access to NTFPs is part of a larger bundle of resource rights and obligations that are determined by local communities. This is also the case in countries with strong central governments such as China, where village councils, with the consent of higher officials, enact NTFP harvest codes of conduct that effectively exclude non-members of their villages from access to locally controlled forests (see Menzies and Li, Chapter 10). Likewise in Senegal, charcoal harvesters are obliged to consult with village chiefs before being granted access to local forests (Ribot, 1998).

In some countries, the devolution of control over forests to local groups has resulted in greater local control over resource access. In Nepal, for example, the government has ceded control over the management, use and sale of NTFPs to local forest user groups, who may in turn create their own NTFP harvest codes and exclude outsiders from using forest resources (Subedi, 1999). In southern Africa, customary – rather than statutory – laws are generally followed and enforced for the use and protection of the marula tree, *Sclerocarya birrea* (Wynberg and Laird, 2007). When land tenure and resource rights are secure, customary laws are strong, local capacity exists to manage the resource base, and commercial pressures on species are not overwhelming, local, customary laws often provide effective access and resource management oversight (Wynberg and Laird, 2007).

Disjuncture with natural resource management policies

Laws and policies regulating the harvest, management and trade of NTFPs are often incompatible with government natural resource laws. For example, most statutory forest policies promote timber management, ecosystem services and ecotourism above NTFP production. Protected areas' management guidelines generally discourage or forbid NTFP harvesting, often causing conflict with local communities that rely on these resources for subsistence and, in some cases, trade (Baird and Dearden, 2003; Jaireth and Smyth, 2003; Dowie, 2005). The cultural and economic importance of forest resources to local communities is typically ignored, and even when conservation law and policy are directed at specific resources, they often backfire. In Brazil, for example, laws forbidding the felling of Brazil nut trees failed to halt deforestation for cattle pasture. Developers simply cleared land of all trees except for the Brazil nut trees, which subsequently died, resulting in landscapes of 'kilometer after kilometer of lone, white, bone-looking, dead Brazil nut trees standing in solitude amid great extension of decaying pasture' (Ortiz, 2002, p69).

Transport and local trade regulations

As NTFPs move from forest to consumer they are subject to a variety of policies and laws. Some countries impose hauling fees for the transport of NTFPs. Middlemen and vendors of NTFPs require a variety of governmental permits, whether they are local permits to operate a street stall, sales permits or retailer licences. In Cameroon (see Laird et al, Chapter 2; Ndoye and Awono, Case Study B; Sunderland et al, Case Study C), the Philippines (Arquiza et al, Chapter 6) and other countries, bribes and other forms of corruption that are manifested along local trade routes have enormous knock-on effects, not only on the livelihoods of harvesters and traders, but also on the quantity of material harvested and transported.

Manufacturing and health regulations

NTFPs destined for processing and manufacturing plants must comply with a suite of regulations including labour laws, tax laws, manufacturing standards and, perhaps most onerous of all, health and safety regulations. As concerns about food safety increase, producers of edible NTFPs may be required to demonstrate more sophisticated documentation procedures for tracing batches of products back to packing facilities or to individual gatherers.

Iqbal (1993) has noted that food and safety legislation is the most formidable and usual obstacle to trade in NTFPs. However, relatively little scholarly attention has been given to the issue. Standards are typically created by governments to ensure the public health and safety of edible products, as well as by companies and trade associations to protect businesses from lawsuits or excessive governmental regulations. The aim of phytosanitary regulations is not only to prevent the contamination of edible products but also to protect the host country from potentially invasive species of animals, plants, fungi and micro-organisms.

Developed and some developing nations often require NTFP exporters to acquire phytosanitary certificates. The phytosanitary requirements of importing countries can be onerous and act as a trade constraint on NTFP producers from developing countries, yet are necessary to ensure biosafety as well as access to these markets (Bürgener, 2007). For example, EU standards for maximum allowable levels of aflatoxin in Brazil nut imports are so stringent that they seriously hamper trade in one of the flagship NTFP species from the Amazon Basin (Newing and Harrop, 2000). Dolan and Humphrey (2003) found that the quality and safety requirements of many UK supermarkets posed serious market access impediments to African producers of vegetables, particularly small-scale producers – a lesson that NTFP producers aiming to supply UK grocers should heed. As interest in good agricultural practices (which place a great emphasis on the safe handling of materials) and good manufacturing processes in the medicinal herb trade grows (Pierce and Laird, 2003), policy-makers will need to take extra steps to make sure that such policies do not exclude producers or products, rearrange commodity chains or exacerbate unequal power dynamics in trade.

Health and sanitation regulations can be arduous for NTFP gatherers in developed nations as well. In March 2005, the Los Angeles County Department of Health Services banned the sale of wild mushrooms at local farmers' markets, citing public safety concerns over the sale of wild harvested species not subjected to the formal state food inspection apparatus (Brown, 2005). Current US Department of Health and Human Services (2004, 3–201.16) policy on wild mushroom sales to the public states that 'mushroom species picked in the wild shall be obtained from sources where each mushroom is individually inspected and found to be safe by an approved mushroom identification expert'. Unfortunately the government has yet to define who may qualify as an approved mushroom expert, leaving gatherers and restaurants wishing to sell wild mushrooms in a legal catch-22 situation. In an effort to forestall government interventions that may curtail the sale of wild edibles to the public, two Vermont gatherers have created their own 'Certification and Identification Form' to accompany each sale of wild products. The form includes a batch number for tracking, to be kept on file by the restaurant for 90 days, as well as safe handling instructions (Schapiro, 2008). Whether such local initiatives will prove satisfactory to state and federal health inspectors in the future remains to be seen.

National laws governing the sale of medicines can have tremendous impacts on the trade of plants. In the USA, most herbal medicines are sold as dietary supplements and are largely unregulated. Manufacturers of herbal medicines need not prove that their products are either safe or effective through clinical trials. This loophole fuels fads for herbal remedies that cause boom and bust cycles for particular plants. In the late 1990s, demand for griffonia (*Griffonia simplicifolia*), an appetite suppressant, skyrocketed in the USA. In the autumn of 1998, medical researchers raised questions about possible contaminants found in samples of griffonia. The American media reported the contamination fears widely and demand for the herb quickly fell off, leaving traders in West Africa with warehouses full of their product but no orders for it (Gbewonyo, 2002). Other herbs such as kava (*Piper methysticum*) and ephedra (*Ephedra sinica*) have undergone similar boom and bust cycles in the US herb market, experiencing rapid sales (due to the lack of government oversight) followed by market crashes spurred on by media enquiries into safety and efficacy.

INTERNATIONALLY TRADED NTFPS

NTFPs bound for international markets face additional legal requirements that govern the trade of species between countries. These include tariffs, taxes and licences, as well as customs and health and sanitation inspections. The documentation required is often substantial and governmental oversight is sometimes poorly coordinated, both of which add to the burden on smaller traders and companies. For example, Bürgener (2007) lists nine specific documents required of NTFP exports from Bolivia, including a commercial invoice, a packing list, a declaration of export, a document of transport, the registration of the company with the Department of Forestry, a certificate of origin and a sanitary certificate from the relevant government agency. Laird et al (Chapter 2) attribute the contraction in the trade of medicinal plant exports from Cameroon to excessive requirements for documents, corruption and taxes. In addition, some exported NTFPs may fall under the remit of international agreements and treaties such as the Convention on International Trade in Endangered Species of Wild Fauna and Flora (CITES), the Convention on Biological Diversity (CBD) and various voluntary labelling schemes (e.g. organic, fair trade and ecological certification).

Tariffs

A tariff is a tax or duty imposed upon imported or exported goods. In the case of import tariffs, a government generally levies little or no tariff on desirable goods that it cannot produce in-country and imposes stiffer tariffs on products that compete with in-country industries. Developed countries typically maintain low tariffs for NTFPs because such products are unavailable in-country and because the cost of overseas labour to procure and process NTFPs is significantly lower than in-country labour. Conversely, developing countries often impose higher tariffs on NTFPs because such imports compete with local jobs and industries. Iqbal (1995) reports that China and India both levy protectionist import tariffs on NTFPs, ranging from 12 to 65 per cent ad valorem for China and 30 to 60 per cent for India. Export tariffs are mainly used by developing countries to capture greater revenue for the state, but they are often counterproductive because they depress prices paid to local collectors and promote illicit trade (Iqbal, 1995; Laird et al, Chapter 2).

Harvest and export bans

Closely related to protectionist tariffs are state-imposed harvest bans and export bans. A number of countries have imposed harvest and export bans on NTFPs as a way to protect overexploited resources destined for overseas markets or to promote the value-added processing of exported products. In 1991, surging demand for *Prunus africana* from the international market prompted the government of Cameroon to impose a harvest ban on the species as a measure to protect the tree from overharvesting. Nevertheless, large quantities of *Prunus* bark were harvested and exported during that period, probably due to corruption. Bürgener (2007) concludes that it is unclear whether the ban was ineffective or whether it stimulated illicit trade. Inconsistent policies on wild *Hoodia* harvesting in Namibia and South Africa similarly led to a spike in trade of the species and facilitated illegal harvesting, with exports channelled through the port with the most permissive policies (Wynberg, Chapter 13). India similarly restricted the harvest and export of gum karaya (*Sterculia* spp.), which, according to Iqbal (1995), only served to spur the substitution of the NTFP by other products.

In 1979, Indonesia imposed an export ban on raw rattan canes in order to encourage more in-country value-added processing; this was later followed by a ban on semi-processed rattan in 1988 (Iqbal, 1993). According to Cahyat (1999), the export ban was largely regressive because it lowered prices paid to rattan harvesters and concentrated profits in the hands of well-capitalized elites in the rattan manufacturing sector. Indonesia's raw rattan export ban was lifted in 1998 after lobbying by the International Monetary Fund. Regulations continue to block the sale of raw rattan by smallholders, however, resulting in a thriving illegal trade (Cahyat, 1999).

Nepal has banned the export of seven wild plants in their unprocessed form, including the well-known medicinal *Rauwolfia serpentine* (Subedi, 1999). We do not know if these export bans have been effective in capturing additional revenue from value-adding processing.

Convention on International Trade in Endangered Species of Wild Fauna and Flora

CITES was introduced in the 1970s as a measure to protect wild species and their habitats from the effects of unregulated demand created by international trade. CITES contains three appendices. Species listed on Appendix I are considered to be threatened with extinction and are all but barred from trade. Appendix II includes species of concern: their harvest and export is permitted, but national governments must monitor trade and provide a CITES export permit with each shipment. Appendix III includes species for which a signatory nation considers cooperation with other countries necessary to prevent unsustainable or illegal trade (Bürgener, 2007).

CITES has a limited impact on NTFPs because most are traded and consumed at the local and regional level. Yet for some high-value, internationally traded species, the impact is great. Compliance with CITES may range from an outright ban on commercial harvesting to the local procurement of a harvesting licence and issuance of a special export permit. The impact of CITES on NTFP trade has been mixed and controversial.

Devil's claw (*Harpagophytum procumbens*) is a medicinal root from southern Africa that has been traded internationally for more than 50 years, with most material exported from Namibia to Germany. Although the trade has been erratic, there has, over the years, been a steady increase in export volumes. Harmful harvesting techniques are one of a growing number of concerns about the status of *Harpagophytum* populations that, combined with the escalation of international trade in the plant, culminated in 2000 in the tabling by Germany of a controversial proposal to list *H. procumbens* and its lookalike *H. zeyheri* on Appendix II of CITES (CITES, 2000). This caused much concern in Namibia, Botswana and South Africa (Kathe et al, 2003a).

One of the immediate effects of the proposal was a decline in market demand – and thus a loss of income for harvesters and exporters – but it also gave renewed impetus to domestication and cultivation efforts, stemming from the perceptions created of an endangered resource (Wynberg, 2006). Although Appendix II allows for 'controlled trade' in a species, rather than imposing an outright ban, successful listing undoubtedly affects trade in wild species, and thus the livelihoods of thousands of harvesters dependent on the resource for income. Lombard and du Plessis (2003) contend that the act of proposing the devil's claw listing on CITES discouraged buyers and investors because it implied a sustainability problem with the resource base. The proposed listing, Lombard and du Plessis (2003) believe, undermined efforts to support the sustainability of devil's claw harvesting and thus worked against the stated goals of CITES.

Prunus africana, an African medicinal bark, was listed on CITES in 1994 due to concerns about overharvesting. Cameroon, a major supplier, instituted a harvest quota and permit system but there were difficulties in enforcing the regulations, identifying *Prunus* material at import and export facilities (because it is traded in many forms, including dried bark, bark extract, powdered form and capsules) and encouraging

Prunus cultivation, while safeguarding against the overharvesting of wild sources (Cunningham et al, 1997). Evidence suggests that an illegal trade in the bark flourished despite the listing. However, this species has received more attention than most NTFPs in Cameroon, and a number of programmes promoting domestication and better management in the wild are having a positive effect in some areas.

A common problem among many CITES-listed species is a lack of harvester and trader involvement in the creation of harvesting standards and of trade and monitoring protocols. Proposals for new listings of species are driven by governments in consultation with national environmental agencies and conservation NGOs. Such consultations often ignore the importance of trade to the livelihoods of local people. The listing of the woodland herb goldenseal (*Hydrastis canadensis*) provoked resentment in the American Herbal Products Association and among goldenseal traders in the USA over what was perceived as a flawed listing process that lacked transparency and failed to give adequate attention to the input of major stakeholders. The CITES community needs to raise awareness about species listings among the public by stressing that a CITES designation is not equivalent to a suggested ban of products. Such information campaigns may be assisted by providing links with certification programmes or other trade mechanisms.

Convention on Biological Diversity

The CBD was signed at the 1992 Rio Earth Summit and, although it neglects issues relating to the governance of NTFPs, a number of provisions are included that may potentially impact NTFP trade. A forest biodiversity thematic programme specifically encompasses NTFPs, albeit peripherally, and five cross-cutting issues – access and benefit sharing, the global strategy for plant conservation, protected areas, traditional knowledge and the sustainable use of biodiversity – are also pertinent to NTFPs (see www.cbd.int).

National governments are the parties to the CBD, but environmental and social NGOs have input into the process. The private sector has had only sporadic involvement in the CBD process, however, and industries involved in the trade of NTFPs – including the botanical, personal care and cosmetic, food and beverage industries – have consistently remained outside the process. For the most part, these sectors are unaware of the new legal and ethical obligations raised by the CBD, particularly those that apply to the development of new products or ingredients from genetic resources – 'bioprospecting' – and the use of traditional knowledge (Ten Kate and Laird, 1999; Laird and Wynberg, 2008). As a result, some companies, such as the US company Pure World Botanicals, which patented pharmaceutical applications of the traditional edible and medicinal root of *Lepidium meyenii* (maca), found only on the Andean central sierra of Peru (Brinckmann, 2007), have been accused of 'biopiracy' (Laird and Wynberg, 2008).

The World Trade Organization

The World Trade Organization (WTO), created in 1995 to replace the Bretton Woods financial and trading regime, has set new rules for global trade and its liberalization. In part this has included the development and aggressive implementation of a suite of new global trade agreements, such as the Agreement on Agriculture and the Agreement on Trade-Related Aspects of Intellectual Property Rights (TRIPS). In some situations, the growing integration of the global economy has provided the opportunity for substantial economic and income growth through NTFP trade, but deregulation has also escalated logging and overharvesting of resources, with detrimental impacts on rural communities and forest habitats (Haque, 1999; Shanley et al, 2002). Trade liberalization may also have negative impacts on NTFP harvesters if prices for raw materials fall and they find themselves in competition with harvesters around the globe producing similar products.

One of the greatest potential impacts of the WTO on NTFP trade relates to TRIPS. Under TRIPS, a global regime has been created for intellectual property rights (IPRs) over biological resources. This has significant implications for member states, who are now obliged to implement minimum IPR standards and to allow patents and other forms of IPRs to enter the realm of agriculture, food production and health care. Many NTFPs are increasingly included in patents and other IPRs, although there is great variation in the way that different countries approach issues of intellectual property. This variability in policy has an effect on national access to health care. Brazilians have relatively good access to medicines because the state made the patenting of pharmaceutical knowledge difficult, facilitated the rights of third parties to use patented knowledge and encouraged the sale of generic drugs. By contrast, Mexico issues lengthy patents for pharmaceuticals and discourages the use of patented knowledge by third parties, thereby hindering access to low-cost, widely available pharmaceuticals (Shadlen, 2007). For further information on TRIPS-related aspects of the WTO and its impact on medicinals, see Dutfield, 2000; WHO, 2001; Dutfield, 2003; Dutfield, 2004; and Finger and Schuler, 2004.

The WTO discourages countries from adopting protectionist measures such as high tariffs. NTFP-producing countries can file complaints of protectionism against nations that have traditionally imposed high tariffs on NTFPs, thereby exposing in-country producers to greater competition from abroad. Product labelling and geographic indications are one potential mechanism to combat overseas competition, because they promote NTFPs produced in specific locales.

Voluntary certification schemes

Voluntary certification schemes are achieving a large market presence in developed nations, particularly in Europe. The Organic Monitor (2008) reports that global sales of organically labelled food and drink reached US$40 billion in 2007, while sales of fair trade products rose by 47 per cent in 2007 and were worth approximately €2.3 billion. As markets for labelled products grow, producers will feel increasing pressure to adhere to certification schemes in order to obtain or maintain market access.

A number of certification and labelling systems potentially apply to NTFPs, including organic certification, fair trade certification and ecological certification. Organic certification focuses primarily on the species level and guarantees that products come from production systems that use little or no pesticide. Fair trade's strength lies in assuring adequate wages and safe working conditions. Ecological certification looks at the sustainability of the ecosystem from which products are derived. Many voluntary certification labels provide a mix of social and environmental standards.

Certified products occupy a niche market in global trade and few NTFPs have been certified to date, so the benefits of labelling NTFPs are unclear. However, the introduction of organic and fair trade certification for products such as rooibos teas has generated distinct and substantial benefits for producer communities (Wynberg, 2006). Certification is a challenging proposition for NTFP producers because it entails a high level of organization, detailed management planning and record-keeping, product-tracing procedures and marketing expertise (Shanley et al, 2002). Certification can also be costly, so producers must carefully weigh the pros and cons of acquiring a label. To date, the most well-known scheme in the international marketplace is the organic label. Newly developed labels for NTFPs include FairWild certification (www.fairwild.org)

Box 14.1 The importance of tracking and statistics for NTFP policy

Many people around the globe depend upon NTFPs for subsistence and income. Dransfield and Manokaran (1994) estimate that 0.7 billion people are involved in the rattan trade alone. Yet the consumption and trade of many NTFPs are rarely reflected in national statistics, such as gross national product, because they largely occur at the local level. If true cost accounting (replacement costs, contingent valuation) were applied to the collection, consumption and trade of NTFPs at the local level, the totals would probably dwarf those of internationally traded NTFPs. NTFP commerce is thus akin to an iceberg. Internationally traded products are highly visible and receive a great deal of attention, yet only comprise a fraction of the total trade in NTFPs. Local and regional markets for NTFPs lie submerged from official view and are difficult to discern, yet make up the bulk of commerce in non-timber forest products globally.

It is challenging to create effective NTFP trade policy and law without solid statistics. Yet who should track NTFP trade statistics? Vantomme (2003, p160) opines that 'monitoring of the resources and evaluation of the economic value of the products for the entire variety of NWFPs in a given country is neither feasible nor desirable'. Rather, Vantomme calls upon governments to choose NTFPs of national relevance (e.g. export products or widely used products) and create better monitoring and product classification systems for their tracking. Such measures would not only improve national accounting of NTFPs in trade, but also provide the data required for the creation and implementation of NTFP policies.

Industries are often more capable than governments when it comes to monitoring NTFPs in commerce. The American Herbal Products Association, a US trade group for herbal products companies, has, since 1997, implemented a survey of raw material suppliers to determine the amounts of select, wild harvested North American plants in trade (AHPA, 2003). A number of problems exist with the AHPA tonnage survey, including the survey response rate, company self-reporting and the fact that data sources are limited to the AHPA membership. However, market data on wild harvested plants, whether compiled by governments or industry, can identify wide swings in market demand, prompting policy or legal interventions to further monitor (or, if warranted, protect) ecologically vulnerable species.

and, for medicinal plants, the International Standard for Sustainable Wild Collection of Medicinal and Aromatic Plants (ISSC-MAP: www.floraweb.de/map-pro/). The Union for Ethical BioTrade (www.ethicalbiotrade.org), a consortium of small and medium-sized enterprises from developing nations, has developed a framework for the ethical and sustainable sourcing of products that addresses conservation, benefit sharing, labour practices and human rights. For further information on NTFP certification, see Nelson et al, 2002; Shanley et al, 2002; Pierce and Laird, 2003; Jain, 2004; Dudley and Stolton, 2006; Maciel, 2007; and Shanley et al, forthcoming. Another interesting avenue for NTFP labelling involves the use of designation of origin, geographic indication or locally developed labels (Parrott et al, 2002; Hayes et al, 2004; Illsley et al, Chapter 8).

CONCLUSION

Policy-makers generally ignore NTFPs because they are a diffuse class of products characterized by economic instability and low profit margins (Homma, 1992). In cases where countries create NTFP trade laws and policies, it is often with the intention of capturing revenue for state coffers. Such laws are often crude and ineffective because governments lack the resources and institutional capacity to implement them effectively (Wynberg and Laird, 2007).

NTFP harvesters are affected most by laws and policies regarding resource access and resource harvesting, including approved methods and quotas. As NTFPs move from harvester to point of sale, and in particular as they are processed and packaged for sale, they are subject to taxation as well as transport, health and labour laws. NTFP researchers and harvesters often complain of the mark-up that occurs between the harvest and final sale of a product as a result of these and other factors (Arquiza et al, Chapter 6). It is true that many harvesters receive a pittance for their goods, but critics often overlook the onerous costs that are borne by middlemen, processors and exporters of NTFPs. NTFP traders incur a great deal of expense in complying with basic laws of commerce as well as the costs of dealing with products that, in many cases, are highly perishable. Government bureaucracies and corruption are additional, significant hurdles to trade in NTFPs.

Governments have a variable track record in creating and enforcing NTFP trade policies. National efforts to protect species from overharvesting often fail to consider the impact of legislation on harvester communities and sometimes result in illegal trade. International law and policy, such as CITES, the CBD and the WTO, and voluntary certification, have also alienated harvesters through the lack of consultation or top-down consumer-driven prescriptions, and thus have not always achieved desirable conservation or trade goals.

For many of the world's NTFP harvesters who depend on wild species for subsistence and income, policy and law are viewed – rightly, in many cases – with suspicion or dread. There are few examples of government policies or laws that specifically aim to support local gatherers (see Kainer et al, 2003, for an exception regarding Brazil's Chico Mendes Law) or, graver yet, support the safety net functions of forests for subsistence users. As long as government intervention is inconsistent and reactive, which seems likely to remain the case for the foreseeable future, NTFP harvesters and traders will have to negotiate numerous layers of local, state and federal laws and policies.

REFERENCES

AHPA (2003) *Tonnage Survey of North American Wild-Harvested Plants, 2000–2001,* American Herbal Products Association, Silver Spring, MD

Baird, I. G. and Dearden, P. (2003) 'Biodiversity conservation and resource tenure regimes: A case study from northeastern Cambodia', *Environmental Management,* vol 32, no 5, pp541–550

Boa, E. (2004) *Wild Edible Fungi: A Global Overview of Their Use and Importance to People,* Non-Wood Forest Products 17, FAO, Rome

Brinckmann, J. (2007) 'Peruvian Maca and allegations of biopiracy', *Herbalgram,* American Botanical Council, vol 75, nos 8–9, pp44–53

Brown, C. (2005) 'Regulating the wild mushroom; L.A. County halts sales at farmers markets. Could restaurants and supermarkets be next?', *Los Angeles Times,* 16 March, F1

Bürgener, M. (2007) *Trade Measures – Tools to Promote the Sustainable Use of NWFP? An Assessment of Trade-Related Instruments Influencing the International Trade in Non-Wood Forest Products and Associated Management and Livelihood Strategies,* Non-Wood Forest Products Working Document no 6, FAO, Rome

Cahyat, A. (1999) 'Improving the rattan trading system in Kalimantan', *Voices from the Forest* (Bulletin of the NTFP Exchange Programme Southeast Asia), vol 1, February, pp14–16

Cameron, G., Pendry, S., Allan, C. and Wu, J. (2004) *Traditional Asian Medicine Identification Guide for Law Enforcers,* TRAFFIC International, London

Ciesla, W. M. (2002) *Non-Wood Forest Products from Temperate Broad-Leaved Trees,* Non-Wood Forest Products 15, FAO, Rome

CITES (2000) Inclusion of *Harpagophytum procumbens* in Appendix II in accordance with Article II 2(a), Prop. 11.60. Available on www.cites.org/eng/cop/11/prop/60.pdf, accessed 7 April 2006

Cunningham, M., Cunningham, A. B. and Schippmann, U. (1997) *Trade in Prunus Africana and the Implementation of CITES,* German Federal Agency for Nature Conservation, Bonn

Dewees, P. and Scherr, S. (1996) *Policies and Markets for Non-Timber Tree Products,* Environment and Production Technology Division Discussion Paper No 16, International Food Policy Research Institute, Washington, DC

Dolan, C. and Humphrey, J. (2003) 'Governance and trade in fresh vegetables: The impact of UK supermarkets on the African horticulture industry', *Journal of Development Studies,* vol 37, no 2, pp147–176

Dowie, M. (2005) 'Conservation refugees: When protecting nature means kicking people out', *Orion,* vol 24, no 6, pp16–27

Dransfield, J. and Manokaran, N. (eds) (1994) *Plant Resources of SE Asia – Rattans,* PROSEA, Bogor, Indonesia

Dudley, N. and Stolton, S. (2006) *Edible Non-Timber Forest Products: Harmonizing FSC and IFOAM Certification,* ISEAL, Oxford

Dutfield, G. (2000) *Intellectual Property Rights, Trade and Biodiversity,* Earthscan, London

Dutfield, G. (2003) *Protecting Traditional Knowledge and Folklore: A Review of Progress in Diplomacy and Policy Formulation*, IPRs and Sustainable Development Issue Paper no 1, UNCTAD-ICTSD, Geneva

Dutfield, G. (2004) *Intellectual Property, Biogenetic Resources and Traditional Knowledge*, Earthscan, London

Finger, M. J. and Schuler, P. (eds) (2004) *Poor People's Knowledge: Promoting Intellectual Property in Developing Countries*, Oxford University Press, Oxford

Gbewonyo, K. (2002) 'Griffonia (*Griffonia simplicifolia*)', in P. Shanley, A. Pierce, S. Laird and A. Guillen (eds) *Tapping the Green Market: Certification and Management of Non-Timber Forest Products*, Earthscan, London, pp200–207

Haque, M.S. (1999) 'The fate of sustainable development under neo-liberal regimes in developing countries', *International Political Science Review*, vol 20, no 2, pp197–218

Hayes, D., Lence, S. and Stoppa, A. (2004) 'Farmer-owned brands?', *Agribusiness*, vol 20, no 3, pp269–285

Homma, A. K. O. (1992) 'The dynamics of extraction in Amazonia: A historical perspective', in D. Nepstad and S. Schwartzman (eds) *Non-Timber Products from Tropical Forests*, The New York Botanical Garden, Bronx, NY, pp23–32

Hunter, I. R. (2003) 'Bamboo resources, uses and trade: The future?', *Journal of Bamboo and Rattan*, vol 2, no 4, pp319–326

Iqbal, M. (1993) *International Trade in Non-Wood Forest Products: An Overview*, FO: Misc/93/11 Working Paper, FAO, Rome

Iqbal, M. (1995) *Trade Restrictions Affecting International Trade in Non-Wood Forest Products*, Non-Wood Forest Products no 8, FAO, Rome

ITTO (1997) 'Bamboo and rattan: Resources for the 21st century?', *Tropical Forest Update*, vol 7, no 4, p13, International Tropical Timber Organization, Yokohama

Jain, P. (2004) *Certifying Certification: Can Certification Secure a Sustainable Future for Medicinal Plants, Harvesters and Consumers in India?* TRAFFIC International, Cambridge

Jaireth, H. and Smyth, D. (eds) (2003) *Innovative Governance: Indigenous Peoples, Local Communities and Protected Areas*, Ane Books, New Delhi

Kainer, K. A., Schmink, M., Pinheiro Leite, A. C. and Silva Fadell, M. J. da (2003) 'Experiments in forest-based development in western Amazonia', *Society and Natural Resources*, vol 16, pp869–886

Kathe, W., Barsch, F. and Honnef, S. (2003a) *Trade in Devil's Claw (*Harpagophytum procumbens*) in Germany – Status, Trends and Certification*, FAO, Rome

Kathe, W., Honnef, S. and Heym, A. (2003b) *Medicinal and Aromatic Plants in Albania, Bosnia-Herzegovina, Bulgaria, Croatia and Romania*, BfN – Skripten 91, Bundesamt für Naturschutz, Bonn

Laird, S. and Wynberg, R. (2008) *Access and benefit sharing in practice: trends in partnerships across sectors*, Volumes I, II and III. CBD Technical Series 38, Secretariat of the Convention on Biological Diversity, Montreal

Lange, D. (1998) *Europe's Medicinal and Aromatic Plants: Their Use, Trade and Conservation*, TRAFFIC International, Cambridge

Lange, D. (2006) 'International trade in medicinal and aromatic plants', in R. Bogers, L. Craker and D. Lange (eds) *Medicinal and Aromatic Plants*, Wageningen UR Frontis Series, vol 17, Springer-Verlag, Dordrecht, Holland, pp155–170

Lange, D. and Schippmann, U. (1997) *Trade Survey of Medicinal Plants in Germany: A Contribution to International Plant Species Conservation*, Bundesamt für Naturschutz, Bonn

Lintu, L. (1995) 'Trade and marketing of non-wood forest products', in *Report of the International Expert Consultation on Non-Wood Forest Products, Yogyakarta, Indonesia, 17–27 January*, FAO, Rome, pp195–222

Lombard, C. and du Plessis, P. (2003) 'The impact of the proposal to list devil's claw on Appendix II of CITES', in S. Oldfield (ed) *The Trade in Wildlife: Regulation for Conservation*, Earthscan, London

Maciel, R. (2007) 'Certificação ambiental: uma estratégia para a conservação da floresta Amazônica', PhD dissertation, Universidade Estadual de Campinas (Unicamp), Campinas, Brazil

McLain, R. J. (2000) 'Controlling the forest understory: Wild mushroom politics in central Oregon', PhD dissertation, College of Forest Resources, University of Washington, Seattle, WA

Moussouris, Y. and Regato, P. (2002) 'Mastic gum (*Pistacia lentiscus*), cork oak (*Quercus suber*), argan (*Argania spinosa*), pine nut (*Pinus pinea*), pine resin (various spp.) and chestnut (*Castanea sativa*)', in P. Shanley, A. Pierce, S. Laird and A. Guillén (eds) *Tapping the Green Market: Certification and Management of Non-Timber Forest Products*, Earthscan, London, pp183–199

Nelson, V., Tallontire, A. and Collinson, C. (2002) 'Assessing the benefits of ethical trade schemes for forest dependent people: Comparative experience from Peru and Ecuador', *International Forestry Review*, vol 4, no 2, pp99–109

Newing, H. and Harrop, S. (2000) 'European health regulations and Brazil nuts: Implications for biodiversity conservation and sustainable rural livelihoods in the Amazon', *Journal of International Wildlife Law and Policy*, vol 3, no 2, pp109–124

Organic Monitor (2008) *European Supermarkets Going Bananas over Fairtrade*, www.organicmonitor.com/r1106.htm, accessed 30 June 2008

Ortiz, E. (2002) 'Brazil nut (*Bertholletia excelsa*)', in P. Shanley, A. Pierce, S. Laird and A. Guillén, *Tapping the Green Market: Certification and Management of Non-Timber Forest Products*, Earthscan, London, pp61–74

Parrott, N., Wilson, N. and Murdoch, J. (2002) 'Spatializing quality: Regional protection and the alternative geography of food', *European Urban and Regional Studies*, vol 9, no 3, pp241–261

Pierce, A. and Laird, S. (2003) 'In search of comprehensive standards for non-timber forest products in the botanicals trade', *International Forestry Review*, vol 5, no 2, pp138–147

Ribot, J. (1998) 'Theorizing access: Forest profits along Senegal's charcoal commodity chain', *Development and Change*, vol 29, pp307–341

Schapiro, D. (2008) 'Foraging: Into the wild', *Edible Green Mountains*, vol 1, no 2, pp22–25

Shadlen, K. (2007) *The Politics of Patents and Drugs in Brazil and Mexico: The Industrial Bases of Health Activism*, Global Development and Environment Institute Working Paper No 07–05, Tufts University, Medford, MA

Shanley, P., Pierce, A., Laird, S. and Guillén, A. (eds) (2002) *Tapping the Green Market: Certification and Management of Non-Timber Forest Products*, Earthscan, London

Shanley, P., Pierce, A., Laird, S. and Robinson, D. (forthcoming) *Beyond Timber: Certification of Non-Timber Forest Products*, CIFOR, Bogor, Indonesia

Subedi, B. (1999) 'Non-timber forest products sub-sector in Nepal: Opportunities and challenges for linking the business with biodiversity conservation', paper presented at the workshop on Natural Resources Management for Enterprise Development in the Himalayas, 19–21 August, Nainital, India

Ten Kate, K. and Laird, S. A. (1999) *The Commercial Use of Biodiversity: Access to Genetic Resources and Benefit-Sharing*, Earthscan, London

US Department of Commerce (1995) *Merchandise Trade – Imports and Exports by Commodity: Ginseng Roots, Cultivated, and Ginseng Roots, Wild*, CD-ROM, National Trade Data Bank – the Export Connection Program, US Trade Information, Washington, DC

US Department of Health and Human Services (2004) *Supplement to the 2001 Food Code*, US Department of Health and Human Services, Washington, DC, www.cfsan.fda.gov/~dms/fc01–3.html, accessed 25 February 2008

Vantomme, P. (2003) 'Compiling statistics on non-wood forest products as policy and decision-making tools at the national level', *International Forestry Review*, vol 5, no 2, pp156–160

WHO (2001) *Report on the Inter-Regional Workshop on Intellectual Property Rights in the Context of Traditional Medicine, Bangkok, Thailand, 6–8 December 2000*, World Health Organization, Geneva

Wynberg, R. (2006) 'Identifying Pro-Poor, Best Practice Models of Commercialisation of Southern African Non-Timber Forest Products', PhD thesis, University of Strathclyde, Glasgow

Wynberg, R. P. and Laird, S. A. (2007) 'Less is often more: Governance of a non-timber forest product, marula (*Sclerocarya birrea* subsp. *caffra*) in Southern Africa', *International Forestry Review*, vol 9, no 1, pp475–490

Chapter 15

The State of NTFP Policy and Law

Sarah A. Laird, Rachel P. Wynberg and Rebecca J. McLain

INTRODUCTION

The case studies presented in this volume indicate that despite wide variations in cultural, economic and political conditions, experiences with NTFP law and policy are remarkably similar around the world, and are characterized by common regulatory features. This finding applies to both developed and developing countries, and includes regions that still have strong traditional and subsistence use of NTFPs and those that may have reduced their dependence on NTFPs, but have recently 'rediscovered' natural products.

Shared characteristics include a tendency to draft inconsistent and poorly coordinated laws in reactive or opportunistic ways. These laws rarely reflect a strategy and often grow from limited understanding by government of the complex ecological, economic and cultural realities of NTFP use, management and trade. Other commonalities are insufficient consultation with harvesters and producers, and under-resourced and ineffective implementation of those laws which do exist. The following is a discussion and synthesis of these and other experiences reported in the preceding chapters.

WHY AND HOW NTFP LAWS AND POLICIES ARE DEVELOPED

NTFP policies and laws are usually a complex, and often confusing, mix of measures developed over time, with poor coherence or coordination. They rarely resemble an overall policy 'framework'. Many policies are enacted as ad hoc responses to a crisis (e.g. perceived overexploitation of a species) or an overly optimistic view of potential tax revenue should 'informal' activities be made more formal. Rarely does regulatory activity follow from a careful and systematic assessment of the range of opportunities and threats associated with species, ecosystems and livelihoods.

As almost all the cases in this volume indicate, a strategic approach to regulating the NTFP sector as a whole is uncommon. A comprehensive policy approach is sometimes developed for individual species with high commercial demand, but this is not always the case. For example, brazil nuts – a pillar of the regional economy – are regulated in Bolivia under a legal system described as 'piecemeal or peripheral' (Chapter 1) and the valuable southern African species *Hoodia* is overseen by a multitude of laws in an ad hoc manner (Chapter 13).

Reactive policy-making

The tendency for NTFP laws to be drafted in response to a real or perceived overharvesting crisis is widespread, especially when use of a species changes from local trade and subsistence to large-scale commercial trade. Reactive policy-making is often an inevitability associated with the NTFP commercial production cycle. As Homma (1992) describes in his widely cited model, this cycle is characterized by four phases. An expansion phase, represented by growth in extraction of the resource, is followed by a period of stabilization, where equilibrium is reached between the supply and demand for the product. Typically, the maximum capacity of wild populations to supply raw material is then reached. If demand continues to increase and supply falls short, prices begin to rise and pressure on wild populations increases. At this point NTFP policies tend to be developed in order to protect the sector, stimulate sustainable production, or protect wild populations (e.g. palm hearts in Brazil, Case Study A; and *Hoodia* from southern Africa, Chapter 13).

A third phase involves shrinkage of the resource base which, combined with the increased cost of harvesting from ever more remote sources, leads to gradual failure of extraction. If technologies are available, prices are high and substitutes or alternative sources of supply are not available, domestication or cultivation begins to take place during the final stabilization phase. In some cases, substitution creates a collapse as seen with the once thriving trade in Amazonian rubber (*Hevea brasiliensis*) in the early 20th century (Chapter 1) and in Finland in the 1990s, when wild lingonberry crop failures shifted industrial demand towards sweeter and cheaper berries from southern and central Europe and towards cultivated cranberries and blueberries from North America (Chapter 12).

The processes of depletion, substitution and domestication vary across species and locations, and are part of a complex array of ecological, political, social and economic circumstances (Neumann and Hirsch, 2000). Alexiades and Shanley (2005) suggest that for many products Homma's (1992) model might be revised to incorporate repeated expansion–stabilisation–decline cycles, and that some production systems actually undergo de-intensification. They also emphasize that most NTFPs are part of multi-species production systems, all of which are dynamic, complex and difficult to represent in a single model.

Booms and busts in NTFP commercial cycles also result from consumer fads, scientific research that supports or undermines markets, and health concerns. In the botanical and herb industry, for example, griffonia (*Griffonia simplicifolia*), kava (*Piper methysticum*), ephedra (*Ephedra sinica*), and cat's claw (*Uncaria tomentosa*) are just a few examples of species that have experienced increased sales in recent decades, followed by market crashes after media reports raised concerns about safety and efficacy (Chapter 14; Alexiades, 2002; Nalvarte Armas and de Jong, 2005). Health concerns associated with raw material supplies in the food sector often trigger reactive policy responses, as in the case of aflatoxins found in Brazil nuts sold in Europe and North America (Chapter 1), with Chinese matsutake harvested in Yunnan and sold in Japan (Chapter 10), and with palm hearts in Brazil and Bolivia (Fantini et al, 2005; Stoian, 2005b).

Despite the risks associated with reactive and iterative NTFP policy-making, such interventions can also have strengths. The *Hoodia* case described in this volume (Chapter 13) demonstrates that such an approach may be necessary to cope with changing conditions, in this case market and trade fluctuations. *Hoodia*'s entry into the weight-control market in 2001 led to a surge in demand for raw material that required southern African governments to respond rapidly by introducing a stringent permit system and, in some cases, prohibiting wild harvesting. A few years later, an increase in the availability of cultivated material reduced pressure on wild populations, and governments responded with a less severe permitting system. The significant changes in *Hoodia* markets, availability and demand necessitated an iterative and flexible approach by government towards permitting and regulation, a situation that is likely to apply to other 'boom–bust' species.

Opportunistic policy-making

Government action is often triggered when politically powerful groups lobby for regulation in order to increase their control over NTFP production and trade. For example, the Rooibos Tea Control Scheme established by the apartheid state of South Africa in 1954 was promoted by and benefited the white farming elite, rather than the mostly 'coloured' farmers who had traditionally gathered rooibos tea from the wild. The scheme was a statutory, one-channel marketing system set up to regulate the production and marketing of indigenous rooibos (*Aspalathus linearis*) tea and to support the sector, including subsidies for affiliated producers, research and the provision of extension services (Hayes, 2000; Wynberg, 2006).

Governments are also quick to act when a species or set of products appears to show great economic promise, part of which they might capture through royalties, taxes or other means. In Cameroon, the government instituted new taxes on medicinal plants in the 1990s in response to a widespread belief that these NTFPs were 'green gold' (Chapter 2). In India, tendu (*Diospyros melanoxylon*) – which provides as much as 74 per cent of Orissa state's total earnings from forests – was nationalized in several states in the 1960s and 1970s due to its high value and the interest of government bodies in benefiting from its trade (Chapter 3). State intervention in the management of devil's claw (*Harpagophytum* spp.) in southern Africa grew alongside increased commercial extraction in the 1960s and 1970s and peaked in the late 1990s along with the trade (Wynberg, 2006).

Information requirements for drafting effective policies

A common problem with NTFP law and policy is limited understanding on the part of policy-makers about the products, people and activities they seek to regulate. Unlike timber or agricultural crops, NTFPs include a broad range of species with extremely different ecologies and cultural and livelihood roles, and equally diverse market chains, end products and consumers (Peters, 1996; Arnold and Ruiz-Perez, 1996; Shanley et al, 2002; Alexiades and Shanley, 2005). For many species there remain enormous gaps in our understanding, including those widely used such as Brazil nuts, devil's claw, and eru (*Gnetum* spp.).

Solid background information is critical to policy-formulation, however. For example, because NTFPs are a diverse group of species, with a wide range of ecological niches, policy-makers cannot assume that intensification of harvesting will have similar impacts in all cases. Marula (*Sclerocarya birrea*) is widespread and common, fruits abundantly and is planted in yards, retained in fields and is usually well managed in the southern African region. These circumstances suggest a resilience that does not require immediate government intervention, but rather calls for monitoring of populations in areas with heavy harvesting rates (Shackleton et al, 2003; Wynberg and Laird, 2007). *Intsia bijuga* in Fiji, on the other hand, is slow-growing, occurs in low densities, is scattered in distribution and does not disperse well – all characteristics that make it vulnerable to overharvesting. In addition, *Intsia bijuga* is experiencing commercial pressure from the tourist trade, new technology has increased harvesting rates, and cultural changes have eroded customary laws and beliefs that hold *Intsia bijuga* to be a sacred species. This combination of factors has led to a sustainability crisis that – unlike the case of marula – requires legislative and policy attention (Chapter 9).

In addition to ecological data, policy-makers must also have access to information on marketing and production chains, the history of NTFP harvest and trade, technological developments that impact harvesting rates and pressure on a resource, and an understanding of broader cultural values that might promote or undermine sustainability (Posey, 1999; Alexiades and Shanley, 2005). If the objectives of policy are as broad and complex as 'sustainability' and 'equity', the information required to draft measures to achieve these objectives will necessarily be complex too.

This said, how can governments adequately understand, and so regulate, the hundreds, and perhaps thousands, of species for which there is little scientific or other information? Information requirements for policy-making exist along a gradient, increasing alongside the need for policy intervention. It is unnecessary and undesirable to regulate most NTFP species; governments should focus their data collection efforts on heavily traded species, and those under threat.

Consultations associated with laws and policies

Consultations with stakeholders are probably the most important way to gather information and to set priorities and objectives for policy. However, in most countries NTFP harvesters and producers are drawn from the least powerful members of society and typically have little say in policy-making (Hecht et al, 1988; Shanley et al, 2002; Shackleton and Shackleton, 2004; Alexiades and Shanley, 2005; Wynberg and Laird, 2007). Because harvesters and producers often belong to marginalized groups and cannot (or sometimes choose not to) participate in organized political action, they are rarely consulted during policy design, and their needs seldom drive the policy-making process. Technical experts and even non-governmental organizations (NGOs) (which may not be representative of producers and harvesters, but can provide important assistance) often have more significant input into the design and drafting process than those directly involved in the harvest or trade of products. The consultations that do take place for NTFP law and policy are often with larger and more powerful business interests.

One reason for the limited involvement of harvesters in the policy process is the dearth of producer organizations or institutional vehicles through which their views and concerns can be expressed, and a lack of organizational capacity to do so. Even in recent decades, Brazil nut measures were drafted and passed in Bolivia without public consultation. It was only in the late 1990s that small Brazil nut producers finally forced their views into the public arena, in part by being better organized (Chapter 1). In the United States, Canada and the United Kingdom, some effort has recently gone into including harvesters, buyers and processors in proposed regulatory reforms, either through the formation of industry-specific task forces, as in the United Kingdom (Chapter 5) and Canada (Chapter 4) or through public hearings, as in the United States (Chapter 11).

In southern Africa, the non-profit trade association PhytoTrade Africa plays an important role in enabling the voice of marginalized producers to be heard. PhytoTrade Africa works to develop a natural products industry that enables poor rural communities to generate income through the sustainable use of indigenous plants. A core component of its work involves lobbying and advocacy to positively influence trade and policy regimes relating to natural products (Phytotrade Africa, 2006).

The few strategic exceptions

A few governments have developed NTFP law and policy in a more strategic manner. This includes undertaking research and building ecological, economic, social and cultural understanding of species, incorporating comprehensive consultations with stakeholders, and developing a strategy for the resulting legal framework.

In the past decade, for example, Namibia has taken a proactive and progressive approach towards NTFP policy and regulation, recognizing that these products provide vital income and livelihoods for communities in an environment characterized by extreme aridity and few economic opportunities (Bennett, 2006; Cole and Nakamhela, 2008; Nott and Wynberg, 2008; Chapter 13). Much of this has been done through the multi-stakeholder Namibian Indigenous Plant Task Team, which promotes collaborative approaches and effective regulation, and facilitates development of the local natural products industry (Nott and Wynberg, 2008).

Finland is also a notable exception to the rule of government neglect for NTFPs. The Finnish government has supported scientific research on wild berries for decades, including studies of their cultural and economic importance, as well as biological and ecological research (Kanga, 1999). At the same time, it has actively promoted berry and mushroom harvesting as an economic activity and cultural practice. Indeed, rather than discouraging harvesting as many countries have done, the government has developed programmes to promote harvesting and related industries. These include a berry crop forecasting system and income-tax relief favourable to harvesters, providing them with the information and incentives they need to participate more effectively in NTFP industries (Chapter 12).

THE POLICIES

Policies and laws that directly address NTFPs

A number of laws and policies directly address NTFPs, often to conserve or sustainably manage resources, and in some cases to improve rural livelihoods or promote broader economic growth in a region. These measures tend to focus on species in commercial trade, form part of national efforts to protect endangered or indigenous species, or regulate international trade under the Convention on International Trade in Endangered Species of Wild Fauna and Flora (CITES). The majority of measures directly addressing NTFPs, however, are found in natural resource law, in particular forestry laws. A range of other measures explicitly regulates aspects of NTFP trade and use, including quality control, safety and efficacy standards, transportation, taxation and trade (Chapter 14).

The inclusion of NTFPs in forestry laws of the 1990s

In most countries, forestry laws historically focused almost exclusively on timber resources and paid limited or no attention to NTFPs. Moreover, the subsistence and commercial value of NTFPs to local communities was totally disregarded when timber management plans were designed and logging operations undertaken. In recent decades, however, NTFPs were incorporated into forestry laws as a response to changing international policy trends. In many cases, this resulted from the direct pressure of international agencies, such as large conservation organizations and finance institutions, including the World Bank, to diversify forest management and make it more sustainable (Chapter 2). As a result, in the 1980s and 1990s, many countries integrated a wider range of objectives into forest policies, including forest health and biodiversity conservation, ecosystem functions and long-term sustainability, as well as broader economic values such as tourism, recreation and NTFPs.

However, initial efforts to address NTFPs in these new forestry laws were poorly formulated and rarely implemented. The scope and definition of the products covered remained unclear, and few specific actions were stipulated (e.g. Fiji Islands, 1992; Republic of Cameroon, 1994; República de Bolivia, 1996a). When actions were prescribed, they usually focused on permits, quotas (often set in arbitrary ways), management plans, and royalties or taxes – an approach lifted directly from the timber sector, and one that proved entirely inappropriate for the diverse, complex and less lucrative NTFP sector.

More usefully, some forestry laws of this generation included NTFPs in timber norms, requiring their consideration in management plans and logging operations in order to minimize negative impacts on locally valuable products. In many countries, the logging of high-value NTFP species for timber has proved their greatest threat. In Brazil in recent years, national and state governments have passed laws prohibiting the logging of high-value NTFP species (Table A.1, Case Study A), and in Bolivia prohibitions on felling Brazil nut trees arrived in 2004 as part of a decree addressing property conflicts (Chapter 1). But the track record for implementing such policies is often poor (e.g. Ortiz, 2002; Chapter 14).

In the past 10–15 years, a number of countries have begun to fine-tune forest policies passed in the 1990s to reflect the socioeconomic, ecological and cultural realities of NTFP use. This has resulted in a number of specific improvements in the ways these products are regulated, including re-thinking the use of costly and complex inventories and management plans for NTFPs, and revising quota and permitting systems (Chapter 1; Chapter 2; Chapter 9; Case Study A). There is still a long way to go, and NTFPs continue to have low priority in most forestry departments, but the trend in several countries is towards greater understanding and better-elaborated regulatory frameworks.

Quality control, safety and efficacy

Quality control and proof of safety and efficacy are increasingly important in developed country markets. This means that NTFP producers may be required to institute sophisticated procedures for tracking materials that end up as botanicals, personal care and cosmetic products, and food and beverages. Food safety legislation has often proved a formidable obstacle to international trade of NTFPs (Chapter 14; Iqbal, 1993; Brown, 2005; Bürgener, 2007). However, governments tend to act quickly when these obstacles arise; unlike environmental and social justice concerns, health concerns often get their attention, and pressure from influential commercial players involved in the trade can be great. For example, in the 1990s when the EU and the USA set maximum acceptable levels of aflatoxins that threatened the Brazil nut trade, the Bolivian government jumped into action, passing a series of measures that created norms for Brazil nut classification, sanitation practices and aflatoxin sampling, drawing upon the Food and Agriculture Organization's Codex Alimentarius (Soldán, 2003, in Chapter 1). These steps allowed the Bolivian government to maintain access to international markets for Brazil nuts.

The exponential increase in trade of *Hoodia* in the past ten years has been fed in part by demand for dozens of non-patented dietary supplements, many of dubious authenticity, containing unsubstantiated quantities of *Hoodia*, and making unfounded claims (Stafford, 2009). Concerns from the US Food and Drug Administration led regulators in South Africa, Namibia and Botswana to introduce permitting procedures to help track trade in the raw material across borders and support initiatives by local industries to monitor quality (Chapter 13).

Transportation

Transportation laws can have direct and indirect impacts on NTFPs. Most significant for all natural resources, including NTFPs, is the opening of previously remote forest areas following road building. More specific to the case of NTFPs is the use of transportation law to monitor trade. The State of Washington in the USA relies heavily on transportation permits as a mechanism for monitoring and tracking the harvesting of floral greens and other NTFPs; these permits also play an important role in identifying thefts of products from state and private land (Chapter 11). In Brazil, a 1993 regulation required a licence to transport any forest product. This included essential oils, medicinal plants and the seedlings, roots, bulbs, vines and leaves of native plants, many of which were not regulated in any other way. Because the law was so broad, and local harvesters and traders could not easily acquire the necessary licence, they could either not participate in commercial trade, or did so illegally. This measure was amended in 2006, in response to these problems (Case Study A).

Taxation, including 'unofficial taxation'

Governments sometimes tax the NTFP trade in order to gain revenue from what is perceived as a lucrative business, but this often negatively impacts the sector. In Cameroon, new taxes instituted in the 1990s on the medicinal plant export business resulted in the near collapse of that sector, and a blossoming of bureaucracy and opportunities for corruption (Chapter 2; Case Study B). In Bushbuckridge, South Africa, the government charges kiaat (*Pterocarpus angolensis* – African or wild teak) harvesters and craftsmen a fee per running metre of wood in order to promote responsible use of this valuable material. In reality, however, reports of harassment and corruption (e.g. government rangers taking wood or issuing incorrect receipts) are common. As a result, craftsmen and harvesters usually choose to bypass the system (Case Study D). Some governments, however, use tax structures as a way of providing incentives to the NTFP sector. In Finland, for example, in order to encourage and support harvesters, and to offer the sector a 'carrot', the government makes picking income exempt from tax (Chapter 12).

'Unofficial' or 'informal taxation' (i.e. bribery) is a very real cost of doing business in many countries. Bribes are tolerated, and even encouraged, by some governments, and they work like any other policy 'stick' to change behaviour. In a number of countries, roadblocks set up by government officials to 'control' the transport of goods from rural to urban areas, and check required documents, bleed profits from traders and have knock-on effects for harvesters (Case Study B; Case Study C; Chapter 6). In The Philippines, one study showed that unofficial payments, or 'SOPS" (standard operating procedures), significantly impact the already meagre NTFP livelihoods of indigenous peoples (Chapter 6).

Bribery can be a good indicator not only of problems with broader governance, but also with NTFP policies and laws. Bureaucratic and confusing NTFP measures can leave communities and government authorities unclear about proper procedures, providing openings for corruption (Chapter 2; Case Study B; Chapter 6). Inappropriate and burdensome measures can also make 'unofficial payments' preferable to following regulations. In The Philippines, harvesters and traders often find it more efficient and cheaper to pay a bribe, than navigate elaborate official management plan and licensing requirements (Chapter 6).

Policies and laws that indirectly impact NTFPs

In addition to laws that explicitly address NTFPs there are a myriad of measures that may not mention the term, and yet impact their use, management and trade as much as, or more than, those that do (Dewees and Scherr, 1996). The high impact of these measures is largely because the role of NTFPs in subsistence and local livelihoods is often poorly understood and rarely considered when drafting other measures. Laws tend to be drafted along sectoral lines that do not take into account other land uses, and the complex and interconnected nature of activities.

Laws and policies with an indirect impact on NTFPs include agricultural poli-cies, land tenure and resource rights, intellectual property, and labour law. In addi-tion, a range of natural resource laws have a significant impact on NTFPs, including the forestry laws discussed above, mining (Chapter 7) and protected area laws that discourage or forbid NTFP harvesting in core areas (e.g. Baird and Dearden, 2003; Jaireth and Smyth, 2003; Dowie, 2005).

Agricultural policies

Agricultural policies can impact NTFPs in a range of ways. They might discourage or promote farming practices that are linked to NTFP harvests and associated livelihoods. For example, in the 1990s an international policy movement identified swidden ('slash and burn') agriculture as a major cause of tropical deforestation. Although this was unproven and controversial, the impact of restricting practices associated with swidden agriculture was significant, including on NTFPs. In the case of the Batak in Palawan, these policy restrictions led to a surge in NTFP harvesting and trade to buy food to supplement low agricultural production (Chapter 7). Agricultural policies can also include subsidies and other incentives to cultivate NTFPs, with both positive and negative impacts on rural livelihoods and species. The cultivation of rooibos tea in South Africa, for example, is promoted by a regulatory framework that encourages the clearing of natural biodiversity for rooibos plantations, and discourages wild collection of this species (Wynberg, 2006).

Agricultural policies can also be a vehicle for land and resource rights reform, with significant consequences for NTFPs. For example, the 1996 Agrarian Reform Law (República de Bolivia, 1996b) in Bolivia initially appeared to have little relevance for the Brazil nut economy, but its impact was dramatic because it sought to resolve the complex and contradictory property rights system of the country (Chapter 1). Agri-cultural policies can also impact NTFPs through their effect on the supply of labour available to harvest products. In Finland, the loss of domestic price supports for agri-cultural products following the country's accession to the EU in 1995 accelerated rural economic restructuring and the out-migration of many rural residents to urban areas. To overcome the resulting labour shortage during the berry season, Finnish berry companies have increasingly turned to the use of immigrant labour, thereby creating further changes in the NTFP economy (Chapter 12).

Land tenure and resource rights

NTFPs are harvested under a wide range of landownership systems, including communal, private, and various tiers of state control, and under different access regimes, from strict prohibitions on use through to open access. Four basic kinds of rights typically underpin such systems: use, transfer, exclusion and enforcement (Neumann and Hirsch, 2000). The many combinations of rights and forms of owner-ship mean that NTFP tenure systems are complex. However, clear land tenure and resource rights are fundamental to the success of any NTFP policy measure seeking equity and sustainability. These rights do not necessarily take the form of government titles, something often not possible in vast rural areas, but there must be a working under-standing between stakeholders. When such understanding is not in place, conflicts over NTFP resources are common (eg Chapter 1; Chapter 2; Chapter 6; Chapter 7).

In some cases, land tenure may be secure, but resource rights are not. In Mexico, most forests are collectively owned, and while local communities have some autonomy in the management of their natural resources, the state sporadically exerts control over their use. For example, agave extraction has been regulated for hundreds of years through local institutions within the *ejido* and indigenous community structure. These have been responsible for regulating access, management practices and the distribution of benefits based on history and traditional knowledge of the species. Norms and agreements are established by general assembly and are continually modified or replaced in a dynamic process that responds to new situations and to tensions of environmental, socioeconomic, cultural or technological origin. Even with such a dynamic and sophisticated system, however, the Environmental Protection Agency now often fines local harvesters when they do not present a legal harvesting permit (Chapter 8).

In Yunnan, China, changing land and resource rights have created opportunities for greater local control and a more effective policy framework for matsutake mushroom harvests. During most of the latter half of the 20th century, China's forests were under state ownership. In the 1980s, however, forests were divided into state, collective and household holdings. In Yunnan, forests under the new tenure arrangements continued to be managed largely for timber until 1998, when logging was banned as a flood prevention measure. These developments coincided with expansion in demand for the region's matsutake, a product that previously had little value and for which rights of tenure and usufruct were in flux. This state of flux and the resulting flexibility in tenure arrangements left space for villages to develop codes of conduct for access to local matsutake grounds and the monitoring of harvest practices. Local regulation has had the added benefit of fostering adaptive management, since villages can adjust to new conditions more quickly and easily than higher levels of government (Chapter 10).

The security of resource rights may also depend on the commercial value of an NTFP. This is illustrated in India, where the state owns all NTFPs and grants usufruct rights for collection, as well as transport and sale. In theory, the state is involved in resource rights in order to protect and benefit collectors, but in practice the distribution of income from these resources is considered highly inequitable, and government is interested only in those species with high commercial value like tendu. Political devolution has recently transferred rights over many NTFPs to local communities, but these are primarily products of low commercial value and the state retains control over more lucrative NTFPs (Chapter 3).

Resource rights are undergoing change alongside broader views of property rights in many developed countries of the North. In Sweden and Finland, for example, the centuries-old principle of 'everyman's right' to harvest wild berries and mushrooms is being tested by the seasonal in-migration of large numbers of non-Nordic pickers, raising public concerns about immigration and tax policies, labour practices and benefit sharing (Chapter 12); in England and Scotland, tension exists between customary rights to roam and the codified versions of those rights (Chapter 5); and in Canada, in a reversal of trends in many other countries, as part of asserting aboriginal rights and title, First Nations are demanding the return of their right to regulate access to NTFPs (Chapter 4).

When intact, customary law can play an important role in ensuring sustainable and equitable use of NTFPs. Arquiza et al (Chapter 6) describe landownership vested in

Philippine communities, each with its own rattan territory, and many with strong customary laws that promote sustainable rattan management. Communities with a poorly defined sense of collective ownership and no traditional institutions tend to have weaker enforcement and manage rattan less sustainably. Similarly, in the case of marula (*Sclerocarya birrea*) in southern Africa, Wynberg and Laird (2007) found that where tenure is secure, customary laws are strong and local capacity exists to manage the resource base and deal with the pressures of commercialization, customary law achieves a balance between sustainable resource use and livelihood needs. However, when customary laws are weak and insecurities persist with land tenure and resource rights, significant conflicts arise around resource management, and government intervention is often necessary. In Fiji, 83 per cent of the total land area is under customary tenure ('native lands') as a result of British colonial policy that prohibited the sale of land to colonial settlers. However, even with secure land tenure and resource rights, dramatic social, cultural, technological, economic and other changes have strained customary and local laws and have led to significant sustainability problems for *Intsia bijuga* (Chapter 9).

In many countries, customary and statutory laws play complementary roles, but it is common for new statutory laws to weaken effective customary systems. In Bolivia, for decades small producers maintained strong de facto control over the resource base through a customary system of 'tree tenure'. Access rights were based on rubber trails and later, when Brazil nuts became important, on access to Brazil nut trees and related infrastructure. All these activities operated in a statutory policy vacuum until 1995. At that time the government superimposed another layer of 'rights' over the region's forests by allocating timber concessions. Conflicts were further exacerbated when efforts to modify the 1996 Agrarian Reform Law to expand the size of land grants to communities also undermined customary tree tenure arrangements. Land reform gave smallholders formal recognition of their tenure rights, but by basing it on control of contiguous territory (allocating each family 500ha), it undermined effective traditional tenure arrangements and access rights based on key resources (once rubber, and now Brazil nut trees) (Chapter 1; Stoian, 2005a).

Intellectual property rights

Policies relating to intellectual property rights (IPRs) can also have a significant impact on NTFP harvest and trade. The Agreement on Trade-Related Aspects of Intellectual Property Rights (TRIPS) of the World Trade Organization has created a global regime for IPRs, the result of which is that many NTFPs are increasingly included in patents and other forms of IPRs (Dutfield, 2002). This has important implications for the broader trade in and use of these products, since IPRs can create barriers against non-affiliated companies entering the market (Gebhardt, 1998). If narrowly applied, IPRs need not restrict the trade or commercialization of products by other companies or groups, but there are a number of cases where this has occurred. For example, the 1997 patenting of active components of *Hoodia* and the specification of a particular extraction technique have directly inhibited trade in *Hoodia* extracts over the past decade (Wynberg et al, 2009; Chapter 13).

The pharmaceutical, crop protection and seed industries, in particular, use patents to protect innovations, and plant breeders' rights (or plant patents in the USA) serve the same function in the horticultural industry. To a lesser extent, patents and other IPRs are also used in industries that rely on whole plant material, such as the botanical medicine and personal care and cosmetic industries. These products contain multiple compounds and therefore do not lend themselves easily to patent protection, but other areas of product development, such as manufacturing and processing techniques, formulations, dosage forms and unique release characteristics, enable IPRs to be secured. IPRS are clearly a complex, difficult and expensive way for small-scale producers to ensure benefits from NTFPs, although trade organizations such as PhytoTrade Africa are using intellectual property tools to protect small producers and enhance their competitiveness.

Increasingly, geographical indications, or appellations of origin, are used as an intellectual property mechanism to protect regional products and the communities associated with them. This is done through labels on products identifying the country, region or locality from which they originate, and that yields the particular qualities or reputation associated with the products (Commission on Intellectual Property Rights, 2002). Because geographical indications are anchored to a region and are a means to identify and market products easily, they can play a role in protecting traditional and cultural practices, as well as local economies associated with non-timber and other products. However, if poorly applied, geographical indications can also result in the disenfranchisement of local groups. For example, the use of geographical indications for *Agave cupreata* in Mexico favoured the development of monoculture plantations, undermined traditional management practices and created a complex and confusing policy environment. Traditional producers are thus unable to benefit from the system, and as Granich et al (Chapter 8) observe, 'the number of regulations and the studies and administrative procedures required make the process of legal extraction of NTFPs difficult and expensive, a great burden to communities and a disincentive to compliance'.

Labour

Labour and related policies such as immigration that directly affect labour supplies can have significant impacts on NTFPs and those whose livelihoods depend on them. These impacts are particularly evident in the case studies from the global North, where many countries have experienced significant rural restructuring in the past two decades. In the north-western USA in the 1990s, for example, floral greens harvesters were transformed from self-employed sole proprietors or micro-firms with relatively independent access to floral greens harvesting sites, to predominantly de facto wage labourers heavily dependent on the floral greens companies not only for access to harvesting sites, but also for the transport needed to get to those sites (Chapter 11). In the UK and Finland rural restructuring has also been accompanied by an influx of immigrants to harvest NTFPs, but most of these have legal authorization to be in those countries and wage labourer conditions analogous to those in the USA have not developed (Chapter 5; Chapter 12).

Insider–outsider conflicts around accessing, harvesting and trading NTFPs are significant and occur consistently around the world, and hence throughout this book. NTFPs are an important, and sometimes the most easily accessed, source of cash for rural communities. 'Outsiders' often enter communities' lands to harvest products without permission, use destructive methods and take more than wild populations can support, disregarding customary laws and controls (Lynch and Alcorn, 1994; Michon, 2005; Wynberg and Laird, 2007; Chapter 2; Chapter 7). This dynamic is played out from northern Europe to South Africa, and from Palawan to Canada to Bolivia. Migrants might harvest for their own use, but most often they exploit an available commercial opportunity, sometimes under contract with companies. The government of Sweden sought to ease tensions between local and migrant harvesters of wild berries by eliminating tax advantages for migrants (Chapter 12). In some cases, however, so-called 'outsiders' have resided in a region for generations (e.g. Chapter 1). Policy-makers must tread carefully when dealing with this potential minefield. Both 'insiders' and 'outsiders' require support, but in very different ways, and measures should take into account, and guard against inflaming, this common form of conflict.

It is also important for policy-makers to consider the many different types of 'labour' involved in the harvest, trade and processing of NTFPs. Harvesters and producers typically receive a small fraction of the final value of NTFPs (e.g. Padoch, 1988; Hersch-Martinez, 1995; King et al, 1999; Biswas and Potts, 2003; Schreckenberg, 2004; Chapter 6). In general, profits from NTFPs increase with greater processing and as the value chain progresses, alongside political power (Southgate et al, 1996; Neumann and Hirsch, 2000; Schreckenberg, 2004; Alexiades and Shanley, 2005; Chapter 1). Existing inequities and power imbalances in the value chain should be understood by policy-makers in order to create laws that benefit all stakeholders, and do not set them against each other.

Common features of NTFP policy and legal frameworks

The tension between broad policy prescriptions and the need to limit the scope of laws

Measures regulating NTFPs must carefully balance a wide range of objectives. These might include the protection of species under threat, the promotion of sustainability, the distribution of greater benefits to harvesters and producers, quality control, the generation of government revenues through taxation, and support for local businesses. A law heavily weighted to serve a single goal and one category of products (e.g. increased tax revenues and commercially traded medicinal plants) might create obstacles for achieving objectives associated with different kinds of NTFPs or stakeholders (e.g. improved livelihoods from local trading or subsistence use of the same species).

As described, the majority of laws that specifically regulate NTFPs do so in response to perceived threats to a species, and the result is often a narrow scope: species-based measures or those regulating a category of products, rather than umbrella measures for a wide range of NTFPs. In some cases, this may be the most effective response. However, this type of measure runs the risk of producing 'unintended consequences' if it lumps locally traded and subsistence NTFPs into a regulatory framework designed for commercially traded species.

There is an inherent tension in the objectives and scope of NTFP laws: on the one hand, there exists a need for broad measures that address a range of species, and on the other measures must be focused to be effective and meaningful, and avoid unintended consequences. How to focus and narrow the scope of laws is a challenge and requires significant understanding. For example, the Brazilian government instituted regulations for a small group of *Euterpe* palms, but the species in this genus have very different ecological, harvesting and economic profiles, and static regulations restricted the ability of small producers to quickly adapt and access new markets (Case Study A).

The tendency towards overwhelming bureaucracy and reporting requirements inappropriate for small-scale producers

NTFP regulations are often unnecessarily bureaucratic. Regulations lifted from industrial timber production that include permitting, fees and management plans have proven unworkable. Even regulations tailored to NTFPs can be cumbersome, and often favour large-scale commercial exploitation over small-scale NTFP harvesters or producers. In one area of Mexico, for example, it is easier to obtain authorization to log timber than to extract mushrooms (Chapter 8). In the Philippines, the Department of Environment and Natural Resources established community-based forest management agreements to allow communities to manage forests for NTFPs, but the bureaucratic obligations that came with these agreements proved insurmountable for most indigenous communities (Chapters 6 and 7). In Cameroon, complex bureaucratic requirements create obstacles for both large- and small-scale traders, and have driven much of the commercial trade in medicinal plants underground (Chapter 2).

Most policies assume communities are literate, have technical skills or funds to pay experts, and can easily find cash to pay for permits. This is rarely the case. Additionally, the logic underlying elaborate regulations eludes most harvesters and producers because they offer little or no benefit in return for increased cost and effort, and open the door to corruption and exploitation at the hands of government officials, and can criminalize traditional harvesting and livelihood activities. Bureaucratic requirements associated with government interventions are unlikely to change, however, and this is an important reason why 'less is often more' when it comes to NTFP regulation (Wynberg and Laird, 2007).

Poor coordination of laws and policies resulting in inconsistency, conflicting mandates and confusion about jurisdiction

NTFP laws and policies tend to be poorly integrated with existing federal, provincial or state laws, and are rarely coordinated with customary law. A comprehensive policy framework for NTFPs that addresses laws and policies acting at different levels requires time, funds, research and comprehensive consultations with stakeholders. This level of investment in NTFP law and policy is extremely rare. The result is legal frameworks that are inconsistent and confusing, and a lack of clarity about which laws and government departments have jurisdiction over these products and activities.

For example, the NTFP policy environment in South Africa is characterized by a plethora of inefficient and sometimes contradictory national and provincial laws. These laws are only sporadically implemented, are often incompatible with each other, and are largely unknown by local communities. The laws then interface with customary systems that have eroded to varying degrees as a result of colonial and apartheid administration, but often offer the most effective regulation for NTFPs (Wymberg and Laird, 2007; Case Study D).

Inconsistent and often underfunded policy implementation

It is difficult to interest governments in effective NTFP law and policy because NTFPs fall into institutional and sectoral 'cracks', are usually part of informal or loosely organized trade, or are consumed for subsistence. Moreover, most producers are politically and economically marginalized and there is little political will to address their needs. When governments do engage with this sector and draft laws, it is common for implementation, monitoring, and compliance to be poor since resources and capacity are rarely allocated to what are perceived as 'minor' products(Tomich, 1996; Wynberg and Laird, 2007; Chapter 2; Chapter 9). In Fiji, for example, the government recently sought to regulate the NTFP sector more effectively through the 2007 National Forest Policy and the Endangered and Protected Species Act of 2002. Despite good intentions, however, implementation has been weak: few traders know of the laws, and monitoring and enforcement is nonexistent (Chapter 9).

Sometimes a lack of implementation results when government departments compete with each other, or their mandates conflict or overlap. As a result, no institution delegates the resources or staff needed to implement NTFP regulations (Antypas et al, 2002). In Cameroon, the 1994 Forestry Law (Republic of Cameroon, 1994) set up an NTFP Sub-Directorate within the then Ministry of Environment and Forests. This new body was provided with a civil servant to oversee activities, but had no budget and extremely limited power compared to the timber interests residing in the same ministry. As a result financial returns from taxes and fees on NTFPs went to other departments and ministries (Chapter 2). It is often the case that NTFP revenue streams, which could strengthen and build capacity within government to effectively regulate and manage NTFPs, are diverted to other, more powerful, entities in government. In the Western Ghats in India, for example, royalties collected on uppage (*Garcinia gummigutta*) go to the state treasury, with no allocation for conservation of the resource, and state efforts focus on policing the movement of material in order to collect royalties, rather than monitoring harvest and trade to ensure sustainability (Chapter 3).

Unimplemented policy measures can be worse than no measures. In some cases they weaken traditional structures that might better promote sustainable management or equity in trade; even cursory government regulation of NTFPs can undermine community institutions and control over resources (Arnold and Ruiz-Pérez, 2001). Confusion, conflict and corruption can also result when laws are unclear or unenforced, making the lives of producers, harvesters, and traders more difficult and encouraging unsustainable harvest of species (Chapter 2; Chapter 6; Case Study B).

THE BROADER CONTEXT: GLOBAL AND REGIONAL TRENDS THAT UNDERLIE AND INFLUENCE NTFP LAW AND POLICY

Seemingly unrelated global and regional economic, social and legal forces can have enormous repercussions in the lives of NTFP harvesters thousands of miles away. This is ever more the case, as the world grows increasingly interconnected and trends move rapidly across societies.

Globalization and trade liberalization

Changes in macroeconomic conditions linked to the processes of globalization have played a role in shaping the content and impacts of policies affecting NTFPs during the past two decades. Since the mid-1970s, the world has experienced the development of capitalist economies in China, the countries of central and eastern Europe, the nations formerly part of the Soviet Union, Vietnam and a number of previously socialist countries in Africa. Simultaneously, advances in communications and transportation technology have facilitated the expansion and intensification of trade networks, so that many NTFPs that were once sold primarily in national or regional markets are now embedded in global exchange networks. Globalization has also affected the flow of people, which in the post-industrial economies of Europe and North America, for example, often results in companies using cheap labour from developing countries for harvesting and processing NTFPs.

In China, market liberalization sparked a thriving trade in matsutake exported to Japan. Villagers in Yunnan have benefited substantially from this trade, although they are vulnerable to declines in Japanese demand, as in 2002 when traces of pesticides were reported in mushrooms (Chapter 10). In contrast, liberalized trade relations between western and eastern Europe damaged the berry sector in Finland because the price of wild berries was substantially reduced. This created serious hardship for many rural residents and businesses in northern and eastern Finland. The Finnish government stepped in to promote harvesting by providing tax incentives for commercial berry harvesters, including immigrants, and implementing liberal immigration policies for seasonal berry pickers from other countries. Russian wild berry exporters, on the other hand, benefited from market liberalization, since they can export berries to the EU market where they can get a better price than at home. However, in Russia, an abundant supply of resources, physical proximity to major export markets and low labour costs have not in themselves proved sufficient for success in global markets; they still require more efficient transport and market infrastructure to get the products to market (Chapter 12).

In many cases, global, regional and local factors combine in unanticipated ways to significantly impact the harvest and trade of NTFPs. For example, in Palawan a combination of changes over the last decade have increased both indigenous peoples' and migrants' dependence upon NTFPs as a source of cash income. These include: the drastic reduction in agricultural production during years of El Niño and La Niña activity and as a result of swidden prohibitions instituted by local governments; the collapse of national and international markets for an important NTFP (copra – dried coconut endocarp); and economic uncertainties associated with the Asian financial crisis (Chapter 7).

Formation of regional economic alliances

Regional economic alliances emerging over the past two decades have substantially affected flows of NTFP products and labour across borders. In the USA such alliances contributed to a radical redistribution of costs and benefits associated with floral greens exchange networks (Chapter 11). The North American Free Trade Agreement (NAFTA), for example, exacerbated the downward slide in prices paid for floral greens, prompting many long-time US harvesters to look for other ways to make a living. At the same time, NAFTA ensured a cheap and plentiful supply of labour from Mexico and Central America, making it possible for a handful of floral greens companies to remain competitive. For many Latino immigrants, NAFTA had a negative push and a positive pull effect, with low corn prices pushing many out of small-scale agriculture or small businesses in Mexico, and the possibility of higher-paying work pulling them into the north-western USA to harvest salal. However, many immigrants who entered the USA illegally had to endure abysmal labour conditions or risk being branded criminals and deported (Chapter 11).

In southern Africa, countries with shared commercial species have increasingly collaborated to design joint policies for management and ensure their effective implementation. However, the complexity and diversity of domestic laws and institutions has meant that governments cannot fully streamline policies. In the case of *Hoodia*, for example, some steps have been taken by southern African countries to collaborate on poaching, trade and the transport of illegally harvested material, but they have not found common ground on the more slippery political issues of benefit sharing and indigenous peoples' rights (Chapter 13).

Rural restructuring in post-industrial societies

In many post-industrial economies, an important consequence of globalization and the formation of regional economic alliances has been massive and widespread restructuring of economies in rural regions. This includes a decline in agriculture, natural resource extraction and associated manufacturing industries. In some countries, such as the USA and Finland, high levels of rural unemployment linked to economic restructuring have caused large numbers of youths and younger families to relocate to urban areas, creating a gap in the labour supply for NTFP harvesting. Seasonal and permanent immigrants are filling these gaps, contributing to tensions between local harvesters and 'outsiders' over access to harvesting sites (Chapter 11; Chapter 12). In British Columbia, Canada, the 'rural flight' phenomenon has been somewhat attenuated by the large proportion of First Nations communities in rural areas reluctant to leave their homes despite high unemployment rates (Chapter 4).

Wider acceptance of indigenous peoples' rights and locally based political organizations

In recent years, NTFP policies have been influenced by the growing political power of indigenous peoples and increased recognition of their land, human, cultural and intellectual property rights. Since the early 1990s, these rights have been articulated through a suite of global instruments and institutions, negotiated texts and processes relating to indigenous peoples and the protection of traditional knowledge, including the Convention on Biological Diversity (CBD) and the United Nations Permanent Forum on Indigenous Issues.

These developments mean that indigenous peoples' rights to harvest NTFPs as part of traditional practices, to control and benefit from access to resources on their territories and to protect the use of their traditional knowledge are now more widely accepted. Non-indigenous communities have also benefited from these developments and from a linked trend towards decentralized governance, or 'devolution' and 'participatory' processes that establish new, or reinvigorate existing, community-based forest governance systems (Case Study A; Chapter 2; Chapter 3; Chapter 6). Related to these developments is the rise of civil society and non-governmental organizations that promote dialogue and political engagement with human rights, social justice and environmental issues (Alexiades and Shanley, 2005).

Devolved, or local governance, could work well for NTFPs given the diverse social, ecological and economic conditions under which they are harvested, used and traded. However, many of these regulatory efforts have not been effectively implemented. Likewise, the rights granted to indigenous peoples are often not recognized in practice, and in the case of NTFPs do not always translate into greater control over resources and improved benefits (Castillo and Castillo, 2009). The 1996 Panchayats Act in India, for example, gave greater authority over NTFPs to tribal groups, but was ambiguous about which forests were included and, with the exception of Orissa state, this measure was largely ignored (Chapter 3). In the Philippines, wider commercial interests such as mining often override the rights of indigenous peoples to use NTFPs and other resources (Chapter 6). Neither have the many laws and regulations that exist to protect human rights and prevent injustice in southern Africa saved the indigenous San peoples from loss of land and natural resources, intellectual property and culture. It has taken a significant process of awareness-raising to enable them to claim and assert their rights to resources, such as those to *Hoodia*, and convert those rights into tangible outcomes (Chapter 13; Chennells et al, 2009). Although the broader legal trend is towards greater rights for indigenous peoples and more local control over resources, including NTFPs, in practice it will take many years for these rights to be realized, and few incentives exist for reluctant governments to cede these powers to local groups.

Broader concepts of conservation that include sustainable use and equity

In recent decades, the field of conservation has moved from a purely protectionist approach to one that incorporates sustainable use and increasingly views equity and social justice as integral to achieving environmental objectives. This has been supported by a suite of new international agreements and processes relating to biodiversity, forests, and climate change. The Convention on Biodiversity (CBD), for example, regulates the commercial use of genetic resources and not NTFPs and other 'biological resources', but its objectives of sustainable and equitable use have influenced national law and international standards for socially responsible business practices (Laird, 1999; Pierce and Laird, 2003; Laird and Wynberg, 2006, 2008; Chapter 13).

A more comprehensive policy approach has emerged that makes room for NTFPs and small-scale producers previously invisible to policy-makers. NTFPs are viewed as important contributors to rural livelihoods, and sometimes as alternatives to more destructive land uses. Interest in the sustainability and equity of the commercial NTFP trade has also grown, including greater attention focused on the distribution of benefits along NTFP value chains. As awareness of the links between social justice, poverty, equity and conservation has grown, so too has awareness of the enormous and diverse role of NTFPs in rural livelihoods.

CONCLUSION

This chapter has described the multiple factors that influence NTFP policy development and implementation, highlighting the remarkable similarities in experiences throughout the world. NTFP policy development is usually reactive or opportunistic, and rarely strategic. Limited information and understanding are key constraints that prevent more effective policy-making, including understanding of the complex and dynamic production systems of which NTFPs are a part. NTFP regulations tend to be inconsistent, unnecessarily bureaucratic, and to operate in an incoherent and conflicting policy environment that provides opportunities for corruption and creates new forms of inequity. A major difficulty in regulating NTFPs is also the need to create laws that are specific enough to be meaningful, and yet broad enough to apply to a range of species and situations.

The tendency for policy-makers to overlook the crucial insights of NTFP producers and traders, many on the economic and political margins, is widespread. All too often governments favour the voices of the politically and economically powerful few, rather than those of the people most directly affected by policy interventions. Governments also tend to support economic activities that generate income they can tax and benefit from, such as mining, logging, oil, or industrial agriculture. It is difficult to attract government support for informal, dispersed activities undertaken by the politically marginal, no matter how superior the economic value or relatively limited the environmental impact of NTFPs.

Although the state of NTFP law and policy is not encouraging, a consistent and important lesson to emerge throughout the world is the value of local and customary law in regulating this complex and diverse group of species, and the need for governments to often 'leave well enough alone' or to intervene minimally. With more careful attention, however, it is possible that recent interest in laws and policies regulating NTFPs will yield more strategic, better-informed and effective policy frameworks. The next and final chapter highlights some of the issues to consider, and information and actions that are required, to achieve this objective.

REFERENCES

Alexiades, M. N. (2002) 'Cat's claw (*U. guianensis* and *U. tomentosa*)', in P. Shanley, A. R. Pierce, S. A. Laird and A. Guillen (eds) *Tapping the Green Market: Certification and Management of Non-Timber Forest Products*, People and Plants Conservation Series, Earthscan, London

Alexiades, M. N. and Shanley, P. (2005) *Forest Products, Livelihoods and Conservation; Case Studies of Non-Timber Forest Product Systems*. Volume 3 – Latin America, CIFOR, Bogor, Indonesia

Antypas, A., McLain, R. J. and Gilden, J. (2002) 'Federal nontimber forest products policy and management', in E. T. Jones, R. J. McLain and J. Weigand (eds) *Nontimber Forest Products in the United States*, University Press of Kansas, Lawrence, KS

Arnold, J. E. M. and Ruiz-Pérez, M. (1996) *Current Issues in Non-Timber Forest Products Research*, Center for International Forestry Research, Bogor, Indonesia

Arnold, J. E. M. and Ruiz-Pérez, M. (2001) 'Can non-timber forest products match tropical forest conservation and development objectives?', *Ecological Economics*, vol 39, no 3, pp437–447

Baird, I. G. and Dearden, P. (2003) 'Biodiversity conservation and resource tenure regimes: A case study from northeast Cambodia', *Environmental Management*, vol 32, no 5, pp541–550

Bennett, B. (2006) *Foreign Direct Investment in South Africa: How Big is Southern Africa's Natural Product Opportunity and What Trade Issues Impede Sectoral Development?*, Regional Trade Facilitation Programme, Pretoria

Biswas, T. and Potts, J. (2003) 'Grounds for action: Looking for a sustainable solution to the coffee crisis', *Bridges*, vol 7, no 2, pp17–18

Brown, H. C. P. (2005) 'Governance of non-wood forest products and community forests in the humid forest zone of Cameroon', PhD thesis, Cornell University, Ithaca, NY

Bürgener, M. (2007) *Trade Measures – Tools to Promote the Sustainable Use of NWFP? An Assessment of Trade-Related Instruments Influencing the International Trade in Non-Wood Forest Products and Associated Management and Livelihood Strategies*, Non-Wood Forest Products Working Document no 6, FAO, Rome

Castillo, R. C. A. and Castillo, F. A. (2009) 'The law is not enough: Protecting indigenous peoples' rights against mining interests in the Philippines', in R. Wynberg, R. Chennells and D. Schroeder (eds) *Indigenous Peoples, Consent and Benefit Sharing: Learning from the San-Hoodia Case*, Springer, Berlin

Chennells, R., Haraseb, V. and Ngakaeaja, M. (2009) 'Speaking for the San: Challenges for representative institutions', in R. Wynberg, R. Chennells and D. Schroeder (eds) *Indigenous Peoples, Consent and Benefit Sharing: Learning from the San-Hoodia Case*, Springer, Berlin

Cole, D. and Nakamhela, U. (2008) 'Review and clarification of procedures for the issuing of permits for the collection and export of biological resources in Namibia', unpublished report, Centre for Research Information Action in Africa and Southern African Development and Consulting (CRIAA SA-DC), Windhoek, Namibia

Commission on Intellectual Property Rights (2002) 'Integrating Intellectual Property Rights and Development Policy', Report of the Commission on Intellectual Property Rights, September, London. www.iprcommission.org/papers/pdfs/final_report/CIPR_Exec_Sumfinal.pdf, accessed 16 May 2008

Dewees, P. A. and Scherr, S. J. (1996) 'Policies and markets for non-timber tree products', EPTD discussion paper no. 16, Environment and Production Technology Division, International Food Policy Research Institute, Washington, DC

Dowie, M. (2005) 'Conservation refugees: When protecting nature means kicking people out', *Orion Magazine*, November/December, pp16–27, abridged version available at www.orionsociety.org/index.php/articles/article/161, accessed 7 June 2009

Dutfield, G. (2002) 'Sharing the benefits of biodiversity: Is there a role for the patent system?', *Journal of World Intellectual Property*, vol 5, no 6, pp899–932

Fantini, A. C., Guries, R. P. and Ribeiro, R. J. (2005) 'Palm heart (*Euterpe edulis* Mart.) in the Brazilian Atlantic rainforest: A vanishing resource', in M. N. Alexiades and P. Shanley (eds) *Forest Products, Livelihoods and conservation; Case studies of non-timber forest product systems.* Volume 3 – Latin America, CIFOR, Bogor, Indonesia, pp83–110

Fiji Islands (1992) *Forest Decree No. 31 of 1992, A Decree Relating to Forest and Forest Produce*, Government of the Sovereign Democratic Republic of Fiji

Gebhardt, M. (1998) 'Sustainable use of biodiversity by the pharmaceutical industry?', *International Journal of Sustainable Development*, vol 1, no 1, pp63–72

Hayes, P. B. (2000) 'Enhancing the competitiveness of the rooibos industry', MPhil thesis, University of Stellenbosch, Stellenbosch, South Africa

Hecht, S. B., Anderson, A. B. and May, P. (1988) 'The subsidy from nature: Shifting cultivation, successional palm forests, and rural development', *Human Organisation*, vol 47, pp25–35

Hersch-Martinez, P. (1995) 'Commercialisation of wild medicinal plants from southwest Puebla, Mexico', *Economic Botany*, vol 49, no 2, pp197–206

Homma, A. K. O. (1992) 'The dynamics of extraction in Amazonia: A historical perspective', *Advances in Economic Botany*, vol 9, pp23–31

Iqbal, M. (1993) *International Trade in Non-Wood Forest Products: An Overview*, FO Misc/93/11 Working Paper, Food and Agriculture Organization, Rome

Jaireth, H. and Smyth, D. (eds) (2003) *Innovative Governance: Indigenous Peoples, Local Communities, and Protected Areas*, World Conservation Union, Ane Books, New Delhi

Kanga, K. (1999) 'Trade of main wild berries in Finland', *Silva Fennica*, vol 33, no 2, pp159–168

King, S. R., Meza, E. N., Carlson, T. J. S., Chinnock, J. A., Moran, K. and Borges, J. R. (1999) 'Issues in the commercialisation of medicinal plants', *HerbalGram*, vol 47, pp46–51

Laird, S. A. (1999) 'The botanical medicine industry', in K. ten Kate and S. A. Laird (eds) *The Commercial Use of Biodiversity: Access to Genetic Resources and Benefit Sharing*, London, Earthscan

Laird, S. A. and Wynberg, R. (2006) *The Commercial Use of Biodiversity: An Update on Current Trends in Demand for Access to Genetic Resources and Benefit-Sharing, and Industry Perspectives on ABS Policy and Implementation*, report prepared for the Convention on Biological Diversity's Ad Hoc Open-Ended Working Group on Access and Benefit-Sharing, Fourth Meeting, Granada, Spain, 30 January–3 February, UNEP/CBD/WGABS/4/INF/5, www.cbd.int/doc/meetings/abs/abswg-04/information/abswg-04-inf-05-en.pdf, accessed 7 June 2009

Laird, S. and Wynberg, R. (2008) 'Access and benefit-sharing in practice: Trends in partnerships across sectors', *CBD Technical Series No. 38*, Secretariat of the Convention on Biological Diversity, Montréal

Lynch, O. J. and Alcorn, J. B. (1994) 'Tenurial rights and community-based conservation', in D. Western and M. Wright (eds) *Natural Connections: Perspectives in Community-Based Conservation*, Island Press, Washington, DC, pp373–392

Michon, G. (2005) 'NTFP development and poverty alleviation: Is the policy context favourable?', in J. L. Pfund and P. Robinson (eds) *Non-Timber Forest Products: Between Poverty Alleviation and Market Forces*, Intercooperation, Switzerland

Nalvarte Armas, W and de Jong, W. (2005) 'Potentials and perspective of cat's claw (*Uncaria tomentosa*)', in M. N. Alexiades and P. Shanley (eds) *Forest Products, Livelihoods and Conservation; Case Studies of Non-Timber Forest Product Systems*, Volume 3 – Latin America, CIFOR, Bogor, Indonesia, pp281–298

Neumann, R. P. and Hirsch, E. (2000) *Commercialisation of Non-Timber Forest Products: Review and Analysis of Research*, Center for International Forestry Research, Bogor, Indonesia

Nott, K. and Wynberg, R. (2008) 'Millenium Challenge Account Namibia Compact. Volume 4: Thematic analysis report – indigenous natural products', Namibia strategic environmental assessment, task order under the Project Development, Project Management, Environmental and General Engineering ID/IQ Contract no. MCC-06–0087-CON-90, Task Order No. 02, ARD, Washington, DC

Ortiz, E. G. (2002) 'Brazil nut', in P. Shanley, A. Pierce, S. Laird and A. Guillen (eds) *Tapping the Green Market: Certification and Management of Non-Timber Forest Products*, Earthscan, London

Padoch, C. (1988) 'The economic importance and marketing of forest and fallow products in the Iquitos region', in W. M. Denevan and C. Padoch (eds) *Swidden Fallow Agroforestry in the Peruvian Amazon: Advances in Economic Botany 5*, New York Botanical Garden, Bronx, NY, pp74–89

Peters, C. M. (1996) *The Ecology and Management of Non-Timber Forest Resources*, World Bank Technical Paper No 322, World Bank, Washington, DC.

Phytotrade Africa (2006) 'Phytotrade Africa's approach', www.phytotradeafrica.com/about/approach.htm, accessed 7 June 2009

Pierce, A. R. and Laird, S. A. (2003) 'In search of comprehensive standards for non-timber forest products in the botanicals trade', *International Forestry Review*, vol 5, no 2, pp138–147

Posey, D. A. (ed) (1999) *The Cultural and Spiritual Value of Biodiversity*, United Nations Environment Programme, Nairobi

Republic of Cameroon (1994) *Law No 94/01 of 20 January 1994 to Lay Down Forestry, Wildlife and Fisheries Regulations*, Yaoundé, Republic of Cameroon

República de Bolivia (1996a) *Ley Forestal* (Forestry Law) No 1700 of 12 July 1996, El Congreso Nacional, La Paz, Bolivia

República de Bolivia (1996b) *Ley del Servicio Nacional de Reforma Agraria* (National Agrarian Reform Law) No 1715 of 18 October 1996, El Congreso Nacional, La Paz, Bolivia

Schreckenberg, K. (2004) 'The contribution of shea butter (*Vitellaria paradoxa* CF Gaertner) to local livelihoods in Benin', in T. Sunderland and O. Ndoye (eds) *Forest Products, Livelihoods and Conservation: Case Studies of Non-Timber Forest Product Systems*. Volume 2 – Africa, Center for International Forestry Research, Bogor, Indonesia

Shackleton, C. M. and Shackleton, S. E. (2004) 'The importance of non-timber forest products in rural livelihood security and as safety nets: A review of evidence from South Africa', *South African Journal of Science*, vol 100, no 11–12, pp658–664

Shackleton, C. M., Botha, J. and Emanuel, P. L. (2003) 'Productivity and abundance of *Sclerocarya birrea* subsp. *Caffra* in and around rural settlements and protected areas of the Bushbuckridge lowveld, South Africa', *Forests, Trees and Livelihoods*, vol 13, pp217–232

Shanley, P., Pierce, A. R., Laird, S. A. and Guillen, A. (2002) *Tapping the Green Market: Certification and Management of Non-Timber Forest Products*, People and Plants Conservation Series, Earthscan, London

Soldán, M. P. (2003) *The Impact of Certification on the Sustainable Use of Brazil Nut (Bertholletia excelsa) in Bolivia*, Non-Wood Forest Products Programme, Food and Agriculture Organization of the United Nations, www.fao.org/forestry/foris/pdf/NWFP/Brazilnuts.pdf, accessed 2 October 2008

Southgate, D., Coles-Ritchie, M. and Salazar-Canelos, P. (1996) 'Can tropical forests be saved by harvesting non-timber products? A case study for Ecuador', in W. L. Adamowicz (ed) *Forestry, Economics and the Environment*, CAB International, Wallingford, UK, pp68–80

Stafford, L. (2009) 'After another cancelled partnership, the future of *Hoodia* remains unclear', *HerbalEGram*, vol 6, no 2, www.cms.herbalgram.org/heg/volume6/02%20February%20/Hoodia_Nixed.html, accessed 30 March 2009

Stoian, D. (2000) 'Variations and dynamics of extractive economies: The rural–urban nexus of non-timber forest use in the Bolivian Amazon', PhD thesis, Albert-Ludwig University, Freiburg, Germany

Stoian, D. (2005a) 'Harvesting windfalls: the Brazil nut (*Bertholletia excelsa*) economy in the Bolivian Amazon', in M. N. Alexiades and P. Shanley (eds) *Forest Products, Livelihoods and Conservation; Case Studies of Non-Timber Forest Product Systems*. Volume 3 – Latin America, CIFOR, Bogor, Indonesia, pp83–110

Stoian, D. (2005b) 'What goes up must come down: The economy of palm heart (*Euterpe precatoria* Mart.) in the Northern Bolivian Amazon', in M. N. Alexiades and P. Shanley (eds) *Forest Products, Livelihoods and Conservation; Case Studies of Non-Timber Forest Product Systems*. Volume 3 – Latin America, CIFOR, Bogor, Indonesia, pp111–134

Tomich, T. P. (1996) 'Market, policies and institutions in NTFP trade: Nothing is perfect', in R. R. B. Leakey, A. B. Temu, M. Melnyk and P. Vantomme (eds) *Domestication and Commercialisation of Non-Timber Forest Products in Agroforestry Systems*, Non-Wood Forest Products no 9, FAO, Rome

Wynberg, R. P. (2006) 'Identifying pro-poor best practice models of commercialisation of southern African non-timber forest products', PhD thesis, Graduate School of Environmental Studies, University of Strathclyde, Glasgow

Wynberg, R. P. and Laird, S. A. (2007) 'Less is often more: Governance of a non-timber forest product, marula (*Sclerocarya birrea* subsp. *caffra*) in southern Africa', *International Forestry Review*, vol 9, no 1, pp475–490

Wynberg, R., Silveston, J. and Lombard, C. (2009) 'Value adding in the southern African natural products sector: How much do patents matter?', in *The Economics of Intellectual Property in South Africa*, World Intellectual Property Organization, Geneva

Chapter 16

Recommendations

Sarah A. Laird, Rachel P. Wynberg and Rebecca J. McLain

The chapters in this book and the NTFP literature in general yield fairly consistent lessons. These range in focus from the ways in which NTFP laws and policies are conceived to how they are drafted and implemented. A few catchphrases emerge repeatedly – 'less is more', 'carrots not sticks', 'leave well enough alone', 'the best-laid plans' – all suggesting a sector that has endured poorly directed and formulated policy. The authors in this book stress the need for better information, simplification, clarity and consistency in NTFP policy frameworks. Below we discuss some of these recommendations.

THE EXTENT OF COMMERCIALIZATION AND THE HETEROGENEITY OF NTFP RESOURCES, MARKETS AND STAKEHOLDERS SHOULD BE REFLECTED IN POLICIES AND LAWS

- The extent of commercialization should have a strong bearing on the nature of regulations. Laws should recognize the different types of NTFP use, including subsistence, local trade, commercial trade and recreation. They should also be sensitive to the scale of activities from local through to global. For example, subsistence use should not be regulated except in cases where there are clear risks of overharvesting, but government attention should be paid to internationally traded industrial-scale NTFPs.
- Policy-makers should anticipate potential, often unpredictable, shifts in market demand due to supply problems, consumer fads, safety and efficacy concerns, and other common disruptions to the NTFP trade. NTFP measures should be flexible and adaptive to accommodate these shifts.
- Market access is as important as market prices for small-scale producers. Policies that support certification and other efforts to set producers apart from competitors are

most effective when the administrative costs of such systems do not exceed their benefits.

- Processors and traders often control NTFP sectors, with small-scale producers having limited power over the commercial trade, including prices. Policy-makers can help reduce monopolistic tendencies in NTFP markets, but should do this in a way that supports all stakeholders along the value chain and does not set them against each other.
- Internationally traded NTFPs cannot easily transform local economies, institutions and management practices in positive ways. In some cases, these NTFPs generate real benefits for local groups, but the greatest and most consistent value for local communities is usually found in subsistence use and local trade of NTFPs.
- Although commercial uses of NTFPs are often based on traditional uses, the relationship between the two grows weaker as commercial demand increases and products move outside the original cultural and geographical context of their use. However, it remains important that traditional knowledge holders provide consent for and benefit from the commercial use of their knowledge, and measures should be instituted to achieve this.

NTFPS ARE PART OF LAND-USE SYSTEMS THAT INCLUDE A RANGE OF ACTIVITIES. REGULATIONS SHOULD REFLECT THESE INTER-CONNECTED PATTERNS OF LAND AND RESOURCE USE

- NTFP laws and policies must take into account the most pressing threats to species and the ecosystems within which they are found. It is often the case that forest degradation and destruction resulting from commercial agriculture, logging, mining and other land uses cause far more damage to NTFP populations than overharvesting.
- Governments should regulate timber and NTFPs in very different ways given the enormous differences in how they are harvested and used, and their role in local economies and cultures. However, timber regulations should minimize the negative impacts of logging on locally and commercially valuable NTFPs.
- NTFPs are part of complex production systems. Prior to drafting regulations, policy-makers should understand the relationship between NTFPs and agriculture, the importance of NTFP harvest timing for subsistence and cash income, and other critical features of these systems.
- Climate change is likely to bring about substantial shifts in the geographic distribution of plant species, including many NTFPs. Climate change mitigation and adaptation strategies and policies thus need to address NTFP harvesting and trade alongside other land-use activities.

POWER AND OTHER SOCIAL RELATIONS MUST BE FACTORED INTO LAW AND POLICY FORMATION

- There are many types of power and social relations manifested in the harvest and trade of NTFPs that help determine whether these activities will be sustainable and equitable, and whether they will support rather than undermine the livelihoods of groups dependent on these resources. So as not to exacerbate existing inequalities, or create new ones, it is vital that the power dynamics between stakeholders be understood prior to policy formulation and implementation.
- Relations between 'insiders' and 'outsiders' are classic points of conflict for NTFPs. The potential for these tensions to arise must be allowed for in policy measures and addressed in consultations with stakeholders. Policy-makers should take great care not to inflame these conflicts with new measures. Where conflict exists, facilitators trained in conflict resolution are likely to be needed to help formulate equitable and viable policies.
- On paper, indigenous peoples increasingly have political power and legal rights to their land, resources, culture and knowledge, but challenges remain. There is a continuing need to assist indigenous peoples to organize, navigate overly bureaucratic NTFP permitting procedures, and assert their rights against more powerful players.
- In many countries, entrenched corruption and abuse of power on the part of governments and their circle of patronage means that new measures will stall. Small producers, who lack political or economic power, can easily lose out if measures are drafted in a way that primarily promotes the interests of the elite.

INFORMATION REQUIREMENTS FOR EFFECTIVE LAWS AND POLICIES SHOULD BE CAREFULLY CONSIDERED BEFORE REGULATIONS ARE DEVELOPED

- Policy-makers require a vast range of information about NTFPs when drafting laws, including: the ecology and management of species, markets for each resource, key stakeholders, the economic and social costs and benefits of use along the value chain, and evolving technologies and harvesting practices. Collecting this information, particularly in countries with severe resource and capacity constraints or where hundreds of NTFPs are used, is difficult or impossible. Capacity-building, and broader research and data-collection efforts should be on-going, but when governments have limited resources they should focus on threatened species and those that are intensively traded.
- The relationship between NTFPs and species conservation should be well-understood before moving into policy formulation. The greatest threats to NTFP populations generally come from degradation or destruction of habitats, but the overharvesting of NTFPs can be a significant problem, as CITES and national

endangered species lists make clear. Policy-makers should, however, be cautious when concluding that overharvesting is the main threat to NTFPs or that concerns about unsustainable sourcing necessarily mean there is a crisis at hand. A tendency on the part of conservation bodies to assume the worst and promote policy interventions has sometimes resulted in conflicts with producer groups who feel that outsiders do not understand the species, trade or local livelihoods dependent upon these products. Measures to conserve and protect species should be informed by consultations with local producers and stakeholders, and supported by a research and data-collection process that allows policy-makers to fully comprehend the products and activities they seek to regulate.

POLICY DEVELOPMENT MUST INCORPORATE COMPREHENSIVE, ONGOING AND ITERATIVE STAKEHOLDER CONSULTATIONS

- Laws and policies should grow from extensive consultations with the full range of affected stakeholders, including harvesters and producers, traders, companies, consumers and government departments. This facilitates the development of more informed and effective policy that reflects real needs and priorities, helping to ensure that policies are widely accepted and implemented. The participation of diverse groups is particularly important for species that are heavily traded and thus involve strong economic interests.
- Intermediary organizations such as producer and harvester groups, trade associations and NGOs should be supported to help strengthen consultations, and ensure these voices are heard in policy processes.

CAPACITY SHOULD BE BUILT IN GOVERNMENT, TRADER AND PRODUCER COMMUNITIES TO ENABLE THE DEVELOPMENT AND IMPLEMENTATION OF EFFECTIVE NTFP POLICIES AND LAWS

- Government capacity to develop and implement NTFP laws and policies is notoriously underfunded and marginalized, due in part to the lack of importance given to these 'minor' forest products. Conflicting or overlapping mandates between departments also mean that it is often not clear who in government has responsibility for these products. Capacity and technical skills should be developed in government departments.
- Producers, traders and their support organizations also need greater capacity to engage with government on the development of effective laws and policies. Creative

approaches should be explored to involve producer communities and traders in monitoring resource use and assisting with policy implementation.

MANY SEEMINGLY UNRELATED AREAS OF LAW CAN SIGNIFICANTLY AFFECT NTFP MANAGEMENT, USE AND TRADE, AND SHOULD BE CONSIDERED WHILE DEVELOPING NTFP POLICY AND LEGAL FRAMEWORKS

- A range of laws directly and indirectly impacts NTFPs, including those regulating natural resources, agriculture, land tenure and resource rights, water, transportation, biodiversity, labour, intellectual property rights and product quality control. Governments should identify the social, economic and environmental effects of such laws on NTFPs when developing a policy framework, and should seek to mitigate the negative impacts of these seemingly unrelated bodies of law.
- Land tenure and resource rights are vital parts of NTFP regulation. However, the many types of rights and ownership, the combinations thereof, and the various layers of NTFP laws create enormously complex systems. It is vital that access and ownership rights to resources and land be clarified when developing regulatory frameworks for NTFPs, particularly for resources with commercial value.
- Policy-makers must understand the impacts of labour relations on the distribution of costs and benefits (social, economic and ecological) associated with NTFP harvesting. Wage labour conditions can have particularly debilitating economic effects on low-income populations, and can set up inequitable trade relations and undermine sustainable harvesting practices.
- Intellectual property rights are powerful tools that can support or restrict NTFP trade. Governments should ensure that these laws provide an enabling environment for traditional knowledge protection and local NTFP industries and producers. Some intellectual property approaches, like geographical indications and trademarks, have the potential to do this, but governments must be careful to build on or complement traditional resource rights, minimize paperwork and avoid duplication of existing laws.
- Policy-makers developing and implementing standards for good manufacturing practices, quality control and food safety need to ensure that they do not, by dint of the high levels of sophistication and reporting required, exclude many producers or products that might otherwise qualify. Efforts should be made to build the capacity of producers to comply with necessary certification, health and safety standards, and to improve their ability to engage and negotiate with standard-setting agencies.

THE IMPACT OF REGIONAL AND INTERNATIONAL POLICIES ON NTFPS MUST BE EXAMINED AS NATIONAL, STATE AND PROVINCIAL NTFP POLICY FRAMEWORKS ARE DEVELOPED

- Regional and international trade agreements can have significant impacts on the distribution of costs and benefits from NTFP harvesting and trade. Policy-makers need to consider how such agreements interact with NTFPs in order to minimize negative, unintended consequences resulting from these agreements.
- Countries that share commercially traded NTFP species should collaborate to develop regional policies for their management, use and trade. This will encourage sustainable use and fair benefit sharing, assist with traceability requirements and give countries a strategic advantage in increasingly competitive markets.
- International treaties such as CITES are important tools to regulate trade in endangered species but need to be used with caution to ensure that trade restrictions are appropriate, targeted and effective, and that the negative effects of regulation on livelihoods are minimized.
- As a result of international policy trends, national, state and provincial policies and laws increasingly require the fair sharing of benefits from the commercial use of biodiversity and associated traditional knowledge. However, these measures are typically not coordinated with laws relating to the bulk trade and use of NTFPs, leading to confusion and ineffectual implementation. Governments should attempt to integrate these bodies of law when developing policy frameworks for NTFPs.

POLICY FRAMEWORKS SHOULD BE STRATEGIC, COMPREHENSIVE AND COORDINATED ACROSS GOVERNMENT DEPARTMENTS

- Policies should be developed strategically. Most NTFP laws are built incrementally and lack an overall strategy or clear objectives. Many are reactive or opportunistic. Incremental approaches may work for some NTFP regulation, given the erratic nature of markets and often uncertain knowledge about resource availability, but they do not offer an effective way to regulate most of these products over time. Policies can be flexible and adaptive while also being strategic.
- Coordination and integration are needed within and across government departments and levels of government. This will help to streamline procedures, minimize bureaucracy and improve policy coherence. Governments should aim to synchronize laws affecting NTFPs, avoid duplication and ensure the mandates of government departments do not overlap.

- Governments should examine NTFP laws with a view to eliminating permits and procedures that are inappropriate and burdensome for small-scale producers and bring no clear management or livelihood benefits.
- Governments and others should be aware that unintended consequences often result from policies regulating NTFPs and from those found outside the sector. Due to the complexity and heterogeneity of these products and associated activities, even when governments make concerted efforts with the best of intentions, NTFP law and policy often do not work out as planned. Policies based on theoretical frameworks and assumptions originating outside a region are particularly likely to lead to unanticipated outcomes when they interact with local political, cultural, economic and ecological conditions. Care should be taken to consider the wide range of issues that converge upon and can distort the effects of NTFP policy and law.

NTFP POLICIES WORK BEST WHEN BASED ON INCENTIVES ('CARROTS') RATHER THAN PENALTIES ('STICKS')

- 'Sticks' are often employed to regulate NTFPs, particularly in a perceived overharvesting crisis, but 'carrots' in the form of incentives and supportive legal frameworks usually work best for this category of products. 'Carrots' might include government support for producer, trade and processing groups; market access and premium prices via certification; tax breaks; and outreach and education on new policies and laws. These contrast with measures that can criminalize traditional harvesting practices or are used as 'sticks' such as permits, quotas, taxes and restrictions on trade or use. In some cases, particularly when there is sudden and high commercial demand, both approaches are necessary.
- Revenue generated by the state from royalties, taxes, or the sale of NTFPs should be channelled to conservation and sustainable management of NTFPs, supporting the sector, and building government capacity on NTFPs, rather than used for purposes not directly related to these resources.

LESS IS OFTEN MORE: NTFP REGULATION SHOULD BE APPROACHED WITH A LIGHT HAND

- One lesson that is emerging around the world is that 'less is often more' when it comes to government regulation of NTFPs. Governments should be encouraged to approach NTFP regulation with a light hand, and in ways that reflect the financial, ecological and social costs and benefits of such actions, the government's implementation capacity and the likelihood of compliance. Regulating lightly will, in turn, reduce bureaucratic procedures and red tape, lessen confusion among harvester communities and eliminate opportunities for bribery and corruption.

- In many cases, governments should 'leave well enough alone'. The first question governments, NGOs and others need to ask is: do we need to regulate? A bias in the fields of conservation and development towards intervention and action often drives the establishment of new laws or government obligations before a solid understanding of the problems and issues they are meant to address is established.

EXISTING CUSTOMARY AND LOCAL LAWS ARE OFTEN BETTER SUITED TO THIS DIVERSE SET OF PRODUCTS AND ACTIVITIES

- Regulators should acknowledge, by adopting a 'less is more' approach, that where land tenure and resource rights are secure, customary laws are still strong, and local capacity exists to manage the resource base and deal with commercial pressures, customary laws often provide a more nuanced approach to regulation, integrating unique local cultural, ecological and economic conditions in ways that better suit this category of products.
- In cases where customary law has broken down to a significant degree, or outside commercial pressure has intensified well beyond the carrying capacity of traditional measures, governments can offer important and necessary complementary levels of regulation, something often requested by local groups. But this must be done in a targeted and informed fashion, and interventions should be crafted to include local-level institutions and management systems, where these are effective.
- Trends towards decentralized and participatory NTFP governance reflect an advance in NTFP regulation throughout the world and should be supported. However, it remains necessary to create coherence in national and regional policies, especially for commercially traded species.
- Governments should explore NTFP policy frameworks that integrate and coordinate customary and statutory governance systems. This requires commitments of time, money, research, and extensive stakeholder consultation.
- The sustainability and equity of NTFP use, management and trade depend upon a myriad of locally specific factors, and are often best addressed by a patchwork of local measures, supported by a streamlined and coherent government framework that sets the floor and intervenes minimally.

NTFP Law and Policy Literature: Lie of the Land and Areas for Further Research

Alan Pierce

INTRODUCTION

The art of finding articles relating to NTFP law and policy requires skills not unlike those needed to locate elusive truffles or ginseng plants in the wild. Researchers may have a general idea of the preferred 'habitat' for NTFP articles – for example, forestry, anthropology and sociology journals – but often, like wild plants and fungi, these articles can be found in a surprising range of habitats, including niches such as websites, white papers and book chapters on resource management, as well as presumptively inhospitable areas like journals relating to law. The breadth of the topic, as well as the different terminologies used by specific research interests, makes compiling an NTFP law and policy bibliography challenging.

The NTFP literature is vast (see von Hagen et al, 1996; Neumann and Hirsch, 2000; Maille, 2001), but relatively little attention has been given to the topic of NTFP law and policy (Wynberg and Laird, 2007). For the purposes of this paper, I use the American Heritage Dictionary's (2000) definition of policy as 'a plan or course of action, as of a government, political party, or business, intended to influence and determine decisions, actions, and other matters'. This review is based upon a survey of more than 150 articles and scores of websites (Pierce, 2009) relating to laws and policies that impact NTFP management, harvest and trade. The review and the bibliography it is based upon are not comprehensive in scope. An exploration of the impact of the Convention on Biological Diversity on NTFPs alone would probably fill several dissertations. Rather its aim is to discuss some of the major legal and policy themes that have been explored to date and to identify research gaps.

Researchers must use a variety of keyword searches to locate NTFP articles, requiring multiple entries and a high degree of trial and error. For example, the topic is broadly defined by a number of monikers including 'non-timber forest products', 'non-wood forest products', 'special forest products', 'minor forest products' and 'secondary forest products'. More in-depth searches may necessitate the use of refined subject headings such as 'medicinal plants', 'wild edibles', 'extractive reserves', 'mushrooms', 'foraging' or 'essential oils'. Many documents can only be located under specific species names or locations. Attempting to link such a diffuse literature to a broad topic like policy presents further difficulties because of the nebulous nature of policy, which itself includes numerous subcategories relating to tenure, resource access, regulations and trade.

I compiled the NTFP law and policy bibliography using three avenues of inquiry.

- Academic search engines (e.g. AGRICOLA, Web of Science, OhioLINK, JSTOR) were used to find published books, journal articles and dissertations.
- The Google search engine was used to locate white papers, conference proceedings, organizations and websites relevant to NTFP law and policy on the Internet.
- NTFP researchers were asked to submit citations of what they considered the most important NTFP policy articles.

The process of asking fellow researchers for leads to find NTFP policy articles was fascinating. Some submitted numerous citations, many seemingly peripheral to policy, while others objected that they knew nothing of policy. Just as there is little agreement over what an NTFP is, there is likewise little consensus on what 'policy' means. For some NTFP practitioners, almost everything has policy implications, while for other field researchers, policy is 'hand waving' that has little to do with serious science.

Reliance upon peer recommendations was essential because of the highly specialized nature of NTFP research. Some individuals spend entire careers dedicated to the analysis of NTFP trade, medicinal plant conservation, common property resources, certification or other niche subjects. Each has its own specialized outlets for publication and dissemination, so insider information is essential for locating some of the more esoteric literature. Although most articles in the bibliography are from published sources, I decided to include grey literature for two important reasons. First, it demonstrates the breadth of organizations working on issues relating to NTFPs. Second, many grey documents relating to NTFP laws and policies are important compendiums that represent prodigious amounts of research into primary sources such as laws, legislative proceedings and departmental memoranda (e.g. de Silva et al, 2001). Such papers are rarely published in peer-reviewed journals because they are generally descriptive in nature and lack theoretical analysis, but they are invaluable to NTFP policy researchers.

THE NTFP LAW AND POLICY LITERATURE

There are at least eight distinguishable macro-themes in the NTFP law and policy literature: tenure and resource access rights; equity, including research ethics, benefit sharing and traditional knowledge; conservation and resource management laws and policies specifically tailored for NTFPs; laws and policies that indirectly impact NTFPs; economic development initiatives that promote NTFP commercialization; NTFP certification; safety and sanitation standards for NTFPs; and trade regulations for internationally marketed NTFPs. These categories are admittedly porous. An article about NTFPs and CITES (the Convention on International Trade in Endangered Species of Wild Fauna and Flora), for example, could logically fit under the heading dedicated to laws and policies that impact NTFPs or be nested in the trade field, while an article that appears under the heading 'conservation and resource management laws and policies specifically for NTFPs' might also include an engaging side discussion about resource access or benefit sharing.

The two themes that have generated the most interest in the literature are, first, the use of NTFP commercialization as a conservation and economic development tool, and, second, benefit sharing and the protection of traditional knowledge. NTFP commercialization has its critics (Browder, 1992; Dove, 1994; Crook and Clapp, 1998) and its proponents (Allegretti, 1990; Shackleton, 2001), but even its champions have tempered their initial optimism about NTFP commercialization as a conservation and development tool. It appears that the success of NTFP commercialization projects, unsurprisingly, is highly dependent upon contextual factors (Marshall et al, 2006) and such projects inevitably create winners and losers within local communities (Rigg, 2006). An offshoot of this literature specifically examines NTFPs and their relationship to poverty. While some researchers claim that NTFPs provide a safety net for rural households, others posit that NTFPs are a poverty trap (see Brown et al, 1998; Shackleton et al, 2002; Angelsen and Wunder, 2003; Paumgarten, 2005). At the global policy level, Kaushal and Melkani (2005) argue that the production, consumption and sale of NTFPs could be central to meeting the United Nations Millennium Development Goals for poverty reduction.

The literature on benefit sharing and the protection of traditional knowledge is evolving rapidly. Much of it focuses on how countries are planning to implement access and benefit-sharing agreements in line with the provisions of the Convention on Biological Diversity (e.g. Wynberg and Laird, 2007; Laird et al, 2008; Taylor and Wynberg, 2008; Kamau and Winter, 2009). Other literature in this area focuses on the pros and cons of bioprospecting by companies from the North in countries of the South that possess great biological wealth (Dutfield, 2002; Greene, 2004; Wynberg et al, 2009). Rosendal (2006) opines that much of the interest in benefit sharing and traditional knowledge is driven by the economic potential of discovering new patentable medicines. The literature in this field is mixed in its emphasis, with many publications focused on the environmental and economic gains from bioprospecting (e.g. Barratt and Lybbert, 2000) and others on the protection of traditional knowledge and on environmental and social justice.

The NTFP literature as a whole has been criticized for inconsistency in its application of research methods and analytical frameworks (Neumann and Hirsch, 2000). Case studies (e.g. Ruiz-Pérez and Byron, 1999; Ruiz-Pérez et al, 2005; Marshall et al, 2006; Belcher and Schreckenberg, 2007) have been widely used as an investigative tool for individual NTFP projects and as a means to compare projects. Surveys, too, have been popular research tools. The NTFP literature is also replete with in-depth qualitative interviews of gatherers, yet most studies of this kind focus on harvester knowledge or livelihoods, with few reporting on gatherer responses to law and policy (some exceptions being McLain et al, 1998; McLain, 2000; Pandit and Thapa, 2003; Larsen et al, 2005). A parallel literature emanates from ethnobiology, where botanical and ecological researchers emphasize the uses of plants and their economic value (Peters et al, 1989; Balick and Mendelsohn, 1992; Grimes et al, 1994; Shanley, 2000). Anthropological researchers have also undertaken in-depth research over many years on people's relationship with their environment (e.g. Posey and Balée, 1989; Peluso, 1992; Redford and Padoch, 1992; Alexiades and Lacaze, 1995; Alexiades, in press). This literature is vast and has a long history, and is not known specifically as 'NTFP literature', but its subject matter is often people's relationships to NTFPs, and it tends to reflect the complexity of this field and have great depth and detail.

A small group of social scientists in the NTFP field have created another type of literature that seeks to 'return results' or exchange knowledge with local communities. This literature takes the form of manuals or illustrated books that allow communities to make informed choices about the ways they use, manage and trade species. In addition to detailed ecological, marketing and socioeconomic data, some of these publications present information on policies that impact NTFPs, or present information that allows communities to engage with the wider economy (e.g. comparing the timber and non-timber values of trees) (for an example, see Shanley and Medina, 2005).

Some scholars (e.g. Peluso, 1992; McLain, 2000) have used power as an analytical lens to examine conflict between NTFP gatherers and the state, particularly with regard to resource access, permit processes and law enforcement. Other scholars have used discourse analysis to parse and categorize the conflicting visions of stakeholders in the formulation and implementation of forest policy (Humphreys, 2004) or applied concepts of participation or democracy to elucidate struggles in policy-making processes (Ribot, 1995; Tobin and Swiderska, 2001; Pattberg, 2005). In general, the NTFP law and policy literature is descriptive, and applications of theoretical frameworks are infrequent and, when used, sometimes not fully explored.

AREAS FOR FURTHER RESEARCH

The NTFP law and policy literature is in its infancy and fractured in approach, perhaps owing to the many disparate policies that impact NTFP management, harvest and trade. As with any young literature, it has its gaps and shortcomings. Below I identify a few areas that are in need of further research.

Researchers note that little is known about the biology, ecology and trade of many locally used species and even a few internationally traded NTFPs. Inventory assessments, valuation studies and investigations of sustainable harvest volumes for target species are still needed. However, the link between such studies and policy formulation and implementation is unclear. In many instances, research results are not transferred to the policy arena. In other cases, research results are bent or distorted for political purposes. For example, inventory data might be used to ease harvest regulations or be cited as evidence for curtailing the harvest of a particular species. In any case, calls for further ecological and economic studies must be balanced by an increase in social science research into subjects such as the impact of laws and policies on access to resources, the consequences of NTFP law and policy implementation (including bribes and corruption), and the impact of law and policy on subsistence use of NTFPs by forest-dependent communities in developed countries (see Emery and Pierce, 2005) as well as in developing countries.

Few studies have examined how NTFP laws and policies are created. In some cases, such studies are nearly impossible, as decisions are often made behind closed doors, with few public records surviving. In transparent forums, however, meeting notes and other public documents, as well as interviews with stakeholders, could provide a fascinating anatomy of how NTFP laws or policies come to life. Such studies could use theories of legitimacy, democracy, discourse analysis, social network analysis and/or the qualitative analysis of in-depth interviews to reveal how actors influence processes leading to the formulation of law or policy.

More scholarly attention needs to be given to the ultimate impact of voluntary codes of conduct (e.g. good agricultural practices, responsible business codes for sourcing herbs, forest management certification standards and wildcrafter guidelines) on law and policy. Do such codes, which originate in the private sector from trade associations, non-governmental organizations and others, ultimately influence legislation and, if so, how? What trajectory do voluntary codes follow as they become governmental law or policy? Why do some codes advance while others fail? And how effectively do stakeholders adhere to such codes?

With respect to NTFP management, relatively few studies have looked at how NTFP gatherers can improve their participation in policy-making and implementation. Gatherers are by no means a homogeneous group, and few papers have looked at their attempts to interface with policy initiatives (one exception being McLain et al, 1998). Even fewer papers have included gatherers and their harvesting techniques in scientific studies (e.g. Peck and Christy, 2006). It would also be worthwhile investigating the impact that different tenure arrangements and management systems have on various NTFP resources, as suggested by Tedder et al (2002).

Vantomme (2003) has discussed the need for better tracking of national and international trade in NTFPs, identifying some major hurdles such as customs product classification codes. This topic is in need of further research, not only to identify volumes and species traded, but to explore existing policies that either facilitate or fetter trade, particularly at the subnational level.

The commodity chain or value chain analysis approach (see Ribot, 1998; Gereffi et al, 2005) is a promising but little-used research methodology in NTFP studies. Increasingly, NTFP researchers are engaging with value chain analysis to understand trade structures, issues of power and benefit sharing within trading relationships, and the impact of trade regulations (e.g. Te Velde et al, 2006; Wynberg, 2006). Commodity or value chain analysis could be particularly useful in examining the repercussions of growing calls for the imposition of quality standards for medicinal plants (e.g. good agricultural practices, which mostly focus on ensuring the safety and proper handling of raw materials) and the ecological certification of NTFPs (to prove sustainability). Such initiatives, while well-intended, are likely to radically transform NTFP market chains, and the commodity chain lens may be an appropriate tool to illuminate such rearrangements. As Ribot (1998) trenchantly observes, the commodity chain approach is essentially a policy tool.

The impact of NTFP labelling and/or certification is still unknown and there are relatively few published case studies on the topic (see Shanley et al, 2008). The number of standards for NTFP harvest and trade is growing – including organic certification, Fairtrade certification, FairWild certification, the International Standard for Sustainable Wild Collection of Medicinal and Aromatic Plants (ISSC-MAP) and the Union for Ethical BioTrade, to name a few – yet the impact of such programmes is unclear. Long-term studies of certified operations would help policy-makers better understand the challenges and benefits of product labelling and certification. Such studies could prove difficult, due to business and certifier reluctance to discuss confidential matters, but more information is needed about why companies choose to become certified, what factors lead to success and what factors produce failure or disillusionment with labelling and certification. Market studies are also needed on the impact of regional labels, designations of origin and voluntary, locally developed labelling and marketing programmes. Do such labelling and marketing programmes offer less expensive, more flexible and more relevant marketing assistance to NTFP producers than certification?

CONCLUSION

The term 'NTFP' was coined in the late 1980s, largely in response to a perceived dominance of timber-centric thinking in forestry, conservation and policy-making circles. 'NTFP' is thus a political construct, promoted largely by private sector conservationists and academics. Unfortunately the term is not well recognized by the political establishments that oversee NTFP management, harvest and trade, specifically ministries of forestry, agriculture, taxation, trade and health. Therefore NTFPs often fall into no-man's-land in policy-making, subject to a variety of laws and policies at the international, national and local levels. Likewise, the term 'NTFP' is not widely recognized across academic disciplines, making the collection of NTFP research articles challenging.

The *NTFP Law and Policy Bibliography* (Pierce, 2009) is by no means a comprehensive collection. Literature searches were limited to modern-era NTFP papers (those published since the late 1980s), omitting important articles from the historical NTFP literature. Funding limitations and a lack of linguistic expertise meant that most of the articles reviewed were in the English language. Although an attempt was made to include literature from all areas of the globe, certain regions (e.g. North America) are better represented than others (e.g. Australia, Oceania). Limiting the literature search to the eight major subtopics listed above has resulted in other policies and laws that impact NTFP management, harvest and trade being overlooked. For example, some NTFPs are harvested by migrant workers whose ability to work is highly dependent on labour and immigration policies (Lynch and McLain, 2003; Chapter 11 of this volume). Likewise, tax policies at the local and regional level may support or undermine NTFP operations, yet few papers cover such matters in detail. Despite these limitations, it is hoped that this review and the bibliography it is based upon provide a useful introduction to the complexities of governance issues surrounding NTFP management, harvest and trade, ranging from international accords to national laws to customary laws.

REFERENCES

Alexiades, M. (ed) (in press) *Mobility and Migration in Indigenous Amazonia: Contemporary Ethno-ecological Perspectives*, Berghahn, Oxford, UK

Alexiades, M. and Lacaze, D. (1995) *'Salud para todos': plantas medicinales y salúd indígena en el Departamento de Madre de Dios, Peru. Un manual práctico*, FENAMAD/Centro de Estudios Rurales Bartolome de las Casas, Cusco, Peru

Allegretti, M. H. (1990) 'Extractive reserves: An alternative for reconciling development and environmental conservation in Amazonia', in A. B. Anderson (ed) *Alternatives to Deforestation: Steps Toward Sustainable Use of the Amazon Rain Forest*, Columbia University Press, New York, NY, pp252–264

American Heritage Dictionary (2000) *The American Heritage Dictionary of the English Language, 4th edition*, Houghton Mifflin, Boston, MA

Angelsen, A. and Wunder, S. (2003) *Exploring the Forest-Poverty Link: Key Concepts, Issues and Research Implications*, CIFOR Occasional Paper 40, CIFOR, Bogor, Indonesia

Balick, M. J. and Mendelsohn, R. (1992) 'Assessing the economic value of traditional medicines from tropical rain forests', *Conservation Biology*, vol 6, no 1, pp128–130

Barratt, C. B. and Lybbert, T. J. (2000) 'Is bioprospecting a viable strategy for conserving tropical ecosystems?', *Ecological Economics*, vol 34, pp293–300

Belcher, B. and Schreckenberg, K. (2007) 'Commercialisation of non-timber forest products: A reality check', *Development Policy Review*, vol 25, no 3, pp355–377

Browder, J. (1992) 'Social and economic constraints on the development of market-oriented extractive reserves in Amazon rain forests', in D. Nepstad and S. Schwartzman (eds) *Non-Timber Products from Tropical Forests: Evaluation of a Conservation and Development Strategy – Advances in Economic Botany*, Volume 9, New York Botanical Garden, Bronx, NY, pp33–42

Brown, R. B., Xu, X. and Toth, J. F. (1998) 'Lifestyle options and economic strategies: Subsistence activities in the Mississippi Delta', *Rural Sociology*, vol 63, pp599–623

Crook, C. and Clapp, R. A. (1998) 'Is market-oriented forest conservation a contradiction in terms?', *Environmental Conservation*, vol 25, no 2, pp131–145

de Silva, S. et al (2001) *Regulation of the Trade in Timber and Non-Timber Forest Products in the Lower Mekong Basin Countries*, IUCN, Gland, Switzerland

Dove, M. R. (1994) 'Marketing the rainforest: "Green" panacea or red herring?', *Asia Pacific Issues, No. 13*, East-West Center, Honolulu, HI

Emery, M. R. and Pierce, A. R. (2005) 'Interrupting the telos: Locating subsistence in contemporary US forests', *Environment and Planning A*, vol 37, pp981–993

Gereffi, G., Humphrey, J. and Sturgeon, T. (2005) 'The governance of global value chains', *Review of International Political Economy*, vol 12, no 1, pp1–27, www.ids.ac.uk/globalvaluechains/publications/govgvcsfinal.pdf, accessed 1 March 2006

Greene, S. (2004) 'Indigenous people incorporated? Culture as politics, culture as property in pharmaceutical bioprospecting', *Current Anthropology*, vol 45, no 2, pp211–237

Grimes, A., Loomis, S., Jahnige, P., Burnham, M., Onthank, K., Alarcón, R., Cuenca, W. P., Martinez, C. C., Neill, D., Balick, M., Bennett, B. and Mendelsohn, R. (1994) 'Valuing the rain forest: The economic values of non-timber forest products in Ecuador', *Ambio*, vol 23, pp405–410

Humphreys, D. (2004) 'Redefining the issues: NGO influence on international forest negotiations', *Global Environmental Politics*, vol 4, no 2, pp51–74

Kamau, E. C. and Winter, G. (2009) *Genetic Resources, Traditional Knowledge and the Law: Solutions for Access and Benefit Sharing*, Earthscan, London

Kaushal, K. K. and Melkani, V. K. (2005) 'India: Achieving the Millennium Development Goals through non-timber forest products', *International Forestry Review*, vol 7, no 2, pp128–134

Larsen, H. O., Smith, P. D. and Olsen, C. S. (2005) 'Nepal's conservation policy options for commercial medicinal plant harvesting: Stakeholder views', *Oryx*, vol 39, no 4, pp435–441

Lynch, K. A. and McLain, R. J. (2003) *Access, Labor, and Wild Floral Greens Management in Western Washington's Forests*, General Technical Report no PNW-GTR-585, US Department of Agriculture, Forest Service, Pacific Northwest Research Station, Portland, OR

Maille, P. (2001) *A Non-Timber Forest Product Bibliography Emphasizing Central Africa*, CARPE, Washington, DC

Marshall, E., Schreckenberg, K. and Newton, A.C. (eds) (2006) *Commercialization of Non-Timber Forest Products: Factors Influencing Success. Lessons Learned from Mexico and Bolivia and Policy Implications for Decision-Makers*, UNEP/WCMC, Cambridge, UK

McLain, R. J. (2000) 'Controlling the forest understory: Wild mushroom politics in central Oregon', PhD dissertation, College of Forest Resources, University of Washington, Seattle, WA

McLain, R. J., Christensen, H. H. and Shannon, M.A. (1998) 'When amateurs are the experts: Amateur mycologists and wild mushroom politics in the Pacific Northwest, USA', *Society and Natural Resources*, vol 11, pp615–626

Neumann, R. and Hirsch, E. (2000) *Commercialisation of Non-Timber Forest Products: Review and Analysis of Research*, CIFOR, Bogor, Indonesia

Pandit, B. and Thapa, G. (2003) 'A tragedy of non-timber forest resources in the mountain commons of Nepal', *Environmental Conservation*, vol 30, no 3, pp283–292

Pattberg, P. (2005) 'The Forest Stewardship Council: Risk and potential of private forest governance', *Journal of Environment and Development*, vol 14, no 3, pp356–374

Paumgarten, F. (2005) 'The role of non-timber forest products as safety-nets: A review of evidence with a focus on South Africa', *GeoJournal*, vol 64, pp189–197

Peck, J. E. and Christy, J. A. (2006) 'Putting the stewardship concept into practice: Commercial moss harvest in Northwestern Oregon, USA', *Forest Ecology and Management*, vol 225, 225–233

Peluso, N. (1992) *Rich Forests, Poor People: Resource Control and Resistance in Java*, University of California Press, Berkeley, CA

Peters, C. M., Gentry, A. H. and Mendelsohn, R. O. (1989) 'Valuation of an Amazonian rainforest', *Nature*, vol 339, pp655–656

Pierce, A. (2009) *NTFP Policy and Law Bibliography*, People and Plants International, Vermont, www. peopleandplants.org

Posey, D. A. and Balée, W. L. (eds) (1989) *Resource Management in Amazonia: Indigenous and Folk Strategies – Advances in Economic Botany*, Volume 7, New York Botanical Garden, Bronx, NY

Redford, K. H. and Padoch, C. (eds) (1992) *Conservation of Neotropical Forests: Working from Traditional Resource Use*, Columbia University Press, New York, NY

Ribot, J. C. (1995) 'From exclusion to participation: Turning Senegal's forestry policy around?', *World Development*, vol 23, no 9, pp1587–1599

Ribot, J. C. (1998) 'Theorizing access: Forest profits along Senegal's charcoal commodity chain', *Development and Change*, vol 29, pp307–341

Rigg, J. D. (2006) 'Forests, marketization, livelihoods and the poor in the Lao PDR', *Land Degradation and Development*, vol 17, pp123–133

Rosendal, G. K. (2006) 'Balancing access and benefit sharing and legal protection of innovations from bioprospecting', *Journal of Environment and Development*, vol 15, no 4, pp428–447

Ruiz-Pérez, M. and Byron, N. (1999) 'A methodology to analyze divergent case studies of nontimber forest products and their development potential', *Forest Science*, vol 45, no 1, pp1–14

Ruiz-Pérez, M., Belcher, B., Achdiawan, A., Alexiades, M., Aubertin, C., Caballero, J., Campbell, B., Cunningham, T., Fantini, A., de Foresta, H., Fernández, C. G., Gautam, K. M., Martinez, P. H., de Jong, W., Kusters, K., Kutty, M. G., López, C., Fu, M., Alfaro, M. A. M., Nair, T. K. R., Ndoye, O., Rai, N., Ricker, M., Schreckenberg, K., Shackleton, S., Shanley, P., Sunderland, T. and Youn, Y. C. (2005) 'Markets drive the specialization strategies of forest peoples', *Ecology and Society*, vol 9, no 2, p4, www.ecologyandsociety.org/vol9/iss2/art4, accessed 15 March 2006

Shackleton, C. M. (2001) 'Re-examining local and market-oriented use of wild species for the conservation of biodiversity', *Environmental Conservation*, vol 28, no 3, pp270–278

Shackleton, C. M., Shackleton, S. E., Ntshudu, M. and Ntzebeza, J. N. (2002) 'The role and value of savanna non-timber forest products to rural households in the Kat River Valley, South Africa', *Journal of Tropical Forest Products*, vol 8, pp45–65

Shanley, P. (2000) 'As the forest falls: The changing use, ecology and value of nontimber forest resources in Eastern Amazonian Caboclo communities', PhD dissertation, University of Kent, Canterbury, UK

Shanley, P. and Medina, G. (eds) (2005) *Frutíferas e plantas uteis na vida Amazônica*, CIFOR/IMAZON, Belém, Brazil

Shanley, P., Pierce, A., Laird, S. and Robinson, D. (2008) *Beyond Timber: Certification of Non-Timber Forest Products*, CIFOR, Bogor, Indonesia

Taylor, M. and Wynberg, R. (2008) 'Regulating access to South Africa's biodiversity and ensuring the fair sharing of benefits from its use', *South African Journal of Environmental Law and Policy*, vol 15, no 2

Tedder, S., Mitchell, D. and Hillyer, A. (2002) *Property Rights in the Sustainable Management of Non-Timber Forest Products*, British Columbia Ministry of Forests/Forest Renewal BC, Victoria, BC

Te Velde, D. W., Rushton, J., Schreckenberg, K., Marshall, E., Edouard, F., Newton, A. and Arancibia, E. (2006) 'Entrepreneurship in value chains of non-timber forest products', *Forest Policy and Economics*, http://quin.unep-wcmc.org/forest/ntfp/cd/4_Journal_publications/a_Entrepreneurship_NTFPs_Te_Velde_et_al_2005.pdf, accessed 20 February 2006

Tobin, B. and Swiderska, K. (2001) *Speaking in Tongues: Indigenous Participation in the Development of a Sui Generis Regime to Protect Traditional Knowledge in Peru*, IIED, London

Vantomme, P. (2003) 'Compiling statistics on non-wood forest products as policy and decision-making tools at the national level', *International Forestry Review*, vol 5, no 2, pp156–160

von Hagen, B., Weigand, J., McLain, R. J., Fight, R. and Christensen, H. (1996) *Conservation and Development of Nontimber Forest Products in the Pacific Northwest: An Annotated Bibliography*, General Technical Report no PNW-GTR-375, US Department of Agriculture, Forest Service, Pacific Northwest Research Station, Portland, OR

Wynberg, R. (2006) 'Identifying pro-poor, best practice models of commercialisation of southern African non-timber forest products', PhD thesis, University of Strathclyde, Glasgow

Wynberg, R. and Laird, S. (2007) 'Less is often more: Governance of a non-timber forest product, marula (*Sclerocarya birrea* subsp. *caffra*) in southern Africa', *International Forestry Review*, vol 9, no 1, pp475–490

Wynberg, R., Chennells, R. and D. Schroeder (eds) (2009) *Indigenous Peoples, Consent and Benefit Sharing: Learning from the San-Hoodia Case*, Springer, Berlin

Index

aboriginal peoples *see* indigenous peoples

açaí (*Euterpe* spp.), Brazil 47, 50, 356

Acre (Brazil), legislation 47

aflatoxins, Brazil nuts 32, 331, 345, 349

African teak *see* kiaat

Agathis resin *see* almaciga resin

Agave cupreata
  access to 214–215, 352
  management of 215–216
  Mexico 8, 213–220
  mezcal production 216–217
  regulations 218–219

agrarian reform, Bolivia 24–34, 42, 351

agricultural policies 351

agriculture, traditional methods 191–193

Alangan Mangyan people, Philippines 163–166, 167, 173

allspice (*Pimienta dioica*), Mexico 207

almaciga resin (*Agathis philippinensis*)
  climate change impacts 194
  conservation projects 192
  gathering techniques 172
  Philippines 7, 155–156, 168–173, 186–188, 190
  prices 169–170

ambé vine (*Philodendron* sp.), Brazil 45

American Herbal Products Association (AHPA) 337

ancestral domains, Philippines 157–162

andiroba (*Carapa guianensis* and *C. paraense*), Brazil 44, 46, 47

appellation of origin, Mexican mezcal 8, 209, 219–220

appetite suppressants, *Hoodia* 10, 312–313

babaçu (*Orbignya phalerata*), Brazil 47

bacaba (*Oenocarpus mapoa*), Brazil 47

bamboo
  India 89, 92
  trade 328

barraqueros, Bolivia 20–28, 30–31, 37–38

Batak people, Philippines 7–8, 183–195

bilberries (*Vaccinium myrtillus*), Finland 290, 299

bioprospecting 321, 335

bluebells (*Hyacinthoides non-scripta*), United Kingdom 142, 143

blueberries (*Vaccinium corymbosum*), Europe 294

Bolivia
  agrarian reforms 24–34, 42, 351
  barraqueros 20–28, 30–31, 37–38
  Brazil nuts 5–6, 15–38, 347
  economy 16–17, 19–24
  export documents 332
  forest management 30–31
  Forestry Law 24, 28–30
  indigenous peoples 26–27
  land transaction 35
  legislation 42
  policy initiatives 31–34
  property rights 20–22, 24, 25–28, 35
  recommendations for Brazil nut sector 34–37
  regional transportation 34
  rubber production 20–23
  structural adjustment policies 23
  Tenure Reform law (INRA Law) 24–28
  tree tenure 21–22, 31, 353

Botswana
  Hoodia spp. 309–324
  legislation 314–319

Brazil
  deforestation 330
  endangered species 43
  extraction and replanting of NTFPs 48–49
  forest management 46
  intellectual property rights 336
  legislation 44–45, 47
  regulations 6, 48
  sustainable use of NTFPs 43–51
  transport permits 49–51, 349–350

Brazilian Institute for the Environment and Renewable Natural Resources (IBAMA) 43–51

Brazil nuts (*Bertholletia excelsa*)
  access control 329
  aflatoxins 32, 331, 345, 349

Bolivia 15–38
Brazil 44, 46
   certification 32–34
   economic importance 16–17, 22–23
   harvest practices 19, 22
   legislation 5–6, 42, 44, 330
   management norms 30–31
   policy process 347
   processing plants 17, 22–23, 32, 33, 38
   stakeholder alliances 37
   trade 328
bribery
   Cameroon 6, 62–63, 74, 331
   Philippines 174, 331, 350
   unofficial taxation 62–63, 65, 350
British Columbia
   aboriginal rights 121
   commercial harvesting 116–118
   community forest agreements 121–122
   First Nations 7, 118, 121, 124–125,
      129–130, 352
   knowledge transfer 124–125
   multi-species management 121–122
   non-commercial NTFPs 118–119
   NTFP policies 6–7, 119–131
   policy gaps 125–128
   private lands 122
   property rights 121
   resource planning 122–123
   single species management 119–120
buriti (*Mauritia flexuosa*), Brazil 47
Bursera spp., Mexico 206
bush mango (*Irvingia* spp.) 6, 60, 77–82
   regulations 79–82
   sustainability and management 78–79
   types of 77–78
   uses of 78
bush plum (*Dacryodes edulis*), Cameroon 58

cacti, Mexico 207
Camedor palm (*Chamaedorea* spp.), Mexico
   206, 207
Cameroon
   bureaucracy 356
   bush mango 6, 77–82
   community forests 60–61
   customary law 55–57, 65
   export documents 332
   forestry laws 57–61, 65
   Gnetum spp. 6, 62, 71–75

government institutions 63–64, 66, 357
land tenure 55, 65
Prunus africana 53, 56, 58, 60, 61, 334–335
quotas and permits 59–60
regulatory framework 6, 63–65
resource rights 55–57
Special Forest Products list 58, 59, 60
taxation 61–63, 65, 345, 350
trade with Nigeria 6, 77
use of NTFPs 53–54
Canada
   floral greens 117, 270
   NTFP policies 113–131
candelilla (*Euphorbia antisyphillitica*), Mexico
   207
CapeNature 317
carnaúba (*Copernicia prunifera*), Brazil 47
carving wood, South Africa 8, 199–203
cascara bark (*Rhamnus purshiana*), British
   Columbia 119
castañales (tree tenure), Bolivia 21–22, 31,
   353
caterpillar fungus (*Cordyceps sinensis*), China
   254
cat's claw (*Uncaria tomentosa* and *U. guianensis*)
   Brazil 44, 47
   safety concerns 344–345
cebolão vine (*Clusia* spp), Brazil 44
cedar foliage (*Thuja plicata*), British
   Columbia 120
cep (*Boletus* spp.)
   Finland 289, 291
   Scotland 137
certification schemes
   Brazil nuts 32–34
   mezcal 220
   Scotland 150–151
   voluntary 336–338
China
   county regulations 251–253
   forest policies 249–251, 260
   matsutake mushrooms 9, 243–261, 352
   multi-tiered NTFP management 9,
      258–261, 352
   trade liberalization 358
   village regulations 253–258, 259, 261
Christmas decorations 120, 265
Cinchona pubescens, Cameroon 60
climate change, Philippines 193–194

cloudberries (*Rubus chamaemorus*), Finland 290, 295, 301

commercialization
British Columbia 116–118
Fiji 232–233
Hoodia spp. 10, 312–314
India 85–86
Mexico 207–208, 217–218
mezcal production 217–218
production cycle 344–345
recommendations 367–368
Scotland 144–145

Commission for Reconciliation, Arbitration and Resolution of Conflicts (CCARC) (Bolivia) 27, 35

community-based forest management agreements (CBFMAs), Philippines 188, 189–191

community development, British Columbia 129–130

community forests
British Columbia 121–122
Cameroon 60–61
China 249–250

community resource management
access rights 330
China 253–258, 259
Fiji 238–240
Mexico 214–216
Philippines 166–167, 171–172, 178–180, 356

conservation
matsutake mushrooms 255–256
Philippines 189–191, 192–193
policy approaches 360–361

consultation, policy-making 346–347, 370

Convention on Biological Diversity (CBD) 2, 209, 333, 335, 360

Convention on International Trade in Endangered Species of Wild Fauna and Flora (CITES) 58, 332–333, 334–335
China 248, 250, 260
Fiji 237–238
Hoodia spp. 314, 317, 321–322

cooperatives, India 95–102, 107

copaíba (*Copaifera* spp.), Brazil 44, 46, 47

copal resin (*Bursera* spp.), Mexico 206–207

copra (*Cocos nucifera*), Fiji 233

cork (*Quercus suber*), trade 328

Council for Scientific and Industrial Research (CSIR) 312–313, 319

Council for Voluntary Forest Certification (CFV) (Bolivia) 32

cowberries *see* lingonberries

cranberries (*Vaccinium* spp.), Europe 290, 294

cuachalalate (*Amphipterygium adstringens*), Mexico 207

customary law
bush mango 79–82
Cameroon 55–57, 65
Fiji 8–9
Philippines 177–179, 352–353
policy-making 374
resource rights 352–353

Democratic Republic of Congo, *Gnetum* spp. 71

devil's claw (*Harpagophytum procumbens*), southern Africa 334, 345

Diospyros egrettarum, Cameroon 60

doradilla (*Selaginella lepidophylla*), Mexico 207

economic alliances 358–359

ecotourism, Canada 125

El Niño 190, 194, 358

ephedra (*Ephedra sinica*), safety concerns 332, 344

eru *see Gnetum* spp.

erva-mate (*Ilex paraguariensis*), Brazil 44

essential oils, trade 328

European Union (EU), wild berry imports 9–10, 294, 298

evergreen huckleberry (*Vaccinium ovatum*), United States 265, 268

everyman's rights, wild berries 295–297, 301, 352

export bans 333–334

exports, wild berries 293–294, 358

fair trade
Brazil nuts 33
certification 336, 337

FairWild certification 337

Fiji
community forest management 238–240
culture 230–232
forest policy 236–237, 357
Intsia bijuga (*vesi*) 8–9, 229–241, 346, 353
land tenure 235–236, 353

legislation 236–238
resource depletion 233–235, 240–241
tourism 232–233
woodcarving 231–233
Finland
    everyman's rights 295–297, 301, 352
    exports 293–294, 358
    government research 347
    harvesting practices 287–293
    labour 9–10, 298–300, 351
    lichen 296
    market access 298, 301–302
    mushrooms 289, 292, 301
    NTFPs collected 288
    taxation 297
    wild berry harvesting 9–10, 287–302
firewood, mezcal production 216
First Nations, British Columbia 7, 118, 121,
        124–125, 129–130, 352
floral greens
    British Columbia 117
    definition 265
    labour force 9, 269–270, 275–280, 354
    legislation 272
    markets 268–270
    Pacific Northwestern United States 9,
        265–283
    poaching 272–275
    tenure system 270–271
    transportation 349
food safety 331–332, 345, 349
forest, definition 4
forest products, definition 1–2
Forestry Commission, United Kingdom 135,
        140, 141
forestry laws 348–349
fungi see mushrooms

geographic indications (GI), Mexican mezcal
        209, 220, 221–222, 354
ginseng (Panax ginseng)
    Russia 297
    trade 328
globalization 358
Gnetum spp. (eru)
    Cameroon 6, 71–75
    informal taxation (bribes) 62, 74
    quotas 73, 75
    regulatory framework 73–75
    types of 72–73

gobernadora (Larrea tridentata), Mexico 207
goldenseal (Hydrastis canadensis), United
        States 335
governance, definition 5
griffonia (Griffonia simplicifolia), safety
        concerns 332, 344
Guano palm (Sabal spp.), Mexico 206–207
gum arabic (Acacia spp.), trade 328
gum karaya (Sterculia spp.), harvest ban 333
gum latex (Manilkara zapota), Mexico 207

harvest bans 333–334
harvesters, consultations 346–347
harvesting practices
    Brazil nuts 19, 22
    Finland 287–293
    matsutake mushrooms 255–256
    rattan 163
    uppage 104–106
Hat palm (Brahea dulcis), Mexico 206–207
health and safety, regulations 331–332,
        349
heno, Mexico 207
honey
    Cameroon 63
    Philippines 158, 173, 178–179, 186
    trade 328
Hoodia Growers Association of Namibia
        (HOGRAN) 318, 319
Hoodia spp.
    access and benefit sharing 309, 321–322
    commercialization 312–314
    ecology 311–312
    harvest bans 333
    intellectual property rights 10, 312–313,
        353–354
    legislation 10, 314–319, 349
    permits 319, 320, 345
    policy frameworks 314–324
    production cycle 322–324, 344
    southern Africa 10, 309–324, 359
    traditional knowledge 313, 318–319
hydroxycitric acid (HCA) 104

India
    central forest region 92–98
    commercialization 85–86
    cooperatives 95–102, 107
    National Forest Policy (1988) 89–90
    NTFP policies 6, 87–107

resource rights 352
rural livelihoods 87, 89
state control of NTFPs 6, 90–91, 107
tendu leaf 92–98, 329, 345
tribal communities 87, 89, 92–98, 107,
    360
uppage 102–106, 107, 357
indigenous peoples
Bolivia 26–27
British Columbia 118, 121, 124–125,
    129–130
India 87, 89, 92–98, 107, 360
Mexico 214–215
Philippines 7–8, 157–180, 183–195
rights of 2, 10, 121, 156–157, 188–191,
    359–360
traditional agriculture 191–193
traditional knowledge 309, 313, 319, 321
Indonesia, rattan export ban 333
informal taxation (bribes), Cameroon 6,
    62–63, 74
information, for policy-making 345–346,
    369–370
Integral Agro-extractive Farmers' Coopera-
    tive of Pando (COINACAPA) (Bolivia)
    33–34, 36
intellectual property rights 10, 312–313, 336,
    353–354, 371
International Standard for Sustainable Wild
    Collection of Medicinal and Aromatic
    Plants (ISSC-MAP) 338
Intsia bijuga (*vesi*)
    commercialization 232–233
    cultural significance 230–232
    depletion of 233–235, 240–241, 346, 353
    Fiji 8–9, 229–241, 346, 353
    see also ipil
'invisible elbows', British Columbia 115,
    125–128
ipê-roxo (*Tabebuia* spp.), Brazil 47
ipil (*Intsia bijuga*), Philippines 192

jaborandi (*Pilocarpus* spp.), Brazil 43
jagube (*Banisteriopsis caapi*), Brazil 44, 47
Japan, matsutake imports 243, 245–246,
    251
jarina (*Phytelephas macrocarpa*), Brazil 47
jatobá (*Hymenaea courbaril*), Brazil 47
joint forest management (JFM), India 89–90

Kabara (Fiji), woodcarving 229–241
Karelia (Russia)
    NTFP policy 297–298
    wild berries 292–293
Karnataka (India)
    Large-scale Adivasi Multi-Purpose Societies
        (LAMPS) 99–102
    uppage 102–106
kava (*Piper methysticum*), safety concerns 332,
    344
kendu (KL) *see* tendu
kiaat (*Pterocarpus angolensis*) (African teak),
    South Africa 8, 199–203, 350
knowledge transfer
    British Columbia 124–125
    see also traditional knowledge

labour issues
    floral greens 9, 269–270, 275–280, 354
    policies 354–355
    wild berries 9–10, 298–300, 352, 355
land tenure
    Cameroon 55, 65
    China 249
    Fiji 235–236
    impacts of 351–353, 371
    Mexico 214–215
    see also property rights
land use, recommendations 368
Lapland (Finland), wild berries 295, 299
Large Area Multi-Purpose Societies/Large-
    Scale Adivasi Multi-Purpose Societies
    (LAMPS), India 91, 97, 99–102
laws and policies 3, 343–361
    agricultural 351
    Bolivia 42
    Brazil 5–6, 43–51
    British Columbia 6–7, 123
    Cameroon 6, 57–61, 65
    common features 355–357
    direct impacts on NTFPs 348–350
    Fiji 8–9, 236–238
    forestry laws 348–349
    Hoodia spp. 10, 314–319
    India 6, 89–90
    indirect impacts on NTFPs 350–355
    Mexico 8, 208–210, 218–222, 226–227
    Philippines 7, 157–162, 173–175, 188–191
    quality control, safety and efficacy 349
    Russia 297–298

Scotland 139–144
southern Africa 314–319
taxation 350
transportation 349–350
United States 272, 275–280, 276–280
see also customary law
lechuguilla (*Agave lechuguilla*), Mexico 207
lichen (*Cladonia*), Finland 296
lingonberries (*Vaccinium vitis-idaea*), Finland
    290, 293, 344

maca (*Lepidium meyenii*), Peru 335
Madhya Pradesh (India), NTFP policies
    95–97, 98
mahogany (*Swietenia mahogani*), Fiji 236
mahua (*Madhuca indica*), India 92, 97
Manila copal *see* almaciga resin
manufacturing, regulations 331–332
markets
    floral greens 268–270
    wild berries 298, 300
marula (*Sclerocarya birrea*), South Africa 330,
    346, 353
mastic gum (*Pistacia lentiscus*), trade 328
matsutake mushrooms (*Tricholoma matsutake*)
    access control 253–254
    China 9, 243–261, 352
    harvesting control 255–256
    history of 244–246
    marketing 257–258
    monitoring and sanctions 256–257
    multi-tiered management 258–260
    protection 247–249, 250
    regulations 249–258
    trade 243–244, 250–253, 257–258, 328,
        358
    village regulations 253–258, 259, 261
    see also pine mushrooms
medicinal plants
    Brazil 46
    Cameroon 345
    commercial cycles 344–345
    intellectual property rights 336
    safety regulations 332, 349
    Scotland 135–136
    trade 327, 329
Mexico
    Agave cupreata 8, 213–220, 352
    commercial NTFPs 207–208
    environmental legislation 209, 218–219

indigenous peoples 214–215
intellectual property rights 336, 354
land tenure 208, 214–215, 352
legislation 208–210, 218–222, 226–227
local institutions 8, 214–217
mezcal production 213–220
mushrooms 207, 210–212
NTFPs used 205–208
mezcal production
    agave management 213–216
    appellation of origin 8, 209, 219–220
    commercialization 217–218
    Mexico 8, 213–220
    regulations 219–220
migrant labour
    Finland and Sweden 9–10, 299–300, 351,
        354–355
    United States 9, 269–270, 275–280
mining, Philippines 175, 194–195
Ministry of Forestry and Wildlife (MINFOF)
    (Cameroon) 63, 64
moss
    Mexico 207
    Scotland 136, 145
    United States 265
mountain pine beetle, British Columbia 124,
    125, 127–128
murmuru (*Astrocaryum* spp.), Brazil 47
mushrooms
    British Columbia 117, 119, 120, 122–
        123
    Finland 289, 292, 301
    gathering rights 141
    knowledge of 146–147
    Mexico 207, 210–212
    Pacific Northwest 280–282
    safety regulations 331–332, 345
    Scotland 136–137, 138–139, 149
    see also matsutake mushrooms

Namibia
    Hoodia spp. 309–324
    laws and policies 314–319, 347
    permits 319
narra (*Pterocarpus indicus*), Philippines 192
National Institute for Agrarian Reform
    (INRA) (Bolivia) 24–28
natural health products 126
natural resource management, policies
    330

Nepal
  access control 330
  export bans 333–334
Nigeria
  bush mango 6, 77–82
  trade with Cameroon 6, 77
njangsang *Ricinodendron heudelottii*, Cameroon 58
non-governmental organizations (NGOs), conservation projects 192–193, 259–260
Non-Timber Forest Products Task Force (NTFP-TF), Philippines 156–157, 163, 168, 171–173, 174–175
non-wood forest products (NWFPs) 4
North American Free Trade Agreement (NAFTA) 270, 359

oak (*Quercus* spp.), Mexico 207
Olympic Peninsula (United States), floral greens 9, 265–283
organic certification 336, 337
Orissa (India), NTFP policies 92–94, 96–97, 345, 360

Pacific Northwest (United States)
  floral greens 265–283
  mushrooms 280–282
Palawan (Philippines)
  almaciga resin 168–173
  indigenous peoples 7–8, 183–195, 358
palm hearts (*Euterpe edulis*), Brazil 43, 44, 50, 344
palm leaves, Mexico 206–207
paraná pine (*Araucaria angustifolia*), Brazil 44, 46
patoá (*Oenocarpus bataua*), Brazil 47
Pausinystalia johimbe, Cameroon 60
pequi (*Caryocar brasiliense*), Brazil 47
Pericopsis elata, Cameroon 58
Pfizer 312–313
pharmaceutical industry 312–313, 336, 353–354
Philippines
  almaciga (*Agathis*) resin 7, 155–156, 158, 168–173, 186–188, 190, 194
  ancestral domain legislation 157–162
  bribery 174, 331, 350
  bureaucracy 356
  climate change 193–194

customary law 177–179, 352–353
environmental conservation 189–191, 192–193
forestry sector 180
honey 158, 173, 178–179, 186
indigenous peoples 7–8, 157–180, 183–195
legislation 157–162, 173–175, 188–191
mining 175, 194–195
overharvesting 175, 190
rattan 7, 155–156, 158, 162–168, 186–187, 194
recommendations 175–180
social equity 175
Strategic Environmental Plan (SEP) 162, 189
Phytopharm 312–313
phytosanitary regulations 331
PhytoTrade Africa 347
pine mushrooms
  British Columbia 120, 122–123
  see also matsutake mushrooms
pine (*Pinus caribaea*), Fiji 236
pine resin (*Pinus* spp.), Mexico 207
pinyon (*Pinus cembroides*), Mexico 207
pita (*Aechmea magdalenae*), Mexico 207
poaching, floral greens 9, 272–275
policies
  consultations 346–347, 370
  definition 5
  implementation 357, 370–371
  incentives 373
  information requirements 345–346
  opportunistic policy-making 345
  reactive policy-making 344–345
  recommendations 367–374
  see also laws and policies
power relations, recommendations 369
Programme of Forest Management in the Bolivian Amazon (PROMAB) 32
property rights
  Bolivia 20–22, 24, 25–28, 35
  British Columbia 121
  see also land tenure
Prunus africana
  Cameroon 53, 56, 58, 60, 61, 334–335
  harvest ban 333
  trade 328, 334–335
Pure World Botanicals 335
pygeum *see Prunus africana*

quality control 349
quotas
    Cameroon 59–60
    Gnetum spp. (eru) 73

rainha/chacrona (*Psychotria viridis*), Brazil
    44, 47
rattan
    climate change impacts 194
    community management 166–167
    export bans 333
    harvesting methods 163
    Philippines 7, 155–156, 158, 162–168,
        186–187
    trade 328
    value chain 163–166
Rauwolfia serpentine, Nepal 333
reduce emissions from deforestation and
        forest degradation (REDD) schemes,
        Brazil nuts 37
reindeer lichen (*Cladonia*), Finland
        296
resource cycle, British Columbia 127
resource planning, British Columbia
        122–123
resource rights
    Cameroon 55–57
    impacts of 351–353
'right to roam' 7, 140
'right of user' 140–141
rooibos tea (*Aspalathus linearis*), South Africa
        337, 345, 351
rosewood (*Aniba rosaeodora*), Brazil 43,
        45
rubber (*Hevea brasiliensis*)
    Bolivia 20–23
    Brazil 44, 46, 47, 344
rural livelihoods, India 87, 89
rural restructuring 359
Russia
    market access 300, 301, 302
    NTFP policy 297–298, 301
    wild berries 290, 292–293, 300, 358

salal (*Gaultheria shallon*)
    British Columbia 117, 122
    Pacific Northwest 265, 268, 269, 273
sal (*Shorea robusta*), India 92, 97, 98
San people, traditional knowledge 10, 309,
        313, 319

Scotland
    access rights 7, 135–151
    accreditation and certification 150–151
    bluebells 142, 143
    commercialization 144–145
    damage to wild plants 141–142
    gatherers 137–139
    gathering practices 144–146, 149–150
    knowledge of NTFPs 146–147
    Land Reform Act (2003) 140, 143–144,
        147, 149
    legislation 139–144
    mechanization 145
    mushrooms 136–137, 138–139, 149
    non-local gatherers 145–146
    NTFP policy 7, 148
    recommendations 148–151
    representation 147, 150
    Scottish Wild Mushroom Code 149
    species protection 142–144
    subsidies 151
    theft 140–141
sea cucumbers, Fiji 233
shifting cultivation, Philippines 191–193
silkworms, Philippines 192
South Africa
    Biodiversity Act (2004) 319, 321
    bureaucracy 356
    customary oversight 330
    Hoodia spp. 10, 309–324
    kiaat wood 8, 199–203, 350
    legislation 8, 314–319, 320
Southern African Hoodia Growers Associa-
        tion (SAHGA) 318, 319, 321
species protection
    British Columbia 119–122
    China 250
    Scotland 142–144
stakeholders, consultations 346–347, 370
Strategic Environmental Plan (SEP), Philip-
        pines 162, 189
structural adjustment policies, Bolivia 23
subsidies, Scotland 151
subsistence communities, Philippines
        183–195
Sweden
    export market 298
    labour 298–300, 355
    wild berries 290, 292

tariffs, trade 333
taxation
    Cameroon 61–63, 65, 350
    Finland 297
    Gnetum spp. (eru) 73, 74
    informal (bribes) 62–63, 74, 350
technology, woodcarving 233
Tee Hilman, Brigitte 141
tendu/kendu (KL) (*Diospyros melanoxylon*)
    (cigarette leaves), India 6, 92–98, 329,
    345
Tibet, matsutake mushrooms 245, 247, 260
titica vine (*Heteropsis* spp.), Brazil 44, 45
tourism
    Canada 125
    Fiji 232–233
trade
    fair trade 33, 336
    harvest and export bans 333–334
    international regulations 332–338
    laws and policies 10, 327–339
    liberalization 358
    local and regional regulations 328–332
    matsutake mushrooms 243–244, 250–253,
        257–258
    tariffs 333
    taxation 61–63, 350
    wild berries 293–294
Trade-Related Aspects of Intellectual Prop-
    erty Rights (TRIPS) 336, 353
traditional knowledge
    British Columbia 118, 124–125
    Mexico 214–216, 221
    Scotland 146–147
    southern Africa 10, 309, 313, 319
transportation
    Brazil 49–51, 349–350
    laws and policies 331, 349–350
tree tenure, Bolivia 21–22, 31, 353
tribal communities *see* indigenous peoples
Tribal Development Cooperative Corpora-
    tion (TDCC) 97–98
Trichocaulon spp., southern Africa 311

Unilever, *Hoodia* use 313
Union for Ethical BioTrade 338
United Kingdom
    NTFP policies 7
    see also Scotland

United States
    floral greens 9, 265–283
    immigrant labour 9, 269–270, 276–280
    mushrooms 280–282
uppage (*Garcinia gummi-gutta*), India
    102–106, 107, 357

vesi see Intsia bijuga
Voacanga africana, Cameroon 60

Washington State (United States)
    floral greens 265–283
    labour regulations 276–280
    legislation 272, 275–280, 329, 349
Western Ghats (India), uppage 102–106,
    107
wicker products 155
wild berries
    access to 295–298, 352
    exports 293–294, 358
    Finland 9–10, 287–302
    labour 9–10, 298–300, 351, 354–355
    markets 298, 301–302
    Russia 290, 292–293
    Sweden 290, 292
    yields 290–291
wild foods, Scotland 7, 135–136, 138
wild teak *see* kiaat
woodcarving
    Fiji 8–9, 229–241
    South Africa 8, 199–203
    technology 233
Working Group of Indigenous Minorities in
    Southern Africa (WIMSA) 321
World Trade Organization (WTO) 335–336
Worldwide Fund for Nature (WWF)
    China 247, 259–260
    Fiji 239

xaxim (*Dicksonia sellowiana*), Brazil 43

yew bark (*Taxus brevefolia*), British Columbia
    119–120
yucca (*Yucca* spp.), Mexico 207
Yunnan Province (China), matsutake mush-
    rooms 9, 243–261, 352

zacaton roots (*Muhlenbergia macroura*),
    Mexico 207